# Physical Meteorology

# Physical Meteorology

Henry G. Houghton

The MIT Press
Cambridge, Massachusetts
London, England

This book was set in Times New Roman by Asco Trade Typesetting Limited, Hong Kong, and printed and bound by The Murray Printing Company in the United States of America.

Library of Congress Cataloging in Publication Data

Houghton, Henry G.
Physical meteorology.

Bibliography: p.
Includes index.
1. Atmospheric physics. I. Title.
QC861.2.H68 1985 551.5 84-12225
ISBN 0-262-08146-6

# Contents

# Preface

This book had its origin in class notes that I prepared and revised many times for use in a course on physical meteorology that I taught for more than 30 years. This book version was almost completely rewritten and contains much material that was not in the class notes. The course in physical meteorology was typically taken by first- or second-year graduate students who had taken several other courses in meteorology. However, a minority of the students was majoring in another physical science or engineering. The only genuine prerequisite is a good undergraduate course in physics, but I believe that some background in meteorology would enhance the value of the book for most students.

Reflecting my predilection, the emphasis is on physical reasoning rather than rigorous mathematical development. The book is intended to occupy a position intermediate between the specialist monographs in fields such as cloud physics and radiative transfer and the introductory books that purport to cover most of meteorology. I hope that it will provide an adequate background in physical meteorology for the meteorologist with a different specialty and that it will also serve as a stepping stone for one aspiring to specialize in an aspect of physical meteorology. The many references are designed in part to aid the second category of readers.

Of necessity all fields of science are subdivided, and meteorology is no exception. The traditional names of the subdisciplines of meteorology are rather inept, and "physical meteorology" is particularly poor, since all of meteorology is based on physics. Alternative names for this branch of meteorology have been used, but I decided to stick to the traditional term.

The choice of topics to be included in a book on physical meteorology is somewhat arbitrary. Almost everyone would agree that cloud physics and the transfer of solar and thermal radiation are two major areas that must be included. Eight of the ten chapters deal with these fields. The two other chapters are on meteorological optics and atmospheric electricity. Topics such as weather radar, the upper atmosphere, meteorological thermodynamics, and even atmospheric dynamics are sometimes included; it seemed preferable to me, however, to retain the cohesion that arises from restricting this book primarily to the two major areas.

My thanks to the many former students who took my course in physical meteorology over a span of some 35 years. Their often penetrating questions helped me to deepen my knowledge and to communicate my thoughts more effectively. I am greatly indebted to Isabelle Kole, who drafted the figures, Virginia Mills, who did most of the typing, the late Patricia Lindberg, who helped with the manuscript, and Jane McNabb, who helped in more ways than I can enumerate here.

Henry G. Houghton

# Physical Meteorology

# 1

# The Atmospheric Aerosol

## 1.1 Introduction

By convention, the atmospheric aerosol consists of the solid and liquid particles more or less suspended in the air, except for cloud and precipitation particles. These particles range in size from clusters of a few molecules up to a radius of a few tens of micrometers. The molecular clusters include the small ions that are formed continuously by cosmic radiation and other ionizing radiations. They and other particles smaller than about $5 \times 10^{-3}$ $\mu$m are found in appreciable numbers only in regions where they are being formed rapidly. Further, these very small particles have little effect on light scattering and do not serve as cloud condensation nuclei. However, the small ions are responsible for the electrical conductivity of the atmosphere. For all other applications, the minimum radius of aerosol particles may be taken as about $5 \times 10^{-3}$ $\mu$m and the maximum as about 20 $\mu$m, a range of nearly four orders of magnitude. Sources of the aerosol are ubiquitous and varied and include wind-raised dust, volcanic eruptions, bubbling of sea water, extraterrestrial dust, man-made pollution, and formation in situ from gas phase reactions, condensation, and deposition. The aerosol particles are removed by coagulation with other particles, sedimentation, participating in the condensation precipitation processes (called rainout), and washout by falling precipitation.

The atmospheric aerosol plays several important roles. Perhaps its most obvious manifestation is haze; indeed, in the absence of an aerosol the visual range would be on the order of 200 km! Of more fundamental importance are the scattering and absorption of solar radiation and the formation of clouds and fog by cloud condensation nuclei (CCN) when the water vapor pressure only slightly exceeds its saturation value. The aerosol particles also capture small ions, greatly reducing their electrical mobility and consequently the electrical conductivity of the air, and modify the space charge and the electric field.

In this chapter an account of the atmospheric aerosol will be presented with emphasis on the aerosol properties that are of most importance in radiative transfer and cloud physics. In a larger sense, study of the atmospheric aerosol is a subdivision of atmospheric chemistry, a long-neglected field that is now undergoing active development.

## 1.2 Methods of Measurement

The information available on the properties of the atmospheric aerosol is dependent in variety, quantity, and accuracy on the available techniques of measurement. The aerosol is a trace substance in the atmosphere; its range of sizes is large, its chemical composition is diverse, and it is subject to substantial variations in space and time. For these and other reasons the measurement of its properties is very difficult and is still in the developmental stage.

The small aerosol particles, less than about 0.5 $\mu$m radius, are the most difficult to measure. For particles smaller than 0.1 $\mu$m radius measurement of their electrical mobility has proved to be the most successful approach. Electrical mobility is the velocity of a charged particle in a unit electric field. There is evidence that most of the charged particles in this size range carry a single unit charge. The resisting force imposed by the air on the motion of the particle is given by the Stokes-Cunningham law, and equating this to the electric force yields the particle radius. Some two-thirds of the particles of radius around 0.1 $\mu$m are charged, but this fraction decreases with particle size. By estimating this factor and the particle density, it is possible to deduce the concentrations of particles as a function of radius. Commerical instruments using this principle are now available.

Submicrometer particles may be precipitated by thermal diffusion and viewed under an electron microscope, but the high vacuum and electron bombardment may destroy or modify some of the particles. An interesting device is the particle centrifuge of Goetz and Preining (1960), which sizes particles larger than 0.1 $\mu$m radius by imposing strong centrifugal forces.

## 1.3 Expansion Counters

An important instrument is the Aitken (1923) counter, in which a sample of air is humidified and then expanded so as to produce condensation on the particles. The resultant drops are large enough to be counted on a slide under a low-power microscope included in the instrument. When operated at the usual expansion ratio of about 1.25, condensation occurs on all particles larger than the small ions. Because particles larger than a few tenths of a micrometer sediment out during the humidification and be-

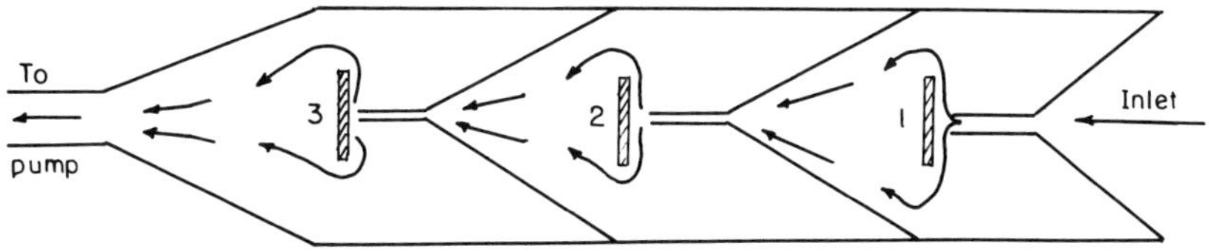


Figure 1.1 Schematic sketch of a three-stage cascade impactor. Numbers identify the three impactor slides, which collect, progressively, particles of decreasing size.

cause particle concentration increases rapidly with decreasing size, the Aitken counter gives a count of particles smaller than about 0.1 $\mu$m radius. More complex and larger expansion chambers permit the user to vary the expansion ratios, leading to counts as a function of the supersaturation. This nucleus spectrum may be interpreted as a size distribution, since size is the most important (though not the only) factor determining the supersaturation required for nucleation. Some expansion instruments provide for a continuous air flow and record the particle concentration by light scattering.

## 1.4 Impaction

Particles larger than 0.3–0.5 $\mu$m radius are easier to capture than smaller ones and may be measured under an optical microscope. Such particles may be caught on spider threads or fine wires exposed to the wind; the larger particles may be impacted onto slides. Collection efficiency depends on air speed, size of the collector, and size of the particles, and the lower cutoff radius increases with the size of the collector and inversely with the air speed.

Small collectors, which are favorable for small particles, collect an inadequate sample of larger particles because of their much smaller concentration. Some of these limitations are ameliorated in a common device called the cascade impactor, described by May (1945). Figure 1.1 illustrates a three-stage cascade impactor, which consists of three nozzles of decreasing diameter and three slides mounted progressively closer to the nozzle exits. The largest particles are deposited on the first slide and progressively smaller ones on the succeeding slides. The instrument is calibrated with particles of known size and density. There is considerable overlap in the size ranges collected in successive stages. Although up to seven stages are found in some versions, three to five are about all that can be used effectively. The slides may be viewed under a microscope to size the particles, but it is more common to use the optical densities of the deposits together with the instrument calibration to get a crude size distribution.

Fairly heavy deposits may be analyzed chemically to yield, information on gross differences in composition with size (Junge, 1954).

An obvious and often used device is the membrane filter through which air is drawn and in which particles are trapped. Larger particles can be observed by immersing the filter in a liquid whose optical index of refraction is the same as that of the filter material. Chemical solutions may be used that react with certain particles (such as NaCl) to produce a characteristic spot. The filter technique is often used to count ice nuclei; the exposed filter is placed in a chamber where the temperature and humidity can be controlled to form ice crystals on the nuclei.

The entrance speed of the air into a filter or cascade impactor should be nearly the same as the free stream velocity to ensure a representative sample. This "isokinetic sampling" is of major importance in the collection of aerosol samples. If the inlet speed is below free stream speed, the smaller particles will be carried around the inlet by the air flow; conversely excess inlet speed enhances the collection of small particles relative to the larger ones.

## 1.5 Optical Methods

A variety of optical techniques has been used to observe the atmospheric aerosol. The newest optical probe is the laser radar, or lidar. A laser pulse is directed into the atmosphere, and the returned back-scattered light is measured. As in microwave radar, "gating" circuitry permits one to observe the return from any desired height. The return pulse includes both the backscatter by the aerosol and Rayleigh molecular scatter; the latter can generally be computed, and the result is often presented as the ratio of the total backscatter to the molecular scatter. Gaseous absorption is avoided by an appropriate choice of laser wavelength. The results are "calibrated" by selecting a height where the aerosol content is negligible. The lidar is particularly useful for detecting aerosol layers up to the middle stratosphere and for providing data on time variations of the aerosol. In its simplest and most used form, the lidar can give particle size and concentration only if the form of the size distribution and the index of refraction are known or assumed. There are some prospects that more information can be obtained from multiple-wavelength lidars and from Raman scattering.

The first optical probing of the aerosol involved scanning a vertical searchlight beam. Here again only the backscatter is observed. Because it is limited more severely than the lidar by stray light sources, this technique is useful only at night.

The integrated effect of the aerosol on solar radiation can be determined by measuring the actual irradiance of the direct solar beam at the surface of the earth. The surface irradiance in an aerosol-free atmosphere can be computed from knowledge of molecular scattering and gaseous absorption (see section 3.29), thus permitting evaluation of the transmissivity due to the aerosol. Turbidity factors can be derived from such observations, which may be made over the entire solar spectrum or for selected wavelength bands. Additional information may be obtained by measuring the diffuse sky light that results from downscatter by both air molecules and the aerosol. Robinson (1962) has estimated the absorption by the aerosol. This requires an assumption about the ratio of the forward to the backward aerosol scatter and is also subject to errors due to multiple scattering. It is one of the very few techniques of remote sensing that yields an estimate of the absorption of solar radiation by the natural aerosol.

In principle, the most satisfactory means for obtaining data on the scattering and absorption of the natural aerosol in the solar spectrum is through the measurement of the up and down hemispheric fluxes as a function of height. The difference between the net fluxes (down minus up) at two levels yields the absorption in the layer. Subtraction of the computed gaseous absorption in the layer then gives the aerosol absorption. The down- and upscatter due to the aerosol may be found in a similar fashion by subtracting the computed molecular scatter. Because of the several subtractions of numbers of similar magnitude, precise measurements are required, and adequate data are scarce. An example of the results that can be achieved is given by Paltridge and Platt (1973), who made an ascent in an appropriately instrumented airplane during a very dense haze. They found that the extra absorption due to the aerosol was about the same as the aerosol backscatter.

Aerosol particles can be counted and sized by observing the light scattered from an intense beam by individual particles. Since the scattered light depends on the optical cross section of the particle, the signal-to-noise ratio of the instrument determines the smallest detectable particle, and this is usually on the order of a few tenths of a micrometer. Interpretation of the results in terms of particle size depends on assumptions regarding the index of refraction. This is further complicated by the usual arrangement in which the scattered light is measured only over a restricted angular span. Absorption by the particle and nonspherical shapes are other sources of uncertainty.

It is simpler to measure the scattered light from all of the aerosol particles in a specific air volume. Instruments called scatterometers are calibrated to yield the scattering coefficient. Because it is not feasible to measure the scattered light very close to the direction of the incident beam, the intense forward scattering by the larger aerosol particles (see section 2.10) is not measured, but is included in the calibration. The calibration can be strictly correct only for an aerosol with a particle size spectrum similar to that of the calibration aerosol.

Information on the stratospheric aerosol may be obtained from quanti-

Table 1.1
Size classes of aerosol particles

| Class | Range of radii ($\mu$m) | Area of interest |
|---|---|---|
| Small ions | $5\text{–}8 \times 10^{-4}$ | Atmospheric electricity |
| Large ions | $1.5 \times 10^{-3}\text{–}1 \times 10^{-1}$ | Atmospheric electricity |
| Aitken particles | $1.5 \times 10^{-3}\text{–}1 \times 10^{-1}$ | Cloud physics |
| Haze particles | 0.1–1 | Optical effects |
| Cloud condensation nuclei | $1 \times 10^{-2}\text{–}20$ | Cloud physics |
| Large condensation nuclei | 0.1–1 | Cloud physics |
| Giant condensation nuclei | 1–20 | Cloud physics |
| Particles containing most of the aerosol mass | $5 \times 10^{-2}\text{–}20$ | Air chemistry |

tative observations of the twilight sky. Spectacular twilights have been observed for a year or more following volcanic eruptions that have injected small particles into the stratosphere. An example of such twilight observations has been given by Volz (1969).

## 1.6 Size Distribution of the Continental Aerosol

Table 1.1 shows several common distributions of aerosol particles by size; it is due primarily to Junge (1963). The radius boundaries are somewhat arbitrary; the overlap if the size ranges reflects the terminologies adopted in different areas of application.

Junge was the first to demonstrate that the aerosol size distribution was continuous over the entire range of sizes by bringing together observations in several ranges obtained by different techniques. The form of the distribution is illustrated in figure 1.2. Logarithmic scales are necessary because of the wide ranges of radius and particle concentrations. Junge selected the ordinate scale so that the area under the curve represents the number of particles per cubic centimeter of air between any two selected radii. The observations were made in Central Europe, and the distribution is therefore that of a continental aerosol. As will appear in section 1.9, a pure oceanic aerosol exhibits a different distribution. Between radii of about 0.1 and 10 $\mu$m, which covers the major part of the aerosol mass, the curve is a straight line represented by

$$\frac{dN}{d(\log r)} = cr^{-\nu}, \tag{1.1}$$

where $c$ is a constant related to the total concentration of aerosol particles

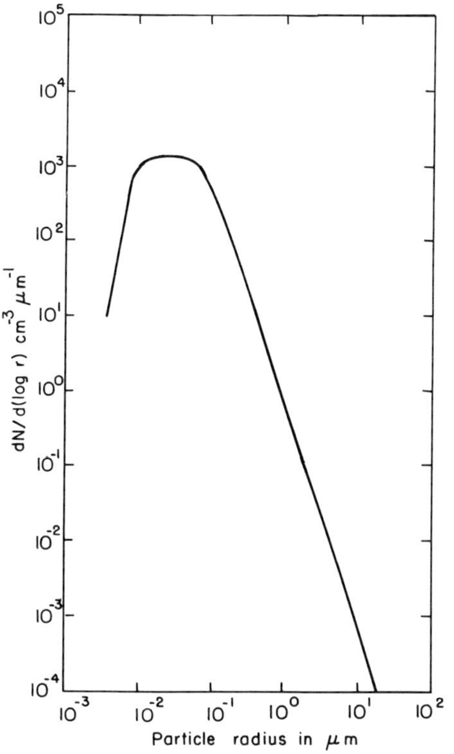


Figure 1.2 Idealized particle size spectrum of a continental aerosol. After Junge (1955).

and $N$ is the total concentration of particles smaller than radius $r$ per unit volume of air. By examining all available data on the size distribution of continental aerosols, Junge concluded that the slopes of the curves in their linear portions vary from about 2.5 to 3.5 and that $v = 3$ is a good average.

Table 1.2 gives particle size (radius) versus concentration for a typical continental aerosol to emphasize further the rapid decrease of concentration with increasing radius.

Subsequent studies have generally confirmed Junge's distributions for a continental aerosol near the surface of the earth. A somewhat larger range of slopes, perhaps from 2 to 4, has been found and also a tendency to an increased slope in the large radius range, that is, fewer micrometer-sized particles than predicted by (1.1). It also appears that the location of the maximum in figure 1.2 depends strongly on the age of the aerosol.

Numerical studies of radiative transfer, heterogeneous nucleation, and similar problems require an analytic form for the aerosol size distribution. The Junge power law has the great virtue of simplicity, but it does not

Table 1.2
Particle concentration versus radius of a continental aerosol[a]

| Radius range ($\mu$m) | Concentration ($cm^{-3}$) |
|---|---|
| <0.01 | 1,600 |
| 0.01–0.032 | 6,800 |
| 0.032–0.10 | 5,800 |
| 0.10–0.32 | 940 |
| 0.32–1.0 | 29 |
| 1.0–3.2 | 0.94 |
| >3.2 | 0.029 |
| Total concentration | 15,170 |

a. After Junge (1963).

replicate the maximum shown in figure 1.2. Two other representations that do include the maximum are that of Deirmendjian (1969) and the zero-order logarithmic distribution, ZOLD, given by Espensheid et al. (1964). The Deirmendjian function is

$$\frac{dN}{d(\log r)} = 2.3026\, ar^{\alpha+1} \exp[(-br^{\gamma})], \tag{1.2}$$

where a, $\alpha$, b and $\gamma$ are constants for a particular aerosol. The ZOLD distribution function is

$$\frac{dN}{dr} A \exp\left[\frac{-\ln^2(r/r_{\rm m})}{2\ln^2\sigma}\right], \tag{1.3}$$

where $A$, $\sigma$, and $r_{\rm m}$, a mean radius, are constants for a given aerosol. Leaving aside the constants $a$ and $A$, which are directly related to the total aerosol concentration, (1.2) has three adjustable constants and (1.3) has two. For a continental haze (haze $L$), Deirmendjian suggests $\sigma = 2$, $b = 15.12$, and $\gamma = 0.5$. Toon and Pollack (1976b) find $\sigma = 2.0$ and $r_m = 0.035$ $\mu$m fits continental size spectra. In addition to a maximum, both (1.2) and (1.3) lead to a slope that increases slowly with increasing radius.

By a simple transformation, figure 1.2 may be converted to a distribution of the particle volume, which, when multiplied by the particle density, becomes the distribution of particles by mass. The volume distribution corresponding to the idealized distribution of figure 1.2 is shown in figure 1.3. The use of a real particle spectrum would only introduce some fluctuations in the curve that would not change its general course. Area under the curve gives the volume contributed by a selected radius interval. It is

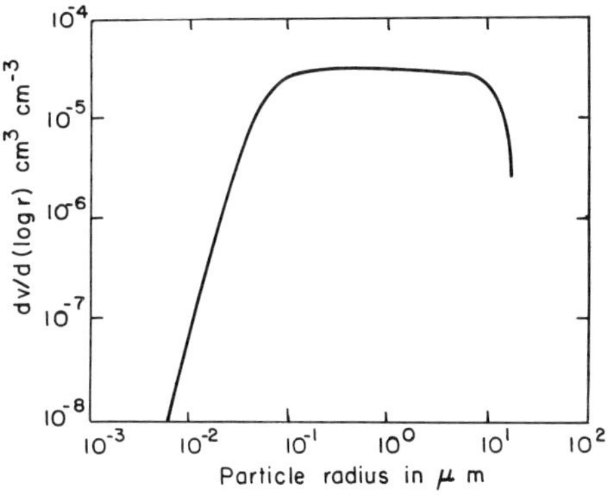


Figure 1.3 Volume distribution of the idealized distribution shown in figure 1.2. After Junge (1957).

evident that most of the volume comes from particles of radii ranging from about $5 \times 10^{-2}$ to 10 $\mu$m, which is also the range of particle radii that determines the optical properties of the aerosol.

The size distributions considered above are based on observations taken at or near the surface over land. The sources of most of the aerosol are also at or near the surface except those formed by gas phase reactions; the latter form many of the very small particles. As the aerosol diffuses up into the atmosphere it is removed from its source and its size distribution will change as a result of radius-dependent decay processes, which will be discussed in section 1.12.

Blifford and Ringer (1969) and Blifford (1970) used an airborne single-stage impactor to determine the aerosol size spectrum as a function of height from about 0.3 to 9.0 km over both land surfaces and the ocean. Of their many sets of data two average curves are reproduced in figure 1.4. The data are limited to radii greater than about 0.15 $\mu$m by the declining collection efficiency of the single-stage impactor. The authors caution that the marked increase of slope for $r < 0.2$ $\mu$m may be an artifact of the corrections for collection efficiency. The upper size spectrum of figure 1.4 was taken at about 2 km, but is generally similar to those taken near the surface. Neglecting the portion for $r < 0.2$ $\mu$m, the slope of this curve is about 2. The size spectrum at 7.6 km shows a marked tendency to flatness in the interval of $r$ from about 0.2 to 1.0 $\mu$m. Many of the other curves given in the reference for the upper troposphere show a broad maximum in this region. It is believed that this results from aging processes that remove both larger and smaller particles.

De Luisi et al. (1976) obtained aerosol size distributions both by impactor and optical measurements to an altitude of 4–5 km. The spectra obtained are similar to those of Blifford, but do not show the variations

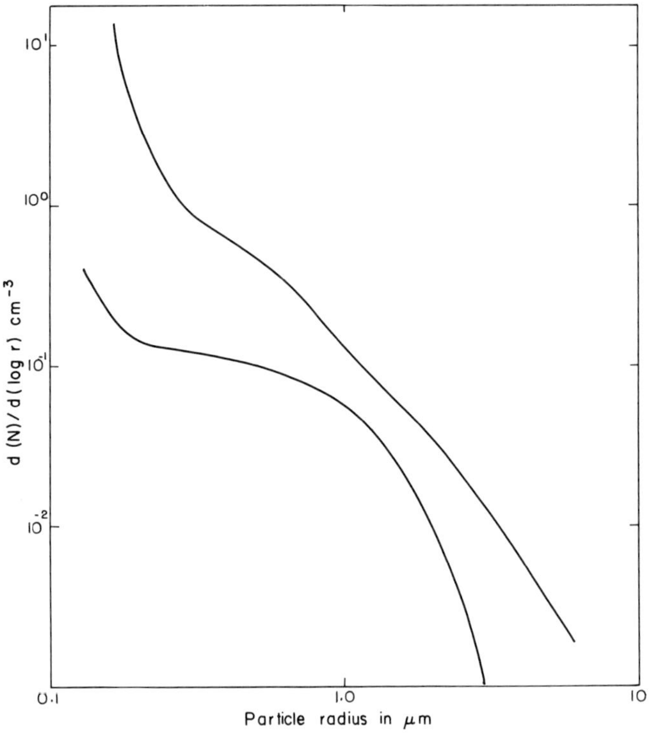


Figure 1.4 Average midtropospheric aerosol size distributions. The upper curve is at about 2 km height, the lower at 7.6 km. After Blifford and Ringer (1969).

with height so clearly as figure 1.4. Discrepancies between the impactor and optical results seem to be due to the fine structure of the vertical distribution of concentration as well as temporal changes. More data of this sort are needed as well as an extension of the impactor results to smaller radii.

Blifford (1970) made measurements over both the continent and the ocean. Characteristic properties of the maritime aerosol were observed only in the first few kilometers over the ocean. In the mid- and upper troposphere the size spectra were found to be much the same over the ocean as over the continent. This suggests that there may exist a rather uniform global aerosol in the upper troposphere. On average, the concentration of particles in the measured range decreases by about an order of magnitude from the unpolluted surface region to the midtroposphere. Above this level the changes with height are smaller and of either sign.

One may conclude that the general form of the size distribution of a continental aerosol near the surface is fairly well represented by figure 1.2, although the location of the maximum is variable and the slope of the curve

Table 1.3
Concentration of Aitken particles[a]

| Region | Average | Average range |
|---|---|---|
| City | 147,000 | 49,100–379,000 |
| Town | 34,300 | 5,900–114,000 |
| Countryside | 9,500 | 1,050–66,500 |
| Mountain (2 km) | 950 | 160–5,300 |
| Islands | 9,200 | 460–43,600 |
| Oceans | 400 | 100–1,000 |

a. In number per cubic centimeter in the air.

for $r > 0.1$ $\mu$m may vary. In the mid- and upper troposphere there appears to be a broad maximum or plateau in the vicinity of 0.1–1 $\mu$m radius. Most important, it appears that the aerosol in the upper troposphere is much the same over the globe, although additional confirmatory observations are needed.

## 1.7 Aitken Nuclei

Many more data are available on the concentration of Aitken nuclei than on any other aspect of the aerosol. Aitken nuclei range in radius from about $1.5 \times 10.^{-3}$ to 0.1 $\mu$m and are of little importance in radiative transfer and of marginal interest in cloud physics. Table 1.3, taken primarily from Landsberg (1938), summarizes a large amount of data taken by numerous observers. Evidently continental regions and, particularly, urban areas are prolific source regions. Some must be formed over the oceans because the concentrations over the southern oceans are not much different from those over the northern oceans.

Rosen et al. (1978), using balloon-borne counters, have obtained a number of vertical profiles of nucleus concentrations up to 30 km in various parts of the world. Some of their results are reproduced in figure 1.5. From the information contained in their paper, I believe that their data are essentially the concentration of Aitken nuclei. Generally, the nucleus concentrations vary by less than an order of magnitude in the troposphere, but fall rapidly above the tropopause and approach a constant value of the order of 100 per milligram of air in the stratosphere.

## 1.8 Variations of the Continental Aerosol

Over the continents the aerosol has both a diurnal and an annual variation. Over the oceans these variations are thought to be small. On a clear day the

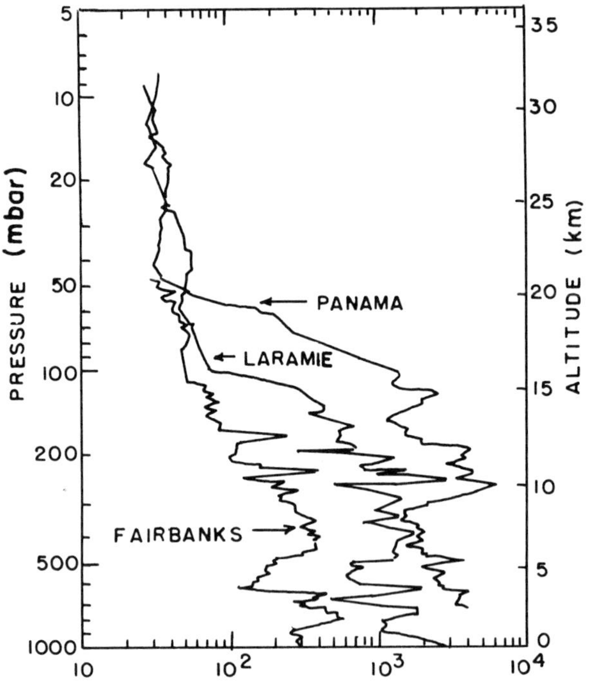


Figure 1.5 Typical vertical profiles of the concentration of Aitken nuclei. After Rosen et al. (1978).

aerosol, as measured by the attentuation of the solar beam, exhibits a maximum in the early morning. This presumably results from photochemical formation of aerosol particles and convective mixing that carries particles of surface origin aloft.

The annual variation of the continental aerosol over the United States has been presented by Flowers et al. (1969) in terms of the Volz turbidity coefficient[1] at a wavelength of 0.50 $\mu$m. They present curves from 29 stations, one of which is reproduced as figure 1.6. This shows the typical summer maximum and winter minimum found at nearly all of the stations. As would be expected, the average turbidity is higher in the eastern portion of the continent than in the western regions.

## 1.9 Size Distribution of the Marine Aerosol

An obvious feature of the marine aerosol is the salt particles that are formed by the bubbling of sea water, which will be described in section

1. The Volz turbidity coefficient, $B$, is defined by $E/E_0 = 10^{-(k_s+k_a+B)m}$, where $k_s$ and $k_a$ are the molecular scattering coefficient and the gaseous absorbtion coefficient and $m$ is the optical air mass.

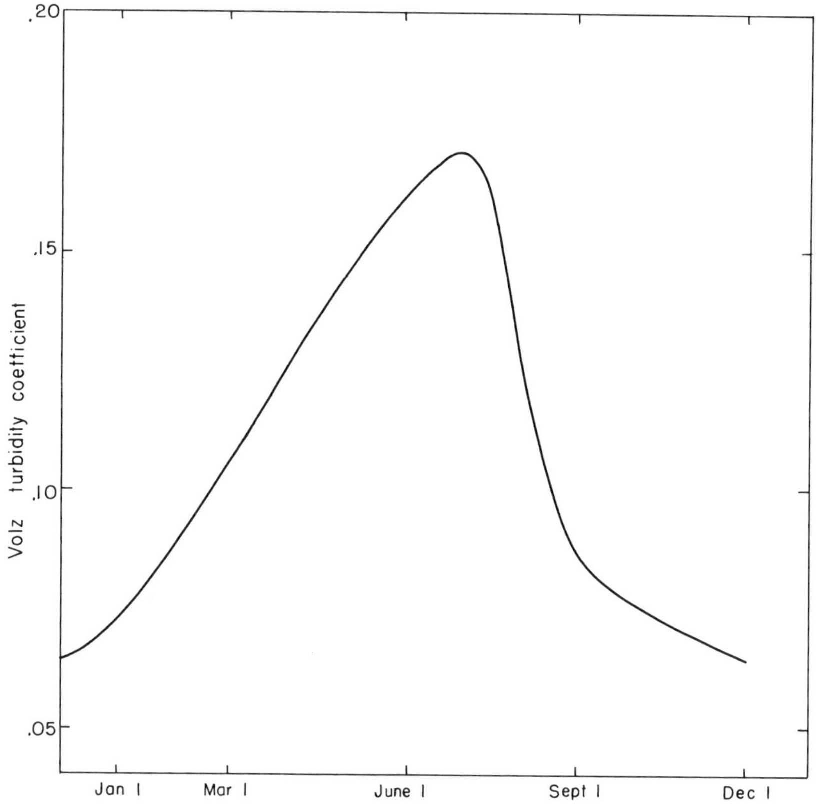

Figure 1.6 Monthly average Volz decadic turbidity coefficients at 0.50 $\mu$m for Blue Hill, Massachusetts. After Flowers et al. (1969).

1.15. The size distribution of sea salt particles with dry radii greater than about 0.5 $\mu$m has been measured by Woodcock (1953) and others as a function of altitude and wind speed. Some of these data are reproduced in figure 1.7. Note that the coordinates are different from those used in previous plots. The particle masses range from $10^0$ to $10^5$ pg (picograms), corresponding approximately to dry salt radii of 0.5–22 $\mu$m. These observations were made at cloud base (640 m). Only at the highest wind speed (34.8 m/sec) does the concentration of these large and giant particles exceed 1 $cm^{-3}$.

As a result of an intensive series of observations using five different instruments in the Atlantic Ocean between 10°S and 60°N, Junge (1972) has deduced the size distribution of the near-surface marine aerosol, shown in figure 1.8. Comparison with figure 1.2 shows that the maximum occurs at a larger radius in the marine aerosol and that there are one to two orders of magnitude fewer Aitken nuclei. Junge has identified five components of

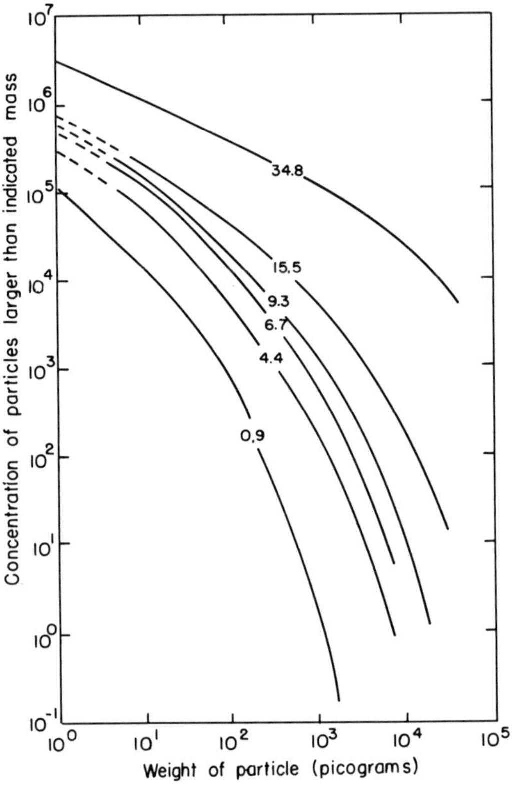


Figure 1.7 Mass distribution of sea salt particles near Hawaii for the indicated wind speeds in m sec$^{-1}$. After Woodcock (1953).

the aerosol. The smallest, composed of particles that have radii less than about $3 \times 10^{-2}$ $\mu$m, must be formed over the sea. Next are the tropospheric background particles, which are the same or similar to, the upper-tropospheric aerosol, which appears to exist worldwide. The Sahara dust, which is only present over the subtropical North Atlantic, is a mineral aerosol raised by the desert winds. Sea salt particles are found ranging in radius from a few tenths of a micrometer to perhaps 20 $\mu$m or more. Finally there is a small concentration of particles of radius greater than 20 $\mu$m, some of which are soluble, while others are not. Those that are soluble may be composed of sea salt. The insoluble particles seemed to be mostly fibers and agglomerates.

The characteristic marine aerosol is found only in the lowest kilometer or so. At higher levels the aerosol is believed to be essentially the same as that of the upper troposphere over the continents.

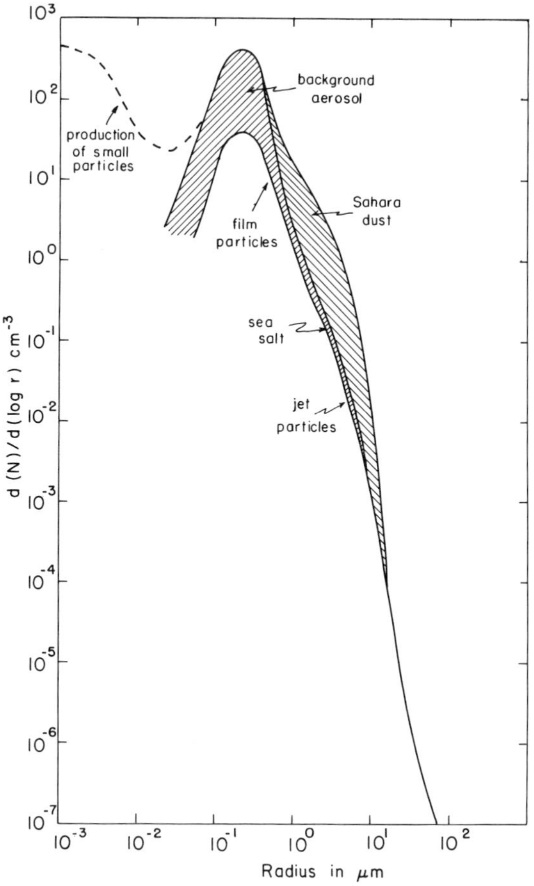


Figure 1.8 Idealized size distribution of an undisturbed marine aerosol. After Junge (1972).

## 1.10 The Stratospheric Aerosol

The presence of an aerosol in the stratosphere (and mesosphere) had been deduced from a variety of remote observations. These included visual observations of twilight phenomena, including the "purple glow," noctilucent clouds just below the mesopause, mother-of-pearl clouds in the stratosphere, observations of dust layers from high-flying aircraft and balloons, and the spectral determination of sodium in the high atmosphere. Renewed interest in the stratospheric aerosol resulted from the in situ observations from balloons and aircraft of Junge et al. (1961) and Junge and Manson (1961). This established the presence of an apparent worldwide aerosol layer centered at 17–20 km with a concentration of

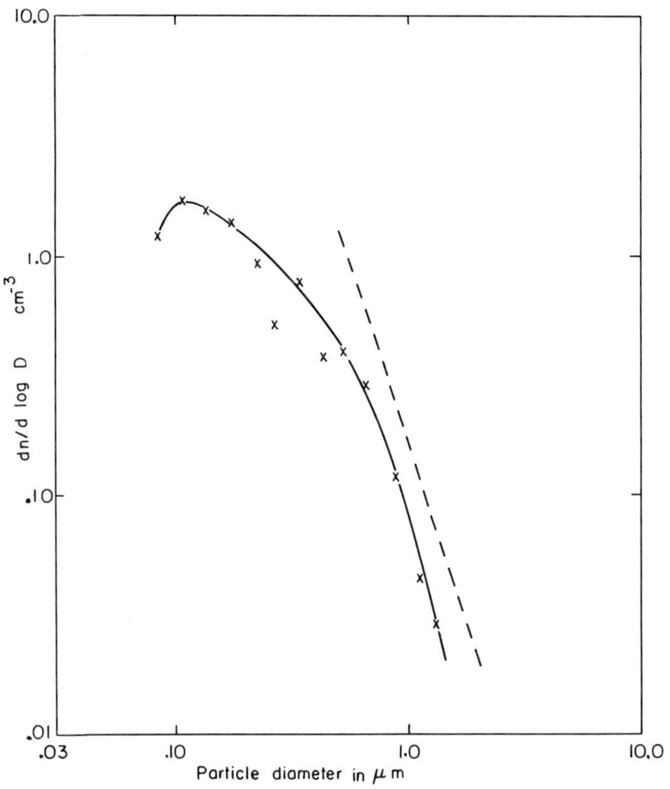


Figure 1.9 Stratospheric particle size distribution at 16–22 km. The dashed line has a slope of 3. After Bigg (1976).

particles of radius 0.1–2 μm of about 0.1 $cm^{-3}$. They found further that sulfur was the dominant element and suggested that much of it was in the form of ammonium sulfate. This layer is called the Junge layer or the Junge sulfate layer, and its discovery stimulated a number of other investigations using searchlights, lidar, and impactor collectors.

The properties of the stratospheric aerosol have been examined with the aid of single-stage impactors and a variety of optical instruments, ranging from those that sense individual particles to those that integrate over a large portion of the spectrum. Both the impactors and the optical instruments are limited to particle radii greater than 0.1–0.2 μm. Further, the concentration of giant particles ($r > 1$ μm) is so small that a representative sample of them is seldom obtained. However, the limited range of particle radii measured includes most of the aerosol mass and largely determines the optical properties of the aerosol. For the typical range of radii, the particle size spectrum is quite similar to that of the tropospheric aerosol. A

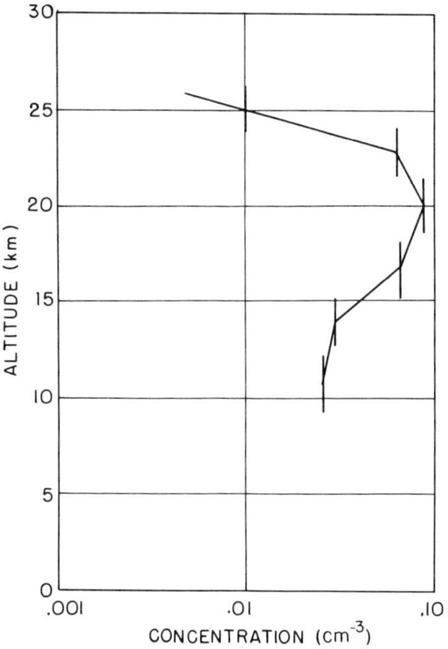


Figure 1.10 Vertical profiles of "large" particles in the lower stratosphere. Vertical bars show altitude increments over which samples were collected. After Junge et al. (1961).

size spectrum obtained by Bigg (1976) from airborne impactors is shown in figure 1.9 as an example.

Because of the difficulty of measurement, the concentration of Aitken particles ($r = 1.5 \times 10^{-3}$–$0.1\ \mu m$) in the stratosphere is not well established. The most promising approach is through the use of diffusion cloud chambers specifically designed for use in the stratosphere. Cadle and Langer (1975) found ten to several hundred Aitken particles per cubic centimeter of air in the lower stratosphere. Note that figure 1.5 shows similar stratospheric concentrations.There Aitken nuclei concentrations may be compared with concentrations of the large particles of 0.1–1.0 $cm^{-3}$. Aitken particle concentrations on the order of 100 $cm^{-3}$ suggests that they are formed in the stratosphere rather than being carried up from the troposphere.

The vertical distribution of the concentration of the large particles is irregular and variable from case to case, but the Junge layer is almost always apparent. Figure 1.10, after Junge et al. (1961), shows a vertical distribution in which the irregularities have been smoothed by their practice of sampling layers about 3 km thick.

Rosen and Laby (1975) and Hofmann et al. (1975) have made soundings of the stratospheric aerosol throughout the year and at 11 Antarctic stations. Their instrument measures the concentration of particles of diameter greater than about 0.3 $\mu$m. They found a simple seasonal variation at midlatitudes with a maximum concentration in winter and a minimum in summer. These variations are found only below 20 km and are highly correlated with the tropopause height. It was found that the Junge layer was present at all latitudes and that it shifted in height with the tropopause height. They found concentrations in the Junge layer of about 0.5–1 $cm^{-3}$, or nearly an order of magnitude larger than in figure 1.10.

Remote optical sensing with lidar is useful for detecting changes in the height and concentration of the Junge layer. As more measurements are accumulated by other methods, the lidar data may be used to deduce particle size and distribution with more confidence.

## 1.11 Noctilucent Clouds

Remote sensing has indicated the presence of aerosol particles at least up to the mesopause. The concentration is evidently very low, and the aerosol here may consist of particles of both stratospheric and extraterrestrial origin. There continues to be some mystery surrounding noctilucent clouds, which are seen with moderate frequency in high latitudes. Particle samples obtained by rockets during a display as well as in the absence of the phenomenon have not settled the many questions. Ice-coated particles may be responsible for these clouds, since there is evidence that they are observed with low-mesopause temperatures at which ice saturation is conceivable.

## 1.12 Factors Affecting Aerosol Size Distributions

Giant particles ($r > \sim 2.0$ $\mu$m) are removed from the atmosphere by sedimentation, and this is thought to be partly responsible for the upper size limit. Very small particles ($r < 0.01$ $\mu$m) in high concentrations coagulate rapidly under the influence of Brownian motion, and this process tends to set the lower size limit. It also leads to the formation of larger particles. The large and giant particles ($r \gtrsim 0.1$ $\mu$m) serve as condensation nuclei, and some are carried down with the precipitation (rainout). Some nine-tenths or more of the cloud drops are not carried down with precipitation but evaporate, releasing aerosol particles. Such a particle is apt to be more heterogeneous than the original nucleus because of collisions of drops, reactions in the drops, and diffusion of small particles and gases to the drops. Some large and giant particles are carried out of the atmosphere by collisions with precipitation particles (washout), a process that seems more efficient in snow than in rain. If all of these processes were quantitatively

Table 1.4
Terminal fall speeds of spherical particles of density 1.5, at four elevations (cm/sec)

| Radius ($\mu$m) | Surface ($\mu m^2$) | 10 km | 20 km | 30 km |
|---|---|---|---|---|
| 1 | 0.010 | 0.027 | 0.046 | 0.133 |
| 2 | 0.076 | 0.100 | 0.139 | 0.311 |
| 5 | 0.463 | 0.584 | 0.666 | 1.11 |
| 10 | 1.84 | 2.29 | 2.53 | 3.32 |
| 20 | 7.09 | 8.90 | 9.59 | 11.0 |
| 50 | 35.9 | 46.1 | 55.6 | 60.1 |
| 70 | 59.2 | 82.1 | 102 | 113 |

understood, as well as the aerosol formation processes, it would be possible to derive the aerosol size distribution as a function of time. The apparently wide applicability of the Junge power law suggested to Friedlander (1961) that it is a self-preserving distribution resulting from the action of some of the processes outlined earlier. Although not all factors affecting the size distribution could be included, tests of his ideas against observed distributions are generally favorable.

Sedimentation of large and giant particles is a simple process, and it can be examined with the aid of the fall speeds of spherical particles of density 1.5 given in table 1.4. This density was selected as intermediate between the large aqueous particles and dry solid particles. These terminal fall speeds were calculated from the formulas given by Beard (1976) for the U.S. Standard Atmosphere of 1962. It is seen that particles larger than about 20 $\mu$m radius will fall through the troposphere in a day or less. The particles fall even faster in the stratosphere. This is usually taken as the reason why particles larger than a few micrometers radius are so rare, except in the immediate vicinity of the ground, when surface particles are lifted by a strong wind. However, other processes that favor the loss of larger particles, such as rainout and washout, must also contribute to the rather sharp upper size limit. It should also be noted that many of the aerosol particles are formed in the atmosphere from gas phase reactions and cannot attain very large sizes.

Evidently very large concentrations of small particles, $r < 0.01$ $\mu$m, are continuously formed in the atmosphere by molecular aggregation, condensation, deposition, gas phase reactions, and photochemical processes. These very small particles are rapidly removed by coagulation under the influence of Brownian motion, and thus larger particles are formed. The basic coagulation equation, as given by Whytlaw-Gray and Patterson

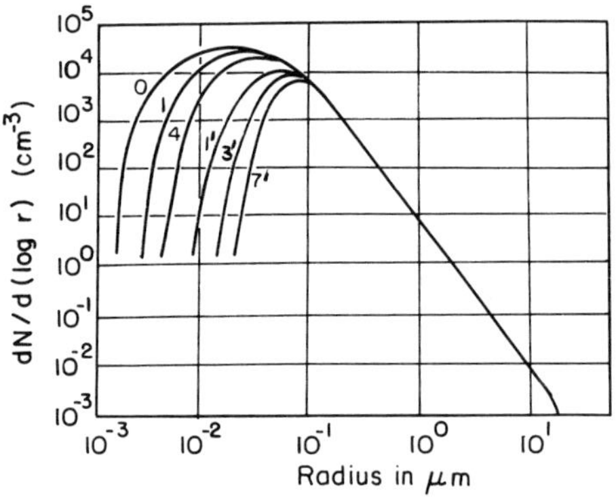


Figure 1.11 Results of Junge's (1963) computations of the effect of coagulation on an aerosol size distribution. Numbers are the elapsed time; unprimed numbers are hours, and primed ones are days.

(1932), is

$$-\frac{d[\ln n(r_1)]}{dt}=\left(\frac{kT}{3\eta}\right)\int_{r_m}^{r_n}(r_1+r_2)\frac{1}{r_1}+\left(\frac{Bl}{r_1^2}+\frac{1}{r_2}+\frac{Bl}{r_2^2}\right)f(r_2)\,dr_2, \quad (1.4)$$

where $n(r_1)$ is the concentration of particles of radius $r_1$, $k$ is the Boltzmann constant, $\eta$ is the viscosity of air, $r_2$ is the population of particles of other sizes as described by the distribution function $f(r_2)$, $l$ is the mean free path, $r_m$ and $r_n$ are the lower and upper limits of the size distribution, and $B = 1.27 + 0.4000 \exp[-(1.10r/l)]$. The integration of (1.4) is time-consuming, especially since it must be repeated for different values of $r_1$. Junge (1963) has performed the integration, with some approximations, with results as depicted in figure 1.11. The effect of coagulation in removing particles of radius $< 10^{-2}$ μm is clearly shown, as well as the near absence of an effect for $r > 10^{-2}$ μm. The maximum in the observed size distributions at $10^{-2}$–$10^{-1}$ μm radius is seemingly explicable by coagulation; the smaller the radius at the maximum, the "younger" is the aerosol.

It would be very helpful to know the rate at which the aerosol is removed from the atmosphere by the various processes noted above. Unfortunately our quantitative knowledge of the residence time of the aerosol is rather limited. For a specific substance, such as sulfur, for which the total world production rate from natural and anthropogenic sources can be estimated, a residence time may be computed from a knowledge of the total sulfur content of the atmosphere. Estimates of residence time of the sulfur compounds in the lower troposphere range from less than a day to a week. Dust clouds raised in Africa can be traced at least to the Caribbean where they arrive in about one to two weeks. Hygroscopic large and giant particles which participate in the condensation-precipitation process would be expected to have shorter residence times than mineral dusts. Average residence times for the tropospheric aerosol have been variously estimated as from about a week to a month or more. The residence time must depend on particle size and composition and on height in the troposphere but quantitative information is lacking. The residence time of the aerosol in the lower stratosphere is thought to be about a year and to increase with height in the stratosphere.

## 1.13 Composition of the Aerosol

In the introduction to this chapter it was noted that the sources of the aerosol include wind-raised dust, bubbling of sea water, volcanic eruptions, extraterrestrial dust, anthropogenic sources, and formation in situ from condensation, sublimation, and gas phase reactions. There are also biological sources, such as seeds, spores, pollens, molds, and terpenes. Nothing approaching a complete assessment of the relative importance of these many sources is available. The yellow sodium flame, seen when almost any noncombustible object is flamed, was once taken as evidence that the sea was the source of most of the aerosol. It is now known that, in general, maritime air has a much smaller aerosol concentration than continental air. Aitken and others showed that high concentrations of condensation nuclei could be produced in carefully cleaned air by inserting a heated platinum wire and inducing gas phase reactions with the aid of ultraviolet light. It is also common knowledge that very high aerosol concentrations occur in urban areas, particularly in the vicinity of industrial plants and home chimneys (see table 1.3).

The measurement of the chemical composition of the aerosol is a difficult task because of the extremely small size of most of the particles and the fact that its total mass concentration is seldom more than 10–100 $\mu$g per cubic meter of air. Coagulation, collisions, and condensation followed by evaporation act to make the particles heterogeneous in composition even if the original particles were composed of pure substances.

A notable step in finding the chemical composition of aerosol particles was taken by Junge (1954). His two-stage cascade impactor deposited large particles ($0.08 < r < 0.8$ $\mu$m) on one slide and giant particles ($0.8 < r < 8$ $\mu$m) on the other. Chemical tests of the deposits were made for $NH_4^+$, $SO_4^{--}$, $Cl^-$, $NO_3^-$, and $Na^+$. Such measurements were made in Frankfurt am Main, West Germany, and at Round Hill on the southeastern coast of Massachusetts. The results are summarized in table 1.5.

Except for one anomalous value at Frankfurt, the concentration of

Table 1.5
Concentration in micrograms per cubic meter of air[a]

| Location | $NH_4^+$ | $SO_4^{--}$ | $SO_4/NH_4$ | $Cl^-$ | $NO_3^-$ | $N_a^+$ |
|---|---|---|---|---|---|---|
| Frankfurt[b] | 1.57 | 4.38 | 2.79 | 0.16 | — | — |
| Frankfurt[b] | 6.60 | 17.40 | 2.64 | 4.00 | — | — |
| Round Hill[b] | 0.72 | 3.00 | 4.17 | 0.02 | 0.05 | 0.10 |
| Round Hill[b] | 0.93 | 6.15 | 6.6 | 0.03 | 0.06 | 0.03 |
| Frankfurt[c] | 0.31 | 2.30 | 7.4 | 0.56 | — | — |
| Frankfurt[c] | 0.70 | 3.40 | 4.8 | 0.50 | — | — |
| Round Hill[c] | 0.14 | 0.91 | 6.5 | 1.17 | 0.87 | 1.61 |
| Round Hill[c] | 0.23 | 1.41 | 6.1 | 0.73 | 0.52 | 0.68 |

a. After Junge (1954).
b. Large particles.
c. Giant particles.

chloride in the large particles is very small. The sulfate is the largest constituent of the giant particles, but chlorides are also important. It may come as a surprise that the chloride concentration at the coastal station (Round Hill) is only about twice that at Frankfurt, but the trajectories of the air sampled at Round Hill were basically continental, with only short excursions over the ocean. Significant concentrations of nitrate were found only in the giant particles.

Woodcock (1953) identified particles he collected over the sea as sea salt (see figure 1.6). Byers et al. (1957) identified giant particles collected in the midtroposphere over the continent as NaCl. These were probably of marine origin. Lodge and Frank (1966) collected Aitken particles with a multistage impactor that removed larger particles and found them to consist of sulfuric acid with varying degrees of ammoniation plus some phosphates. Others have also found that Aitken particles contain sulfates, but this subject needs further examination. It is known, for example, that polluted air contains tremendous concentrations of Aitken nuclei, and it seems unlikely that they are all composed of sulfates. It has also been suggested that aerosol particles may be formed from terpenes by photochemical processes.

Nitrates are found in aerosol samples (see table 1.5) and in rainwater. Nitric acid vapor is found in the troposphere and, in higher concentration, in the stratosphere. The source of the nitrates is presumably the oxides of nitrogen that are introduced both naturally and by man. Nitrates are hygroscopic and should serve nearly as well as sulfates or chlorides as condensation nuclei.

A significant fraction, estimated as 10–20%, of the tropospheric aerosol is surface dust raised by the wind. Such particles range in size from tenths to tens of micrometers and follow a Junge power law distribution with an exponent of around −2. It is not possible to catalog all the many other substances found in small concentrations in the atmospheric aerosol. Great volcanic eruptions are of particular interest, since they inject large quantities of particulates and gases into the stratosphere, where they persist for from one to several years. Such clouds produce spectacular sunsets and measurably reduce the incoming solar radiation. One hypothesis of climatic change attributes it to variations in volcanism. Spores, molds, and seeds are seasonal contributors to the natural aerosol. Meteoritic spherules composed of iron oxides have been found in deep-sea sediments and in precipitation, but their concentration in the atmosphere is vanishingly small. The principal source of extra terrestrial particles is the zodiacal dust. Finally, there is the multitude of man-made pollutants, which includes a wide variety of substances in gaseous and particulate forms. Attention has generally been focused on only certain components, such as soot, fly ash, lead, acid mists, and the like. Anthropogenic gaseous pollutants, such as sulfur and nitrogen oxides and exhaust products of internal combustion engines, are involved in photochemical processes that lead to the formation of aerosol particles.

## 1.14 Composition of the Stratospheric Aerosol

Junge identified sulfate as a principal component of the stratospheric aerosol and speculated that it was in the form of ammonium sulfate. Further observations have confirmed sulfate as a principal ingredient, but have suggested that the sulfate is often in the form of sulfuric acid, as well as ammonium persulfate and ammonium sulfate. These results are based on laboratory studies of single-stage impactor slides under the electron microscope. Different collector surfaces and different methods of handling the slides between collection and examination are probably responsible for a lack of complete agreement. Thus Bigg (1976) concludes that the particles are composed of concentrated sulfuric acid in liquid form; he also found insoluble, electron opaque particles that were probably of volcanic origin. On the other hand, Farlow et al. (1977) found the particles to be a somewhat volatile slurry of crystalline material in a liquid matrix. They observed insoluble granules in about one-third of the particles. The granules contained sulfur and sodium, suggesting a non-volcanic source. The crystalline component of the slurry has not been analyzed.

Tentatively, the stratospheric Aitken particles are also composed of sulfates. They are almost certainly formed in situ by a gas phase reaction. The process of formation of the large sulfate particles is not known. It seems more likely that they are nucleated by the Aitken sulfate particles

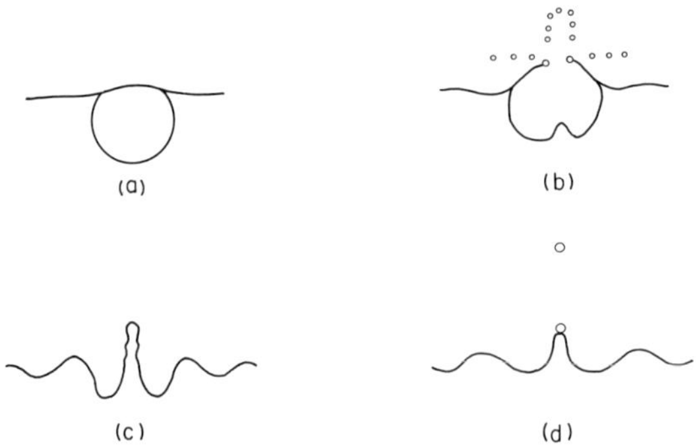


Figure 1.12 Time sequence of a bubble bursting at the sea surface: (a) a bubble pushing surface up and film beginning to form; (b) film breaks, forming toroidal ring, which breaks into film drops, some of which move horizontally, while others are caught in the ring vortex formed by the puff of air and move upward; (c) central jet, formed by surface energy, shows Rayleigh instability; (d) jet breaks into drops, which are projected vertically.

than by the insoluble granules. Clearly there is a need for further studies of the chemical composition and the growth mechanisms of stratospheric aerosol particles.

## 1.15 Formation of Salt Particles from the Ocean

The mechanisms by which salt particles are formed at the sea surface have been studied in some detail in the laboratory. Spray produced by the mechanical disruption of breaking waves is composed of drops much too large to become airborne. From the work of Kientzler et al. (1954), Mason (1954), and Blanchard (1963), it is known that small drops of saltwater are ejected into the air by the bursting of bubbles at the sea surface. Copious bubbles are formed by breaking waves (hence the white of whitecaps), and some may be formed by rain falling on the sea and probably also by biological processes. Some of the mechanisms involved in the formation of droplets by bursting bubbles are illustrated schematically in figure 1.12. In this figure, the bubble has begun to deform the surface and to form a film. In (b) the film, now fully formed, has burst, releasing the so-called film droplets. These drops are formed by the breaking of the toroidal ring of fluid that is gathered up at the edge of the expanding hole. Some of the film drops are projected nearly horizontally, while others are caught in the vortex ring formed by the puff of air released by the bursting of the film; these drops are carried up and have a better chance of becoming airborne than the others. Also shown in figure 1.12b is the beginning of the liquid jet from the bottom of the bubble, driven by the release of the surface energy of the bubble. This jet is shown more fully developed in (c), where the

Table 1.6
Diameter of first and second jet drops versus bubble diameter

| Bubble diameter ($\mu$m) | 10 | 20 | 50 | 100 | 200 | 500 | 1,000 |
|---|---|---|---|---|---|---|---|
| Drop diameter ($\mu$m) | 2.2 | 3.3 | 6 | 10 | 18 | 47 | 115 |

Table 1.7
Ejection velocity of the top jet drop

| Bubble diameter ($\mu$m) | 70 | 200 | 400 | 1,000 | 2,000 |
|---|---|---|---|---|---|
| Ejection velocity (m/sec) | 60 | 25 | 20 | 9 | 5 |

Rayleigh instability has caused the jet to "neck-in." The final phase is shown in (d), where the first jet drop has been ejected and the second is about to follow.

The film drops are thought to be only a few micrometers in diameter and are much smaller than the jet drops. The number of film drops formed depends strongly on the bubble size. Day (1964) found that a bubble of 0.5 mm diameter forms about 5 film drops, while a 1.0-mm bubble may form 20. Bubbles smaller than 100–300 $\mu$m diameter do not seem to form film drops. The vortex that carries film drops away from the surface of the ocean has not been observed with bubbles smaller than 1 mm diameter. Limited observations of the size distribution of bubbles in a surf zone suggest that there are very few bubbles larger than 200 $\mu$m diameter. If this is confirmed, then in view of the previous statements, it can be concluded that film drops play only a minor role in the injection of seawater drops into the atmosphere.

The size of the jet drops is directly related to the size of the bubbles, as shown in table 1.6, which is based on Blanchard's (1963) measurements. These are the diameters of the seawater drops as they are formed. Those that become airborne evaporate to a smaller size depending on the relative humidity of the air. At about 75% relative humidity, the drops become a saturated solution, and their diameters are reduced to about one-half of their initial diameters.

The ejection velocities of the top jet drops are surprisingly large and vary inversely with the bubble size, as illustrated in table 1.7, which is based on Blanchard (1963). There is considerable scatter in the observed values, due probably to differences in the complex bubble-bursting process from bubble to bubble, and table 1.7 should be considered only an approximation.

The height to which the jet drops are projected, along with the near-surface air turbulence and the settling speed of the drops, largely governs the fraction of the jet drops that becomes airborne. The distance a drop can be projected through air declines rapidly with decreasing drop size, more

Table 1.8
Drop ejection heights for the first and second jet drops

| Bubble diameter (mm) | 0.2 | 0.4 | 0.6 | 1.0 | 1.4 | 2.0 | 3.0 |
|---|---|---|---|---|---|---|---|
| Top drop height (mm) | 5 | 20 | 40 | 110 | 150 | 185 | 160 |
| Second drop height (mm) | 5 | 20 | 35 | 53 | 40 | 20 | 7 |

than compensating for the increase in ejection velocity with decreasing drop size. Observed ejection heights for the first and second jet drops from Blanchard (1963) are given in table 1.8. The third, fourth,... jet drops reach decreasing heights and are generally considered unlikely to become airborne. It is notable that the first and second drops from the smaller bubbles reach about the same heights. It is of interest that jet drop ejection is readily observable over a glass of any carbonated beverage.

Jet drops formed by bubbles of the observed size range seem capable of producing the salt particles found by Woodcock (see figure 1.7). According to Junge (see figure 1.8) the marine atmosphere also contains smaller salt particles than could result from bubble film drops. It is rather uncertain that there is a sufficient number of bubbles of diameter greater than 100–300 $\mu$m, required to produce film drops, to account for the small salt particles observed by Junge.

On the basis of a few counts of bubble concentrations in a surf zone and an estimate that 1–2‰ of the world ocean is covered by whitecaps, on the average, it has been estimated that the mean rate of bubbling is around 1 $cm^{-2} sec^{-1}$. This number is probably uncertain by at least an order of magnitude. Only a small, but unknown, fraction of the bubbles yield viable atmospheric salt particles. By combining such estimates with other assumptions, it has been crudely estimated that the bubbling process injects some $10^9$–$10^{10}$ tons of salt a year into the atmosphere, where it is subsequently removed by sedimentation, rainout and washout.

Blanchard (1963) has pointed out that the jet is formed from the surface layer of the water. He found that jet drops from bubbles of 2 mm diameter carried away an organic film placed on the water. He also found that the jet drops carry an electric charge and suggested that this may contribute significantly to the global atmospheric charging mechanism.

## 1.16 The Sulfate Aerosol

The pervasiveness of sulfate as the dominant soluble component of the large aerosol particles in both the troposphere and the stratosphere makes its origin a matter of interest. The initial sources are primarily sulfur dioxide and hydrogen sulfide. Sulfur dioxide may be oxidized after absorption into water drops or by a photochemical gas phase reaction possibly

Table 1.9
Annual global budget of sulfur in units of $10^6$ metric tons of $SO_4$

| | |
|---|---|
| **Sources** | |
| Decay processes (mostly $H_2S$) | 250 |
| Sea salt (from bubbling) | 140 |
| Anthropogenic sources | 150 |
| Volcanoes | 10 |
| Total | 550 |
| **Sinks** | |
| Brought down in precipitation | 475 |
| "Dry" deposition on earth | 75 |
| Total | 550 |

involving the hydroxyl radical. The details of the oxidation of hydrogen sulfide are not well known, but it apparently proceeds fairly rapidly in the atmosphere.

There are both natural and man-made sources of $SO_2$ and $H_2S$. Anthropogenic sulfur is released almost entirely as $SO_2$, mostly from the burning of sulfur-containing fossil fuels, with some contributions from nonferrous smelters and other industrial processes. These sources are found predominantly in the middle latitudes of the Northern Hemisphere, in the major urban areas. Natural sources include soils and marine marshes and flats that yield $H_2S$ under anaerobic conditions and organic sulfur compounds under aerobic conditions, the sulfate content of the marine aerosol produced by bubbling (seawater has a $SO_4/Cl$ ratio of about 0.14) and volcanoes, which may emit $SO_2$, $H_2S$, and other sulfur compounds.

It is relatively simple to determine a reasonably reliable figure for the annual anthropogenic production of sulfates, but the natural sources are difficult to evaluate. Table 1.9 of the annual global sulfate budget is taken from Kellogg et al. (1972), with a modified amount for volcanoes based on Stoiber and Jepsen (1973). These estimates indicate that humanity is contributing somewhat more than a quarter of the total, and projections suggest that this may approach one-half by the turn of the century. In the Northern Hemisphere, humanity contributes nearly 40% of the total, and in urban areas the anthropogenic sources are dominant.

Although there is still some uncertainty about the reaction rates, it appears that $H_2S$ in the atmosphere is oxidized to $SO_2$ in a matter of hours and that $SO_2$ is oxidized to $SO_3$, which nearly immediately becomes $H_2SO_4$, in hours to a very few days. Ammonia is an atmospheric constituent that is released by the decay of organic matter and from fertilizer. It reacts readily with $H_2SO_4$ to form $(NH_4)_2SO_4$, the reaction probably being limited by the supply of $NH_3$.

Nitrates are found in atmospheric particulates and in rain. The nitrates presumably arise from some or all of the oxides of nitrogen (written collectively as $NO_x$), but these reactions are poorly known. The nitrates are often acidic, as are the sulfates. It is not uncommon to find acidic rain tens to hundreds of kilometers downwind of industrial areas. This has apparently led to ecologically important changes in the acidity of unbuffered lakes. It is also the sulfuric and nitric acids from industrial sources that cause the erosion of stone buildings and statues.

Brief mention is made here of photochemical smog (for example, in Los Angeles) because of the great public and scientific interest in this urban phenomenon. Photochemical smog is the result of a photochemical reaction chain in which oxides of nitrogen ($NO_x$), reactive hydrocarbons, and solar ultraviolet radiation lead to the formation of ozone ($O_3$), nitrogen dioxide ($NO_2$), and peroxyacetyl nitrate (PAN). The $NO_2$ is a brownish gas that contributes to the characteristic smog color. The reactions summarized so briefly here are very complex and not understood in all their details and, particularly, the interrelation between photochemical smog and sulfate smog. The smog particulates, which cause the reduction in visibility, have both a complex composition and origin.

## 1.17 Attenuation of Radiation by the Aerosol

The aerosol scatters and absorbs solar radiation and thereby significantly reduces the sunlight reaching the surface. The aerosol also modifies the transfer of thermal radiation in the atmosphere. Consideration of these processes is deferred to chapters 4 and 5. For preliminary orientation some of the results of Elterman (1966) and Elterman et al. (1969) are shown in figure 1.13. The extinction coefficient is defined here by $\tau = \exp(-k_e L)$, where $k_e$ is the extinction coefficient and $\tau$ is the fraction of the radiation transmitted over a path of length $L$.

The smooth curve in figure 1.13 is the extinction coefficient of pure air that is due to molecular scattering. It provides a useful comparison with the other two curves, which are for the aerosol alone, and is to be added to them if the total extinction coefficient of the cloud-free atmosphere is desired. Note that the Junge sulfate layer is apparent in both aerosol curves, although it is naturally more marked in an individual profile than in the average of 105 profiles.

The data on which these curves are based were obtained by scanning a nearly vertical searchlight beam with a ground-based detector. All of these soundings were taken in New Mexico on moonless nights to avoid stray light. Somewhat similar results have been obtained at other locations using a pulsed laser instead of a searchlight.

The data of figure 1.13 in the upper troposphere and in the stratosphere may be used as a first approximation to the attenuation of solar radiation

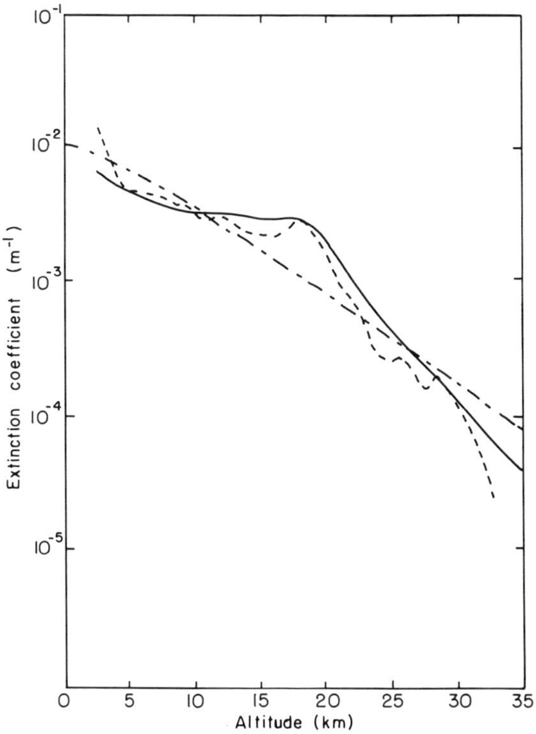


Figure 1.13 Aerosol extinction coefficients as a function of altitude. The solid curve is the average of 105 profiles, while the dashed curve is an individual profile (for 8 May 1964). The extinction coefficient for aerosol-free air (the dash-dot curve) must be added to the aerosol extinction coefficient if the total extinction is desired. All curves are for wavelength = 55$\mu$m. After Elterman (1966).

by the natural aerosol. They are strictly valid only for a wavelength of 0.55 $\mu$m, which lies at about the center of the visible spectrum. This wavelength was chosen to avoid absorption by atmospheric gases, but it also eliminates most of the absorption by the aerosol, which is usually found in the infrared.

## 1.18 Secular Changes of the Aerosol

It has been asserted that the aerosol content of the troposphere has increased since the 1940s and that this increase is continuing. It has been argued that such a presumed worldwide increase in the aerosol will result in reduced solar heating and a colder climate. The assumed increase in the aerosol is poorly documented because there are very few places in the world where solar radiation has been monitored for a long period by unchanged techniques of adequate accuracy. It has been variously estimated that humanity's activities now contribute some 10–15% of the global mass of the aerosol. However, the anthropogenic part of the aerosol has a highly nonuniform distribution over the globe, reflecting its sources in urban areas and intensively tilled soil. There can be no doubt that humanity's production of aerosol particles has increased dramatically over the past century. The rather short residence time of the aerosol suggests that its increase is concentrated in, and a few hundred kilometers downwind of, urban areas.

Cobb (1973) reported on changes in the electrical conductivity of the air over the oceans that have occurred since the cruises of the R/V *Carneigie* (1910–1920). The electrical conductivity of the air changes inversely with the concentration of small aerosol particles that capture the small ions. Over most of the world ocean little change in conductivity was found, implying no significant change in the surface aerosol in 50 years. Notable exceptions were the western North Atlantic, where the conductivity had dropped 80%, and a 20% reduction in the western North Pacific and in a portion of the Indian Ocean. These areas are downwind of continental source regions of anthropogenic aerosols.

Ellis and Pueschel (1973) have analyzed the data collected at Mauna Loa Observatory (Hawaii) on the transmissivity of the atmosphere above this 3,400-m mountain for the period 1958–1970. The effects of the eruption of Mount Agung are apparent, but the transmissivity had returned to the preeruption value in 1970–1971. The period of record is too short to detect a small long-period trend, but the data do not support a rapid increase in the upper-tropospheric aerosol.

Using the same Mauna Loa data, but only to 1967, Peterson and Bryson (1968) interpreted their results as most probably indicating a secular increase in the aerosol. Careful analysis of a variety of pertinent data covering 50 years or more by Rosen et al. (1973) and Dyer (1974) failed to show any clear evidence of a secular trend in the aerosol. The most prominent effects are those due to volcanic eruptions, and this may obscure small long-term trends.

Ludwig et al. (1970) showed that the airborne particulate matter in 58 U.S. urban sites decreased about 19% from 1957 and 1967, presumably reflecting pollution control efforts. At the same time the particulate loading in 20 U.S. nonurban sites increased about 40%.

One may conclude that there is as yet no very persuasive evidence of a worldwide increase in the atmospheric aerosol. There has certainly been an increase over the past century or so in the aerosol within, and downwind of, major urban areas, but this trend may have been arrested by efforts to control airborne pollution at its source. The pollution in urban and adjacent areas has doubtless had an effect on the weather and climate of these regions. It still remains to determine these effects quantitatively so that they may be compared with the effects of other aspects of urbanization, such as heat and water vapor pollution and changes in surface roughness and evapotranspiration. More definitive answers to these important questions will be forthcoming as the period of careful observations lengthens.

## Problems

1.
What are the chemical compositions of the more common atmospheric aerosol particles?

2.
What processes are responsible for producing particles small enough to remain nearly in suspension in the atmosphere? Why do they not continue to accumulate?

3.
What are the possible sources and compositions of noctilucent cloud particles?

4.
Develop an analytical expression for the ejection height of jet drops from bursting bubbles. How do the results compare with Blanchard's (1963) observed results?

5.
How does one tell a noctilucent cloud from a thin cirrostratus cloud?

# 2

# Scattering in the Atmosphere

## 2.1 Introduction

When light passes through a nonhomogeneous medium, some of it is scattered; that is, it is diverted from the initial direction of propagation. In general, the phase and the polarization of the scattered light differ from those of the incident beam, but there is no change in wavelength involved in the scattering process.[1] The nonhomogeneities that cause scattering in the atmosphere are particles of various kinds, ranging from the air molecules themselves to the aerosol and hydrometeors that populate the natural atmosphere. Scattering particles are those that have an index of refraction different from that of the homogeneous medium in which they are imbedded; in the case of the atmosphere the index of the medium (air) is nearly unity as compared with that of the particles.

When a beam of parallel light passes through the atmosphere, both scattering and absorption generally occur. Both of these processes remove radiation from the parallel beam, and their combined effect is called extinction. The absorption may occur in absorbing gases in the atmosphere and/or in the particles. The extinction coefficient is the sum of the scattering and absorption coefficients.

In the definition adopted here, all radiation that is diverted from the initial direction of propagation, no matter how small the scattering angle, is counted as scattered. This has some practical consequences that will be discussed in section 2.7. The definition encompasses not only scattering by particles small compared with the wavelength but also by particles large enough to permit the use of geometric optics. Thus reflection, refraction,

1. The Raman effect, sometimes called Raman scattering, involves a quantum transition, and the "scattered" light is of a different wavelength. This process is not important in radiative transfer, but may be used as a measurement technique.

and diffraction caused by raindrops and ice crystals and leading to such phenomena as rainbows, halos, coronas, and glories are included in the definition of scattering.

The scattering particles in the atmosphere are in relative motion and are far enough apart so that interference between the radiation scattered by different particles does not occur. This is called independent or incoherent scattering and permits the total scattering to be obtained by summing the scattering from the individual particles. An example of coherent scattering is a crystal at very low temperature, in which the atoms are fixed in a regular lattice. In such a case no scattering in the usual sense occurs, but simply a retardation of the phase velocity.

Scattering is important in a number of scientific fields, such as astronomy, astrophysics, colloid and aerosol chemistry, and radar. As a result, a very large literature on it has developed. The mathematical representation of scattering is rather complex and specialized. Fortunately van de Hulst (1957) has assembled all of the basic theory and many of the applications in an excellent book that should be consulted by anyone who wishes to go beyond the simplified introductory treatment to be given here.

When long paths are involved through a cloud of scatterers, multiple scattering becomes important. By this is meant light that is scattered more than once. This problem rapidly becomes unmanageable by simple iterative procedures.

## 2.2 Exponential Law

Scattering in a medium containing a considerable number of particles is described by an exponential law, often called Beer's law or Bouguer's law. Consider a parallel beam of radiation passing through a scattering medium as indicated in figure 2.1. $E$ is the incident irradiance (radiant flux per unit cross section), and $dE$ is the irradiance scattered out of the parallel beam in the distance $dy$. The basic premise is that the fractional depletion is proportional to the length of the path in the scattering medium, or that

$$dE/E = -k_s dy, \tag{2.1}$$

where the coefficient of proportionality $k_s$ is called the scattering coefficient. In general, $k_s$ varies with the path length $y$ and often with the wavelength. The integrated form of (2.1) is

$$E_\lambda/E_{0\lambda} = \exp\left[-\int_0^y k_\lambda\, dy\right], \tag{2.2}$$

and a second integration over wavelength is often required. Here $E_{0\lambda}$ is the irradiance at $y = 0$ and $E_\lambda$ is the irradiance at $y = y$.

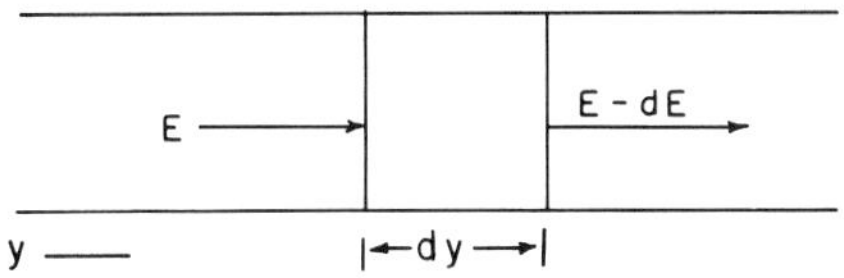


Figure 2.1 Scattering in a parallel beam of radiation.

Equation (2.2) is equally applicable to the depletion of a parallel beam by gaseous absorption. In that case $k$ is called the absorption coefficient, which will be written $k_a$ to distinguish it from the scattering coefficient $k_s$. In gaseous absorption, it is often preferable to measure $y$ not as a distance but as the mass of the absorber per unit cross section. In this case the dimension of $k_a$ is inverse mass. When both absorption and scattering occur simultaneously, $k$ is called the extinction coefficient $k_e$, and $k_e = k_s + k_a$ so long as $y$ is measured in the same units for each.

Because of the exponential form of (2.2) it is not permissible to use a linear average $k$, when $k$ varies rapidly with either $y$ or $\lambda$. This has consequences that are explored in chapter 4.

## 2.3 Small-Particle Scattering

When the scatterers are small compared with the wavelength, the scattering is relatively simpler than in other cases. Also, this was the case treated by Rayleigh (1871), who developed the first true theory of scattering in order to explain the blue of the sky. Because of his pioneering work, scattering by particles that are small compared with the wavelength is almost universally called Rayleigh scattering. Although not in the form originally given by Rayleigh, the scattering coefficient for Rayleigh scattering for optically isotropic spheres in a medium of unit refractive index is

$$k_s = N\pi r^2 \frac{128\pi^4 r^4}{3\lambda^4}\left(\frac{n^2-1}{n^2+2}\right)^2, \tag{2.3}$$

where $N$ is the concentration of the particles, $r$ is their radius, $\lambda$ is the wavelength, and $n$ is the index of refraction of the particles.Note that the quantity $N\pi r^2$ is the total geometric cross section of the scatterers in a unit volume. The balance of the right-hand side of (2.3) is called the scattering efficiency factor; that is, it is the nondimensional factor that, when multiplied by the geometric cross section, yields the scattering cross section. Thus the scattering efficiency factor $Q_{sca}$ is

$$Q_{sca} = \frac{128\pi^4 r^4}{3\lambda^4}\left(\frac{n^2-1}{n^2+2}\right)^2. \tag{2.4}$$

Table 2.1
Indices of refraction of water and ice[a]

| Substance | Spectral region | $n$ | $[(n^2 - 1)/(n^2 + 2)]^2$ |
|---|---|---|---|
| Water | Visible | 1.333 | 0.042 |
| Ice | Visible | 1.309 | 0.037 |
| Water | Microwave | 8.7 | 0.93 |
| Ice | Microwave | 1.75 | 0.17 |

a. After J. C. Johnson (1954).

These expressions contain the ratio of the radius to the wavelength to the fourth power. It is usual to define a size parameter $x = 2\pi r/\lambda$. This yields

$$Q_{\text{sca}} = \frac{8}{3}x^4\left(\frac{n^2 - 1}{n^2 + 2}\right)^2. \tag{2.5}$$

For particles of a given size and index of refraction, these equations show that the scattering coefficient is inversely proportional to the fourth power of the wavelength. It follows that blue light is scattered much more than red. This is the basic cause of the blue of the sky and the reddening of the sun near the horizon. Any complete explanation of the color of the sky must consider multiple scattering, reflection from the surface of the earth, absorption, and non-Rayleigh scattering. It must be remembered that Rayleigh scattering and the above equations apply only when the particles are small compared with the wavelength. For visible light this means that the particle radius should be less than about 0.05 $\mu$m. The "dry air scattering" discussed in section 3.15 is Rayleigh scattering, but the "water vapor scattering," section 3.16, is not. Rayleigh scattering also applies to the scattering of radar beams by raindrops, where the wavelength is typically 3–10 cm and the radius is a few millimeters. The index of refraction of water and of ice are different at optical and radar wavelengths, as indicated in table 2.1

Equations (2.3)–(2.5) give the total scattered radiation irrespective of angular distribution and polarization. This is adequate when all that is desired is the loss of radiation in the initial parallel beam, but in any study of the scattered radiation both its angular distribution and polarization are usually desired. The Mie intensity functions, $i_1$ and $i_2$, for Rayleigh scattering of natural light are given as a function of the scattering angle $\phi$, measured from the direction of the incident radiation by

$$i_1 = \frac{Nr^2}{2}\left(\frac{2\pi r}{\lambda}\right)^4\left(\frac{n^2 - 1}{n^2 + 2}\right)^2, \tag{2.6}$$

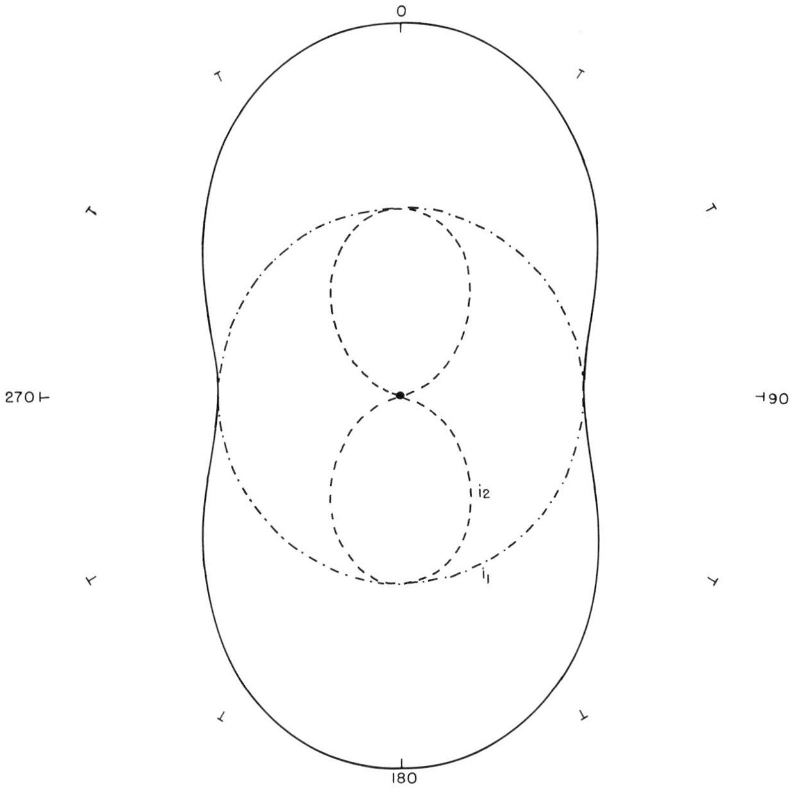


Figure 2.2 Polar scattering diagram or the phase function, for Rayleigh scattering: – · – · – ·, $i_1$; – – –, $i_2$; ———, $i_1 + i_2$.

$$i_2 = \frac{Nr^2}{2}\left(\frac{2\pi r}{\lambda}\right)^4\left(\frac{n^2-1}{n^2+2}\right)^2\cos^2\phi, \tag{2.7}$$

The intensity function $i_1$ is polarized in a plane perpendicular to the plane defined by the incident ray and the scattered ray, and $i_2$ is polarized in this latter plane (the plane of the paper in figure 2.2). The sum of the intensity functions is related to the radiant intensity $I$ (radiant flux per steradian) by

$$I = \frac{\lambda^2 E_0}{8\pi^2}(i_1 + i_2), \tag{2.8}$$

where $E_0$ is the incident irradiance.

Equations (2.6) and (2.7) and their sum are plotted in figure 2.2. The polarization ratio, $(i_1 - i_2)/(i_1 + i_2)$, is unity on the 90°–270° axis (complete polarization), and is zero on the 0°–180° axis. Note that the angular scattering pattern, known as the phase function, is symmetric about the

270°–90° axis. This means that half the scattered light is in the forward hemisphere and half in the backward hemisphere. In this sense, the Rayleigh phase function is symmetric about any axis between the 90°–270° axis and the 0°–180° axis.

As can be demonstrated with a polarization analyzer, skylight is partially but not completely polarized in a direction normal to the solar beam. This is because of the presence of non-Rayleigh scattering, multiple scattering, and the anisotropy of the air molecules, which are the principal Rayleigh scatterers. In agreement with the simple theory, , there is no polarization toward the antisolar point. However, there are often several other small regions in the sky, called neutral points, where there is also no polarization. The more common of these neutral points are the Arago, 15°–25° above the antisolar point, the Babinet, 15°–25° above the sun, and the Brewster, 15°–20° below the sun. The positions of these neutral points vary with the solar zenith angle and with atmospheric turbidity (due to the aerosol). Their explanation is uncertain in detail, but they must result from multiple scattering and non-Rayleigh scattering.

## 2.4 Scattering by Air Molecules

As already noted, the principal Rayleigh scatterers in the atmosphere are the air molecules themselves. The equations given earlier involve the radius of the scatterer and its index of refraction, quantities that are not readily definable for individual molecules. The approach to this problem lies in a consideration of the refractive index of air as due to an ensemble of molecules. A complete discussion of this subject is given by van de Hulst (1957), and only a brief outline will be given here.

Consider a plane electromagnetic wave passing through a layer of air. The permanent dipoles in the air molecules will be excited and will oscillate at the frequency of the incident wave. These oscillating dipoles will reradiate in all directions, and it is this reradiation at the imposed frequency that constitutes scattering. At any instant the molecules are randomly distributed in space; their spacing is large compared with their size; and their positions also change rapidly with time. For these reasons there is no interference between the wave packets scattered by the individual molecules, and the scattering is incoherent. However, there is an important exception to this statement exactly in the direction of propagation of the incident electromagnetic radiation The scattered radiation is initiated when the incident wave front reaches a molecule. Thus the incident wave imposes coherence on the scattered radiation in the direction of propagation and only in this direction. As a result, there is interference in this direction, which leads to a phase lag and hence to a reduction in the wave speed. By definition, the bulk index of refraction of the air (or of any medium) is the ratio of the speed of light in vacuo to the speed in the air.

This clearly makes it possible to relate the forward (0°) scattering to the index of refraction. In turn, the forward scattering is uniquely related to the scattering at all other angles. In this way, the scattering from an ensemble of molecules may be related to the gross or macroscopic index of refraction of the gas, a quantity that can be directly measured. The analytic expression of this relation is

$$(n_0^2 - 1) = 4\pi N_0 r^3 \frac{n^2 - 1}{n^2 + 2}, \tag{2.9}$$

where $n_0$ is the bulk index of refraction of the air and $N_0$ is the number of molecules in a unit volume of the air, both at standard pressure and temperature, and $r$ and $n$ are, respectively, the effective radius and index of refraction of a molecule. Substitution of (2.9) in (2.3) yields

$$k_s = \frac{8}{3} \frac{\pi^3 N}{N_0^2 \lambda^4} (n_0^2 - 1)^2, \tag{2.10}$$

where $N$ is the number of molecules per unit volume at an arbitrary pressure and temperature. This is essentially in the form derived by Rayleigh in his explanation of the blue of the sky. More recently it has been found that molecules are not isotropic scatterers. That is, the molecules and the dipoles they contain are not spherically symmetric. In an ensemble of molecules in thermal motion the effect of the anisotropy is to increase slightly the total light scattered and to modify the polarization so that there is no longer complete polarization on the 90°–270° axis. Equation (2.10) is then modified to

$$k_s = \frac{8\pi^3 N}{3\lambda^4 N_0^2} (n_0^2 - 1)^2 \frac{6 + 3\rho_n}{6 - 7\rho_n}, \tag{2.11}$$

where $\rho_n$ is the depolarization factor, which Hoyt (1977) sets at 0.014 for air. The correction term for the anisotropy is then 1.024.

The index of refraction of air in equations (2.9)–(2.11) is at the standard condition of 1013.3 mbar and 15°C. Numerical values are given in table 9.1 and range from 1.0002861 at 0.35 $\mu$m to 1.0002754 at 0.75 $\mu$m. $N_0$, the concentration of air molecules in a cubic centimeter of dry air, is $2.547 \times 10^{19}$. Equation (2.11) may be used to compute the transmittance of the atmosphere due to molecular scattering. Numerical results are presented in table 3.7.

## 2.5 Scattering by Large Particles

If the scatterers are very large compared with the wavelength $x$ greater than about 100, the scattering may be closely approximated by means of the simple ray-tracing techniques of geometric optics. Involved are external and internal reflections, refraction, and edge diffraction. The effects of internal absorption may also be included. Further reference to this case will be made in sections 2.10 and 2.11.

Many of the more interesting scattering processes result from particles that are neither small nor large compared with the wavelength, say $x = 1 - 20$. Further, in many regions of the spectrum the particles absorb as well as scatter. The complete and general solution to this problem for spherical particles was worked out by Gustav Mie (1908) to solve a problem in the scattering of light by gold sols.

This problem requires the solution of the vector wave equation in spherical coordinates for electromagnetic waves as specified by Maxwell's equations. Both interior (within the drop) and exterior solutions must be found and fitted at the interface. What is finally wanted is the external far-field solution. Since, in general, the particles absorb as well as scatter, Mie gave expressions for the scattering cross section, the absorption cross section, and the extinction cross section. These are given both as a function of the scattering angle (the phase function) and integrated over a sphere to yield the total cross sections. As a sample, the Mie equation for the extinction efficiency factor for spheres of index of refraction $n$ in a medium of unit index is

$$
\begin{aligned}
Q_{\text{ext}} &= \frac{2}{x^2} \sum_{u=1}^{\infty} (2u+1)\,\mathrm{Re}(a_u + b_u), \\
a_u &= \frac{\psi_u'(nx)\,\psi_u(x) - n\psi_u(nx)\,\psi_n'(x)}{\psi_u'(nx)\,\zeta_u(x) - n\psi_u(nx)\,\zeta_n'(x)}, \\
b_u &= \frac{n\psi_u'(nx)\,\psi_n(x) - \psi_u(nx)\,\psi_u'(x)}{n\psi_u'(nx)\,\zeta_n(x) - \psi_u(nx)\,\zeta'(x)}, \qquad (2.12) \\
\psi_u(x) &= (\pi x/2)^{1/2} J_{u+0.5}(x), \\
\zeta(x) &= (\pi x/2)^{1/2} [J_{u+0.5}(x) + i(-1)^u J_{-u-0.5}(x)],
\end{aligned}
$$

$J$ stands for the Bessel function of the first kind, and the primes denote first derivatives with respect to the argument. In general, the index of refraction is a complex number and Re in (2.12) stands for "real part of."

Equation (2.12) defies anything except numerical evaluation. The infinite series converges more slowly as $x$ increases, and a high-speed electronic computer is necessary for $x$ greater than 5–10. A number of asymptotic and approximate expressions have been derived that are useful in special circumstances. Many of these are discussed in van de Hulst's book. With the advent of the computer many machine solutions of the Mie equations

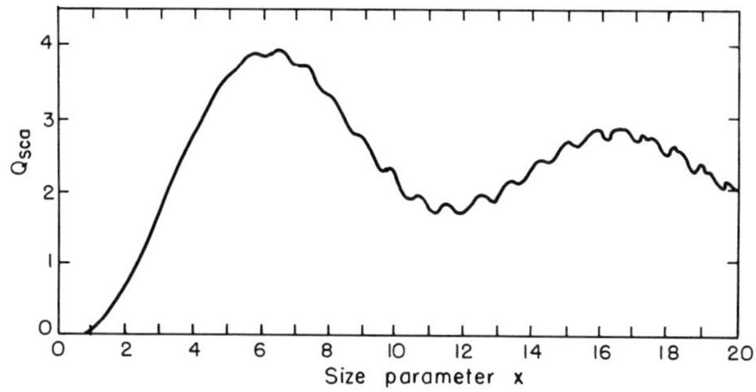


Figure 2.3 The scattering efficiency factor $Q_{sca}$ as a function of the size parameter, showing the ripples. After Penndorf (1956).

have been undertaken. Unfortunately most such tabulations have appeared in reports rather than in the open literature. Fairly complete results are now available for the scattering phase functions and efficiency factors for spheres of real indices of refraction of $\sim 1\text{–}2$ and for $x < \sim 125$. The situation is less satisfactory for absorbing spheres because an extremely large number of computations would be required to embrace the many possible combinations of the real and imaginary parts of the index of refraction.

An unexpected and bothersome feature of the numberical results of (2.12) is the appearance of "ripples" on the curves of $Q_{sca}$ versus $x$. This is illustrated by figure 2.3 taken from Penndorf (1956). These ripples are due to the optical interference between the edge rays (which graze the surface) and radiation from the surface waves on the sphere. The ripples become more pronounced as $x$ or $n$, or both, becomes larger. They are of no practical importance, since the slightest deviation from a monodisperse suspension will smooth them out. To obtain the mean curve, which is more useful, it is first necessary to compute for small enough increments of $x$ to define the ripples. This greatly increases the number of computations.

A smoothed curve for $n = 1.33$ is shown in figure 2.4. This is of particular interest in the atmosphere, since it corresponds to water drops in air in the visible spectrum. It has been shown by van de Hulst that curves for other indices of refraction, no larger than about 2.0, are very similar in general form to that presented here for index 1.33. This curve may be used for another index of refraction reasonably well by multiplying the scale of abscissas by $1/3(n-1)$; thus the first maximum occurs at $x = 4$ for $n = 1.5$.

The Rayleigh region is included, but is lost in the extreme lower corner, where $x$ is less than about 0.6. If $r$ is considered constant, so that the variation with $\lambda$ can be investigated, the abscissas become an inverse wavelength (or a frequency) scale. In the region to the left of the first maximum, $Q_{sca}$ increases with decreasing wavelength, although at a lesser

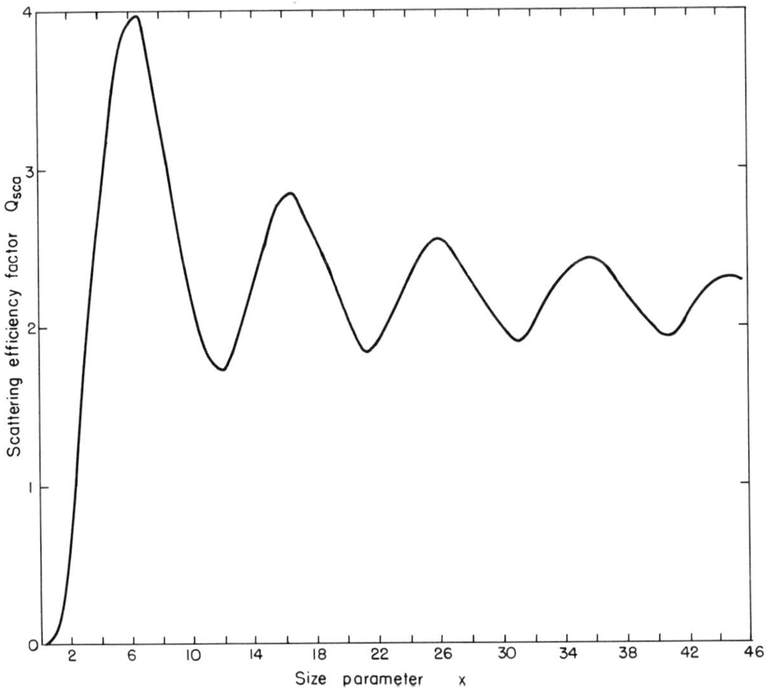


Figure 2.4 Smoothed curve of the scattering efficiency factor as a function of the size parameter. Drawn from data given by Penndorf (1956).

rate than the Rayleigh inverse fourth-power law. For visible light, particles of radius about 0.05–0.5 $\mu$m fall in this region, and the scattered light will be blue and the transmitted red, but not as intensely as for Rayleigh scattering. The particle size that yields the maximum scattering efficiency factor for a given mass of suspended material will be found to correspond to a value of $x$ just short of the first maximum. This was taken into account by Irving Langmuir in the design of the oil-fog generators used for screening smokes in World War II. Between $x = 6$ and $x = 12$, the slope of the curve is reversed. In this region, the scattered light would be red and the transmitted light blue. The rarely seen phenomenon of a blue moon has been explained on this basis. There are similar oscillations of decreasing amplitude for larger values of $x$. These are of little practical significance because the percentage range of $x$ and hence of $\lambda$ for these regions of positive and negative slope is so small compared with the wavelength range of visible light. Further, the amplitudes of the variations of $Q_{sca}$ are small. Finally, it is most unusual for all of the particles in any natural suspension to be of the same size (monodisperse). A range of particle sizes will evidently smooth the oscillations of $Q_{sca}$ and render them indistinguishable.

## 2.6 Physical Interpretation of Mie Scattering

The infinite series in (2.12) may be considered as an expansion of the scattered light in terms of the electric and magnetic dipoles, quadripoles, octopoles,.... Thus the coefficients $a_1, a_2, \ldots, a_u$ represent the contributions of the electric multipoles, while the coefficients $b_1, b_2, \ldots, b_u$ similarly represent the effects of the magnetic multipoles. Rayleigh scattering is represented by the coefficient $a_1$ for the electric dipole.

The major maxima and minima may be considered to be the interference pattern resulting from the radiation diffracted by the edge of the sphere and that transmitted through the sphere. These two components contain most of the scattered radiation. The phase lag of a ray passing through a diameter of the sphere is easily shown to be $2 \times (n - 1)$ rad. The maxima or the minima occur at intervals of $2\pi$ and hence for an index of refraction of 4/3, at intervals in $x$ of $3\pi$, or about 9.4. This corresponds well, but not exactly, with figure 2.4. Note that this phase lag leads to the abscissa scaling factor for figure 2.4 referred to earlier. The physical interpretation given here is based in part on concepts from wave and geometric optics and therefore cannot lead to exact correspondence with the results of the Mie equation.

## 2.7 Numerical Values of $Q_{sca}$

Smoothed values of $Q_{sca}$ are given in table 2.2, which also contains $Q_{sca}$ from the Rayleigh equation for $x$ from 0.1 to 1.0.

Penndorf (1956) has provided similar tables for $n = 1.40$, 1.44, 1.486, and 1.50. Even at $x = 120$, $Q_{sca}$ is still significantly different from the limiting value of 2 for very large $x$. The first part of table 2.2 shows the divergence of the Rayleigh solution from the Mie as $x$ increases. For a path transmissivity of 0.8 and $x = 0.6$, the use of the Rayleigh expression leads to an underestimate of the transmissivity of about 1%. This error increases with $x$ and with decreased transmissivity and is 12% for $x = 1$ and a transmissivity of 0.5.

As $x$ becomes very large, $Q_{sca}$ approaches two. This is the realm of geometric optics where the problem may be discussed in the more familiar terms of reflection, refraction, and diffraction. From this point of view it is, at first, surprising that $Q_{sca}$ is two rather than unity. It can easily be seen that, except for the infinitesimal fraction that traverses a sphere along the 0°–180° axis, all of the radiation geometrically intercepted suffers a change in direction by reflection and refraction and is therefore scattered, thus explaining half of the predicted magnitude of $Q_{sca}$. The other half is the radiation diffracted by the "edge" of the drop. Babinet's principle states that a circular opaque disk forms the same diffraction pattern (except for phase) as a hole of the same radius in an opaque screen. This can be

Table 2.2
Smoothed values of $Q_{sca}$ versus $x$ for $n = 1.33$[a]

| $x$ | $Q_{sca}$ Rayleigh | $Q_{sca}$ Mie |
|---|---|---|
| 0.1 | $1.110 \times 10^{-5}$ | $1.109 \times 10^{-5}$ |
| 0.2 | $1.776 \times 10^{-4}$ | $1.770 \times 10^{-4}$ |
| 0.3 | $8.990 \times 10^{-4}$ | $8.925 \times 10^{-4}$ |
| 0.4 | $2.841 \times 10^{-3}$ | $2.802 \times 10^{-3}$ |
| 0.5 | $6.937 \times 10^{-3}$ | $6.773 \times 10^{-3}$ |
| 0.6 | $1.438 \times 10^{-2}$ | $1.385 \times 10^{-2}$ |
| 0.7 | $2.665 \times 10^{-2}$ | $2.515 \times 10^{-2}$ |
| 0.8 | $4.546 \times 10^{-2}$ | $4.176 \times 10^{-2}$ |
| 0.9 | $7.282 \times 10^{-2}$ | $6.454 \times 10^{-2}$ |
| 1.0 | $11.10 \times 10^{-2}$ | $9.392 \times 10^{-2}$ |

| $x$ | $Q_{sca}$ | $x$ | $Q_{sca}$ | $x$ | $Q_{sca}$ | $x$ | $Q_{sca}$ |
|---|---|---|---|---|---|---|---|
| 1.5 | 0.32 | 10.5 | 1.95 | 19.5 | 2.13 | 28.5 | 2.20 |
| 2.0 | 0.71 | 11.0 | 1.82 | 20.0 | 2.03 | 29.0 | 2.12 |
| 2.5 | 1.21 | 11.5 | 1.74 | 20.5 | 1.95 | 29.5 | 2.05 |
| 3.0 | 1.75 | 12.0 | 1.73 | 21.0 | 1.88 | 30.0 | 1.98 |
| 3.5 | 2.27 | 12.5 | 1.84 | 21.5 | 1.85 | 31.0 | 1.91 |
| 4.0 | 2.82 | 13.0 | 1.97 | 22.0 | 1.91 | 32.0 | 1.99 |
| 4.5 | 3.22 | 13.5 | 2.14 | 22.5 | 2.00 | 33.0 | 2.19 |
| 5.0 | 3.59 | 14.0 | 2.31 | 23.0 | 2.10 | 34.0 | 2.34 |
| 5.5 | 3.83 | 14.5 | 2.49 | 23.5 | 2.21 | 35.0 | 2.24 |
| 6.0 | 3.92 | 15.0 | 2.64 | 24.0 | 2.31 | 36.0 | 2.42 |
| 6.5 | 3.98 | 15.5 | 2.75 | 24.5 | 2.41 | 37.0 | 2.34 |
| 7.0 | 3.81 | 16.0 | 2.83 | 25.0 | 2.48 | 38.0 | 2.20 |
| 7.5 | 3.59 | 16.5 | 2.84 | 25.5 | 2.54 | 39.0 | 2.07 |
| 8.0 | 3.32 | 17.0 | 2.76 | 26.0 | 2.56 | 40.0 | 1.96 |
| 8.5 | 3.02 | 17.5 | 2.68 | 26.5 | 2.52 | 41.0 | 1.93 |
| 9.0 | 2.70 | 18.0 | 2.57 | 27.0 | 2.46 | 42.0 | 2.08 |
| 9.5 | 2.38 | 18.5 | 2.43 | 27.5 | 2.40 | 43.0 | 2.22 |
| 10.0 | 2.14 | 19.0 | 2.26 | 28.0 | 2.30 | 44.0 | 2.29 |
| | | | | | | 45.0 | 2.32 |

a. From Penndorf (1956).

Table 2.3
Ratio of apparent $Q_{sca}$ to Mie $Q_{sca}$ as a function of $x$

| | Half-angle of acceptance cone | | | | |
|---|---|---|---|---|---|
| $x$ | 0.5° | 1.0° | 2° | 3° | 5° |
| 20 | 0.996 | 0.983 | 0.936 | 0.873 | 0.744 |
| 30 | 0.990 | 0.966 | 0.881 | 0.782 | 0.672 |
| 40 | 0.984 | 0.942 | 0.815 | 0.707 | 0.658 |
| 60 | 0.967 | 0.881 | 0.696 | 0.630 | — |
| 80 | 0.944 | 0.812 | 0.606 | 0.588 | — |
| 100 | 0.910 | 0.728 | — | — | — |
| 200 | 0.750 | — | — | — | — |

experimentally verified. Fraunhofer diffraction theory shows that all rays passing through a hole, except for the axial ray, are deviated and hence scattered. It therefore follows that the diffraction cross section of a drop is equal to the geometric cross section. The edge-diffracted light forms the corona discussed in section 9.13. As indicated there, the diffraction pattern of a drop is not quite identical to that of an opaque disk, but these differences are unimportant here when all that is needed is the total diffraction cross section.

Most of the edge-diffracted light is contained in the maximum centered on the forward direction ($\phi = 0$). The angular width of this maximum is defined by the first minimum. This has certain practical consequences. All real instruments, such as the normal incidence solar pyrheliometer and the transmissometer used to measure the visibility, have a finite angle of view, or acceptance cone. Such instruments will accept as part of the undeviated beam that portion of the scattered radiation within their acceptance cones.

The magnitude of this effect can be expressed as the ratio of the apparent value of the scattering efficiency factor to the value derived from the Mie equations. This ratio is evidently dependent on the size parameter $x$. Table 2.3, taken from Gumprecht and Sliepcevich (1953), gives values of the ratio as a function of $x$ for several half angles of the acceptance cone.

The size parameters of the optically important aerosol particles range from roughly 1 to 20 in the solar spectrum. Thus an instrument with a half-angle of about 2° or less should exclude most of the scattered light. In a fog or cloud the size parameters are typically 50–200, and even the best-collimated instrument will accept a substantial portion of the forescattered light.

Also note that at the large-size parameters typical of fog or cloud drops, $Q_{sca}$ is about two and is nearly independent of $x$. Thus both the scattered

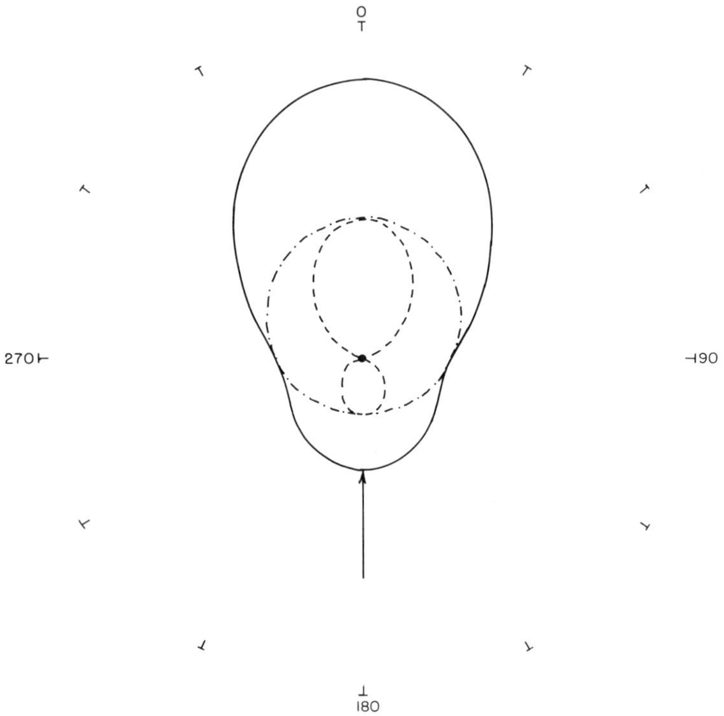


Figure 2.5A Phase function for spheres of refractive index 1.33 and size parameter 1.00: –·–·–·, $i_1$; – – –, $i_2$; ———, $i_1 + i_2$.

and transmitted light in fog or clouds are white, as can be verified by noting the white appearance of clouds and of the sun seen through a thin cloud. Colored "fog lights" will not penetrate fog better than white lights, and in fact, the usual colored lens simply reduces the light. Colored signal lights may be advantageous simply because of their color contrast against a general background of white light, but this is another matter.

## 2.8 Phase Functions for Mie Scattering

As illustrated in 2.2 Rayleigh scattering is symmetric about a plane normal to the incident radiation; that is, the radiation scattered into the forward hemisphere is a mirror image of that scattered into the backward hemisphere. As $x$ increases into the regime of Mie scattering, the angular scattering pattern rapidly becomes asymmetric, in such a way that more radiation is scattered into the forward than into the backward hemisphere. The forward lobe of the scattering pattern also becomes narrower. This behavior is illustrated in figures 2.5A and 2.5B, in which the patterns for

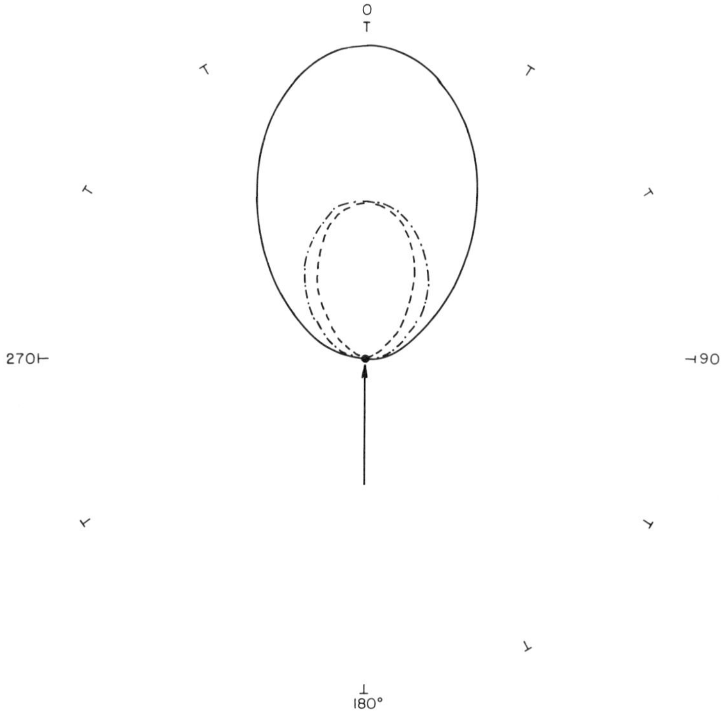

Figure 2.5B As in figure 2.5A, except that the size parameter is 2.00.

$x = 1$ and $x = 2$, for an index of 1.33, are sketched. (Note that the scales of the two figures are not the same.) It is also apparent that the polarization ratio is smaller for $x = 2$ than for $x = 1$, at least in the forward lobe. Although it cannot be seen in the figure, the polarization is not quite complete at 90° for $x = 1$, and this is even more marked for $x = 2$.

It might be anticipated that the trends illustrated in figure 2.5 would continue as $x$ becomes larger. This is true only to a limited extent; the forward lobe does continue to become narrower with increasing $x$, but the backward lobe does not continue to decrease in comparison with the forward lobe. The most striking change is the fine structure that appears at larger values of $x$. This is illustrated in a different type of plot in figure 2.6, which shows the complex situation at $x = 10$ more clearly than a polar plot. The rapid fluctuations of $i_1$ and $i_2$ with the scattering angle lead to a similarly varying polarization ratio. One might have anticipated from the results for $x = 1$ and $x = 2$ that the polarization ratio would simply approach zero at all angles as $x$ becomes large, but this is not the case. The fine structure varies very rapidly with $x$, so that it can be observed experimentally only in a highly monodisperse cloud. La Mer was able to demon-

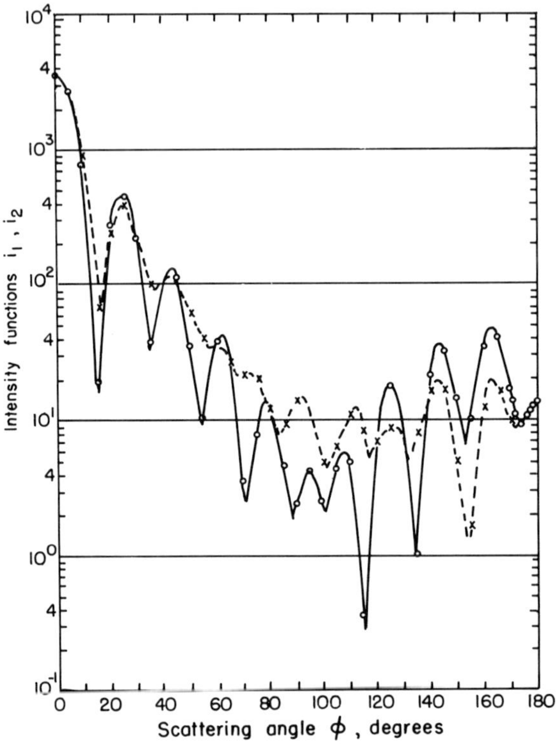


Figure 2.6 Scattering phase function for an index of refraction of 1.33 and size parameter 10. After Penndorf (1961): ○——○, $i_1$, × × ×, $i_2$.

strate these maxima and minima in some beautiful experiments, and he called the effect the higher-order Tyndall spectrum. The effects cannot be of significance in any natural aerosols because they are so poly-disperse.

Of more significance is the ratio of the scattering in the forward hemisphere to that in the backward hemisphere. This ratio, together with the scattering efficiency factors for the forward and backward hemispheres, is shown in figure 2.7. These results were obtained by the numerical integration of the machine computations of Penndorf[2] of $i_1$ and $i_2$ for every 5°. Integrations were performed only for intervals of 0.5 in $x$ from $x = 0$ to $x = 5$ and for integral values of $x$ from 5 to 16. These intervals are not small enough to resolve the "ripples" referred to earlier, and the plotted results have been smoothed as well as possible with the aid of the smoothed values of the total scattering area coefficients of table 2.2. Notable is the extremely rapid rise of the ratio of fore- to backscattering beginning near $x = 1$ and

2. R. B. Penndorf kindly loaned me his computer printouts of $i_1$ and $i_2$.

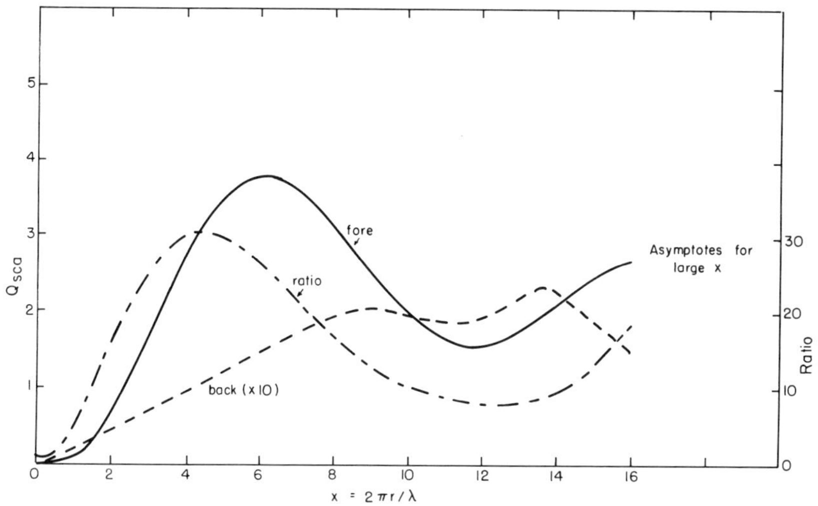


Figure 2.7 Scattering efficiency factors for the forward and backward hemispheres and their ratio: the smoothed index of refraction = 1.33; ———, forward hemisphere; – – –, backward hemisphere ( × 10); – · – · – ·, ratio of forward to backward.

reaching a maximum near $x = 4$. Also interesting is the rather broad minimum centered on $x = 12$. These results have a bearing on the contribution of the upward scattering by the atmospheric aerosol to the planetary albedo.

The optically important portion of the aerosol spectrum is in the radius range from 0.1 to 1.0 $\mu$m. For a wavelength of 0.5 $\mu$m, the corresponding range of $x$ is about 1–12. This encompasses the maximum of the fore to back ratio and a portion of the minimum shown in figure 2.7. It should be noted that figure 2.7 is for an index of refraction of 1.33 and that some shift of the curves can be expected for another index, such as 1.5, often used for the aerosol.

The fore- and back scattering in figure 2.7 are defined with respect to the axis of the incident solar beam. In studies of the albedo we want the up- and downscatter with reference to the local vertical. The ratio of down- to upscatter will evidently depend on the solar zenith angle as well as on $x$. Figure 2.8 shows this ratio as a function of solar zenith angle for a few values of $x$, including the geometric optics case, $x \gg 1$. Note that the ratio converges to one at $\phi = 90°$ in all cases.

## 2.9 The Asymmetry Parameter

The asymmetry of the Mie phase functions can be expressed in terms of a single number, called the asymmetry parameter. This is defined as

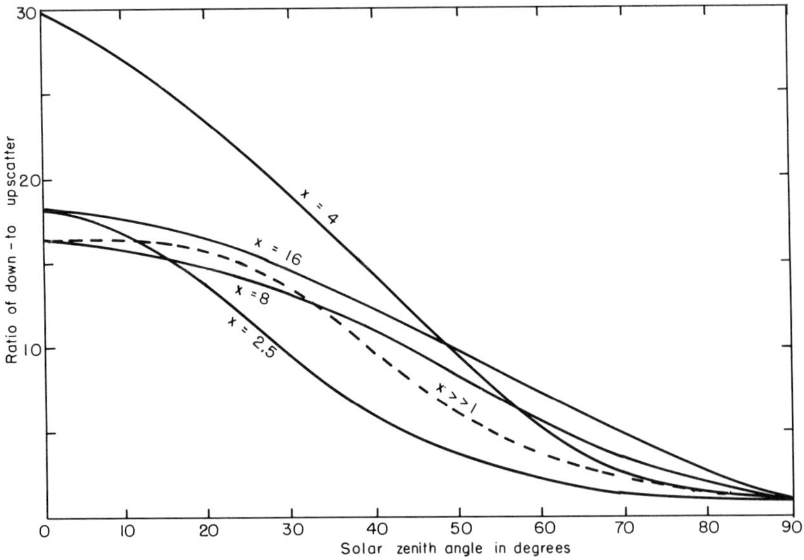


Figure 2.8 Ratio of forward to backward scatter as a function of the solar zenith angle for selected values of the size parameter.

$$<\cos\phi> = \frac{1}{4\pi}\int_{4\pi} P\cos\phi\, d\Omega, \tag{2.13}$$

where $\phi$ is the scattering angle measured from the forward direction, $\Omega$ is an element of solid angle, and $P$ is the phase function, normalized by

$$\frac{1}{4\pi}\int_{4\pi} P\, d\Omega = 1. \tag{2.14}$$

$<\cos\phi>$ may also be expressed in an infinite series as a function of the coefficients $a_u$ and $b_u$ of (2.12).

The asymmetry parameter varies between 1 and $-1$. It is zero for isotropic and Rayleigh scattering and increases with $x$, denoting increasing forward scattering. In the limit for large $x$, $<\cos\phi>$ is about 0.87. For scatterng of solar radiation by cloud drops, $<\cos\phi>$ varies only from about 0.80 to 0.86 for realistic drop size spectra. A negative value of the parameter means that the backward scattering lobe is larger than the forward scattering lobe. The asymmetry parameter for a real index of refraction of 1.33 is shown in figure 2.9 after Hansen and Travis (1974). The ripples have been smoothed out to emphasize the general course of $<\cos\phi>$. Note that the positions of the first maximum and the first minimum are about the same as those of the curve of hemispheric fore-

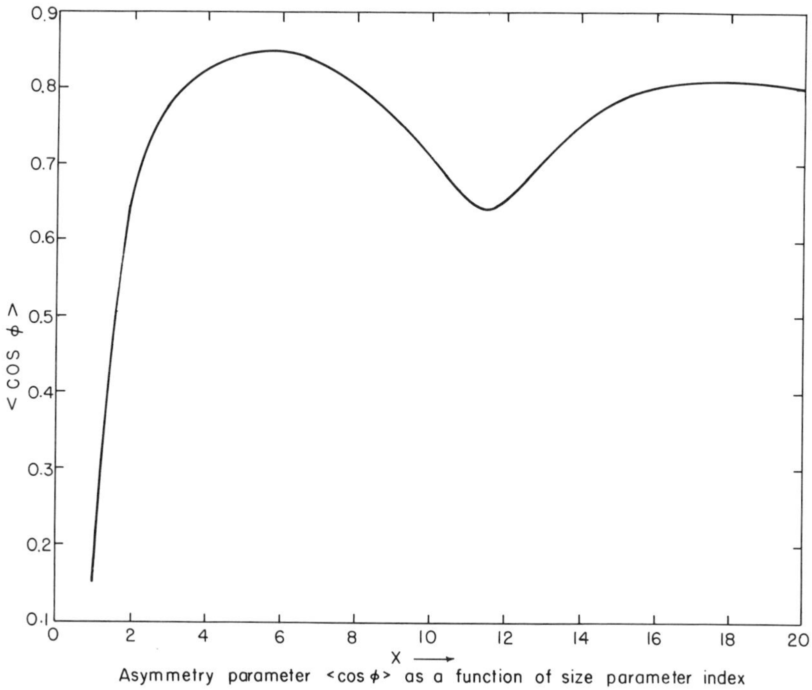


Figure 2.9 The asymmetry parameter $\langle \cos \phi \rangle$ for an index of refraction of 1.33. After Hansen and Travis (1974).

scatter in figure 2.7. This should come as no surprise, since both curves are measures of the asymmetry of the scattering.

## 2.10 Scattering by Large Spheres

In the geometric optics case the scattering phase function may be computed by ray tracing more easily than from the Mie equations. Rather than show this pattern graphically, the data are given in table 2.4 for 10° increments. The values have been normalized, so that $(i_1 + i_2) = 1.0$ at 0° and are therefore in relative rather than absolute units. The relatively large value at 140° represents the rainbow, which cannot be resolved by 10° increments. See section 9.7 for the theory of the rainbow.

## 2.11 Absorbing Scatterers

Absorption and scattering may occur simultaneously in two ways. First, nonabsorbing scatterers may be imbedded in an absorbing medium. The second and more complex situation is when the scatterers are also absorbers.

Table 2.4
Angular distribution of large drop scattering[a]

| Degrees | $i_1$ | $i_2$ | $i_1 + i_2$ | Degrees | $i_1$ | $i_2$ | $i_1 + i_2$ |
|---|---|---|---|---|---|---|---|
| 0 | 0.500 | 0.500 | 1.00 | 100 | 0.0014 | 0.0002 | 0.0016 |
| 10 | 0.426 | 0.428 | 0.854 | 110 | 0.0012 | 0.0003 | 0.0015 |
| 20 | 0.261 | 0.271 | 0.532 | 120 | 0.0017 | 0.0009 | 0.0026 |
| 30 | 0.137 | 0.149 | 0.286 | 130 | 0.0008 | 0.0005 | 0.0013 |
| 40 | 0.066 | 0.078 | 0.144 | 140 | 0.0313 | 0.0033 | 0.0348 |
| 50 | 0.030 | 0.039 | 0.069 | 150 | 0.0088 | 0.0061 | 0.0149 |
| 60 | 0.012 | 0.015 | 0.027 | 160 | 0.0041 | 0.0035 | 0.0076 |
| 70 | 0.0042 | 0.0034 | 0.0076 | 170 | 0.0032 | 0.0029 | 0.0061 |
| 80 | 0.0021 | 0 | 0.0021 | 180 | 0.0030 | 0.0030 | 0.0060 |
| 90 | 0.0016 | 0.0001 | 0.0017 | | | | |

a. Index of refraction = 1.33.

It is easy to visualize what happens when the scatterers are very large compared with the wavelength. Some of the radiation intercepted by a drop passes through the interior of the drop and suffers refraction and internal reflection. If the drop is nonabsorbing, all of this radiation emerges from the drop as part of the scattered radiation. If the drop is an absorber, some of the radiation that enters the drop is absorbed and does not leave as scattered radiation. Thus the total radiation scattered and absorbed is the same as the radiation scattered by a nonabsorbing drop of the same size; the greater the absorption, the less is the scattering. In the limit for very large $x$ the extinction efficiency factor always approaches two. However, for an absorbing sphere, the limiting value of the scattering efficiency factor is a function of the complex index of refraction. As formally proved by Chylek (1975), this limiting $Q_{\mathrm{sca}}$ is given by

$$Q_{\mathrm{sca}}(x \to \infty) = 1 + \left| \frac{m-1}{m+1} \right|^2, \tag{2.15}$$

where $m$ is the complex of refraction.

It is generally true that the extinction coefficient is the sum of the scattering and absorption coefficients, but in the Mie region the scattering and absorption efficiency factors must be computed from the relevant Mie equations. An absorbing sphere has a complex index of refraction of the form $n - in'$, where $n$ is the ordinary optical index of refraction and $n'$ is related to the absorption coefficient of the substance of which the particle

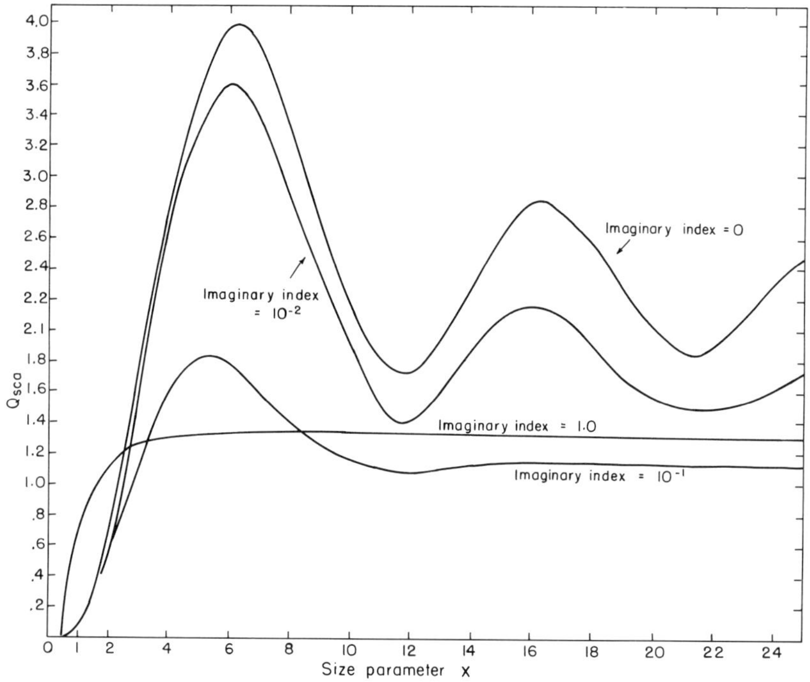


Figure 2.10 Scattering efficiency factor as a function of the size parameter for spheres of real index of refraction of 1.33 and for imaginary indices of refraction as marked. After Plass (1966).

is composed by

$$n' = k_a \lambda / 4\pi, \qquad (2.16)$$

where $k_a$ is the absorption coefficient of the material in $cm^{-1}$ at the wavelength $\lambda$ in cm. The absorption coefficients of most natural substances vary rather rapidly with $\lambda$. For this reason the extinction efficiency factor or any of the other Mie results must be computed separately for each wavelength, using the appropriate value of $k_a$.

Plass (1966) gives the absorption and scattering efficiencies for several values of the imaginary part of the index of refraction. Some of his results for the real part of the index equal to 1.33 are given in figures 2.10 and 2.11.[3] The upper curve in figure 2.10 is for the nonabsorbing case and is

---

3. In the interest of clarity, the ripples have been omitted from the curves in figure 2.9 for $n' = 0$ and $10^{-2}$. The ripples do not appear for $n' > 10^{-1}$.

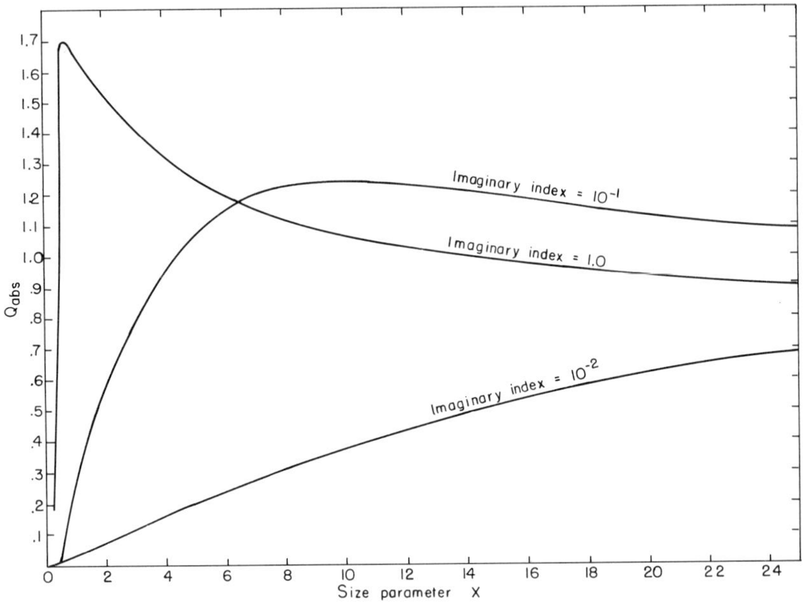


Figure 2.11 Absorption efficiency factor as a function of the size parameter for spheres of real index of refraction 1.33 and for imaginary indices of refraction as marked. After Plass (1966).

therefore the same as figure 2.4. As $n'$ increases the scattering efficiency factor decreases and the major oscillations are damped and nearly disappear at $n' = 1.0$.[4]

## 2.12 Absorbing Rayleigh Scatterers

As shown by van de Hulst (1957), (2.5) applies also to the more general case in which the particles absorb as well as scatter. The absorption efficiency factor is

$$Q_{\text{abs}} = -4x \,\text{Im}\frac{m^2 - 1}{m^2 + 2}, \tag{2.17}$$

where Im stands for, "the imaginary part of" and $m$ is the complex index of refraction. Since $Q_{\text{sca}}$ varies with $x^4$, while $Q_{\text{abs}}$ is linear in $x$, the absorption cross section will become large compared with the scattering cross section for sufficiently small values of $x$. As examples, for a real index of 1.33 and

4. Some workers express the absorption in terms of the single scattering albedo, which is the ratio of the absorption coefficient to the extinction coefficient.

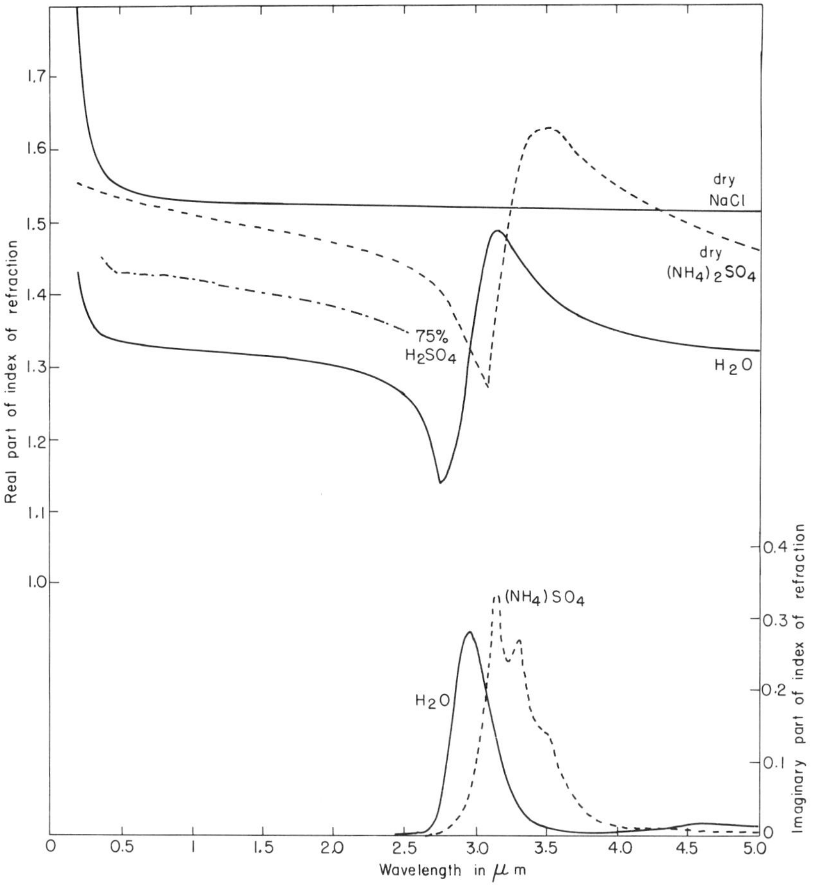


Figure 2.12 Real and imaginary parts of the indices of refraction of several substances in the solar spectrum. The imaginary parts of the indices of sulfuric acid and dry sodium chloride are too small to show on the scale. After Downing and Williams (1975) for water, data of Palmer and Williams (1975) are used for sulfuric acid; the real indices of the dry salts are taken from Toon et al. (1976).

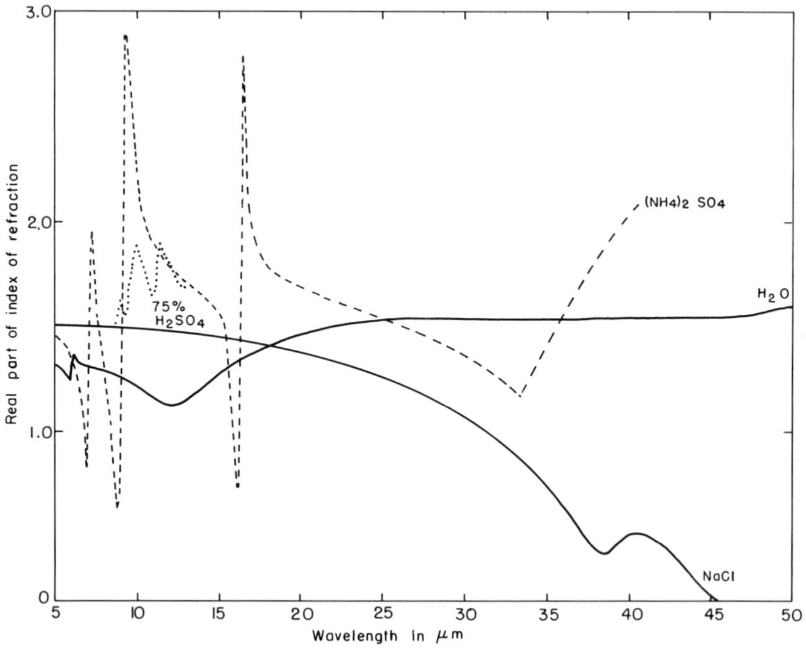


Figure 2.13A Real parts of the indices of refraction of the same substances as in figure 2.12 but in the thermal radiation spectrum: ———, $H_2O$; · · ·, 75% $H_2SO_4$; – – –, dry $(NH_4)_2SO_4$; ———, dry NaCl. Downing and Williams (1975) are the source of the real indices of water; the indices for sulfuric acid are taken from Remsberg (1973); the real indices for the two dry salts were taken from the work of Toon et al. (1976).

an imaginary index of $10^{-3}$, $Q_{abs}$ and $Q_{sca}$ are equal at $x = 0.27$; for the same real index and an imaginary index of $10^{-2}$, the two efficiency factors are equal at $x = 0.59$. Since these values of $x$ are near or above the upper limit for Rayleigh scattering, absorption will dominate the extinction cross section of most natural particles that are small enough to fall in the Rayleigh domain.

## 2.13 Complex Indices of Refraction of the Aerosol

The atmospheric aerosol scatters and absorbs both solar radiation and thermal radiation. The quantitative effects of the aerosol on radiative transfer and the radiation budget of the earth-atmosphere system depend on knowledge of the absorptive and scattering properties of the aerosol particles. The typical size parameters of aerosol particles fall in the Mie region, and computations from the Mie equation depend on a knowledge of the complex indices of refraction of the particles.

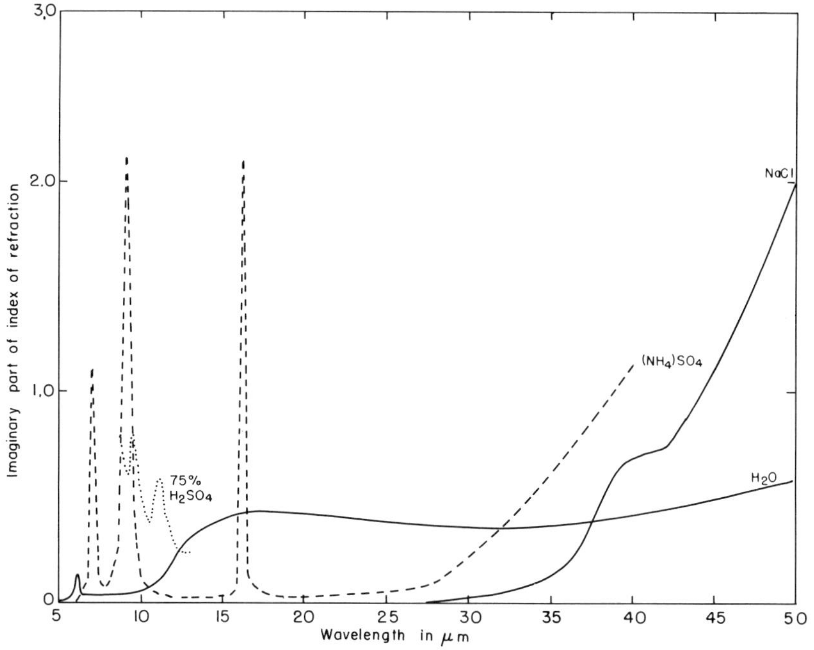


Figure 2.13B As in figure 2.13A, except that the imaginary parts of the indices of refraction are given where available.

One approach to evaluating the complex indices of refraction of aerosols is to measure the indices of pure substances that are known to be common constituents of the natural aerosol. Examples are given in figure 2.12, 2.13A, and 2.13B, in which the complex indices of $H_2O$, NaCl, $(NH_4)_2SO_4$, and $H_2SO_4$ are given as a function of wavelength in both the solar spectrum and in the thermal radiation spectrum. Since there is very little solar radiation beyond 2.5 $\mu$m, there is little absorption of solar radiation by these species. The situation in the thermal radiation spectrum is quite complex, reflecting the similar complexity of the far-infrared absorption spectra. The plotted data for NaCl and $(NH_4)_2SO_4$ are for dry salts; if they were in aqueous solution, some of the features of the curve for water would be apparent, but curves for aqueous solutions cannot be obtained by simple combinations of those for dry salt and water.

A few measurements of the complex indices of the natural aerosol have been reported, and many more will undoubtedly be made. The available data do not permit the construction of spectral curves like those of figure 2.12 and 2.13, and the great variability in the composition of the natural aerosol would make a single spectral curve of doubtful value. All that seems justified now is to note that the imaginary part of the index seems to range from about 0.005 to 0.02 in the solar spectrum and from about 0.1 to 0.5 in the thermal infrared. One notable exception is a value of about 0.5 for carbonaceous aerosols (for example, soot) in both the solar and thermal infrared regions. The real part of the indices seem to range mostly from 1.5 to 1.6 in the solar region and 1.6 to 1.8 in the thermal infrared, but data are sparse.

## 2.14 Small Reflecting Spheres

If the spheres are perfectly reflecting, no radiation enters the sphere, and consequently there is no absorption. The phase function for this case has a large backward lobe and a small forward one. It might be thought that metallic spheres would be nearly perfectly reflecting, but this is true only in the infrared and microwave regions. Small metallic spheres typically absorb more than they scatter in the visible spectrum. Mie undertook his fundamental theoretical solution in seeking an explanation of the colors exhibited by various gold sols; the transmitted light was red, purple or blue, depending on the size of the particles. He found that these colors resulted from large variations of the indices of refraction with wavelength rather than from just the changes in particle size.

In the microwave region at typical wavelengths of the order of 10 cm, raindrops are nearly perfect reflectors. The back-scattered radiation ($\phi = 180°$) is therefore a maximum. Radars used for detecting precipitation are much more effective than would be the case if the phase function had the large forward lobe typical of the visible spectrum.

## 2.15 Multiple Scattering

In a deep scattering layer, such as most clouds and the typical atmospheric aerosol, radiation is scattered more than once. Multiple scattering so rapidly changes the angular distribution of the scattered light and the radiative transfer that they can no longer be adequately described by single scattering. It has been suggested by van de Hulst that single scattering is adequate when the optical depth[5] of the scattering layer is 0.1 or less, that second-order scattering is necessary for optical depths of 0.1–0.3, and that a more general treatment of multiple scattering is necessary for larger optical depths. Single-scattering computations may be adequate for somewhat larger optical depths when attention is focused on the transmitted beam, but not when the properties of the scattered radiation are the subject of enquiry. For example, it has been found that single-scattering

5. The optical depth is the product of the extinction coefficient and the path length or the natural logarithm of the reciprocal of the transmissivity over the path.

models are reasonably satisfactory when used to estimate the solar radiation at the surface at moderate values of the solar zenith angle.

It is easy enough to understand the physics of multiple scattering and the computational problems that it brings. The first scatter distributes the scattered light in accordance with the scattering phase function. The second scatter again redistributes the scattered radiation as dictated by the phase function. At the same time some of the incident radiation remains unscattered. If the scattering is isotropic, it is possible to treat the process analytically, since the scattered radiation remains isotropic regardless of the order of the scattering and the transfer of isotropic radiation is analogous to molecular diffusion. In the more general anisotropic case no such simplification is possible. One can imagine setting up a procedure in the typical strong forward lobe case in which the continuous phase function is represented by a finite number of angular intervals. This can be carried through relatively easily to the second scatter, but rapidly becomes ponderous as the number of scatters increases.

Over the years a number of methods have been developed for computing radiative transfer by multiple scattering; this development has recently become more rapid due to the availability of high-speed digital computers. The first model, now called the two-stream model, was introduced by Schuster (1905) for an astrophysical problem, but has since been used to compute terrestrial cloud albedos. The basic assumption of this method is that both the upward and the downward fluxes are diffuse but, in general, not equal. This permits limited use of the phase function to specify the fractions of the scattered radiation in the up and down directions. If $A$ is the upward flux and $B$ the downward,

$$dA = -k_s\beta(A - B)\,dz, \qquad dB = -k_s\beta(A - B)\,dz, \tag{2.18}$$

where $k_s$ is the scattering coefficient, $\beta$ is the fraction of the light that is scattered into the backward hemisphere, and $z$ is height, here taken as positive down. When applied to clouds in the solar spectrum, the fraction $\beta$ is usually taken from the phase function in the large-size parameter region, where it does not vary with wavelength or droplet radius. Reflection from the underlying surface and also absorption may be included in (2.18). Other two-stream models have been given by Coakley and Chylek (1975), including one that shows the variation of cloud albedo with solar zenith angle.

Of the several more accurate models of radiative transfer with multiple scattering, only two of the more commonly used methods will be briefly described here. The algebraic and computational problems are sufficiently complex to be of interest primarily to those who wish to make such computations. Those who desire a more complete treatment are referred to the review paper by Hansen and Travis (1974), which also contains an extensive list of references.

## 2.16 The Doubling Method

This is also called the adding method. The basic thesis of this method is that if the reflectivity and the transmissivity are known for each of two layers, the reflectivity and transmissivity of the combined layer may be obtained by computing the successive reflections back and forth between the two superimposed layers. Usually not more than five iterations are necessary. If the two layers are identical, the transmissivity and reflectivity of a thick homogeneous layer may be calculated quickly in a geometric or doubling fashion. The initial layers may be thin enough so that their properties can be quickly deduced from single scattering. Alternatively, somewhat thicker layers may be used if their optical properties are known from other methods, such as a model incorporating secondary scattering. Various simplifications are often introduced, the most common being the replacement of the continuous-scattering phase function by a finite number of point values.

The doubling method has been found to yield results that closely duplicate those obtained by other methods. It may be made as accurate as one wishes, subject only to limitations on computer time. It is generally limited to plane-parallel atmospheres and is not as flexible in dealing with variable optical properties as the Monte Carlo method.

## 2.17 The Monte Carlo Method

In this method, multiple scattering is considered to be a stochastic process. One follows large number of photons injected at the top of the scattering layer as they progress through the layer. The phase function is, in effect, the probability density function for scattering at a given angle. Absorption in the gas and in the droplets may be readily included, as well as changes with height of the scatterers or other properties. It is the only method in which it would be relatively easy to consider non-plane-parallel geometry. The disadvantage of the method is the large number of photons that must be followed to obtain stable statistics. The statistical fluctuations decrease in magnitude only with the square root of the total number of photons introduced. Various biasing procedures have been introduced to avoid having too many photons contributing to an easily defined part of the radiation field; biasing may introduce errors unless great care is used.

## 2.18 Some Further Comments on Multiple Scattering

Mention should be made of the method developed by Chandrasekhar (1960). This involves integrations of the equation of radiative transfer with optical depth. The algebra and the computational problems are complicated, but Coulson et al. (1960) have published extensive tables that simplify the computations. Unfortunately it has not been possible to apply this method to asymmetric scattering, but only to isotropic or Rayleigh scattering. It has led to comprehensive results of the angular distribution and polarization of skylight in a Rayleigh atmosphere.

The methods just discussed briefly, the Monte Carlo method, in particular, seem capable of describing radiative transfer in a multiple-scattering atmosphere to any desired accuracy, given the microoptical properties of the scattering layer. Some of the latter are poorly known, and all are observed only during research flights by balloons and aircraft carrying special instrumentation. As things stand, if it is desired to establish statistics on the reflectivity and absorptivity of clouds and aerosol layers, it is easier and more accurate to measure these "whole cloud" properties directly than to measure the microphysical optical properties for insertion in one of the models discussed here. Nevertheless, the models yield an important insight into the physics that underlies the "whole cloud" optical properties.

Perhaps a few general comments on the effects of mutiple scattering will be useful. If the scattering phase function is anisotropic, multiple scatters will make the scattered radiation more and more isotropic. From some calculations on clouds it appears that isotropy is approached rather slowly with the increasing order of scattering. On the other hand, a few orders of scattering are sufficient to obscure the bright region around the direction to the source. Assuming a plane top surface of the scattering layer, the albedo of the layer will increase with the optical depth of the layer. This will be more pronounced for asymmetric Mie scatterers than for Rayleigh scattering. It is also evident that the albedo of the scattering layer will increase with the angle of incidence of the source if the scattering is asymmetric. It is often forgotten that the apparent cloud albedo increases with the albedo of the underlying surface; this is quite important when the surface is snow covered.

## 2.19 Nonspherical Particles

It has been tacitly assumed in this chapter that the scatterers are spherical. In general, only particles composed entirely or partially of liquids will be spherical. However, this class of particles is very common in the troposphere in the form of cloud drops and cloud condensation nuclei. Stratospheric particles composed of more or less ammoniated sulfuric acid are presumably spherical. There are also many nonspherical particles in the atmosphere, such as surface dust, soot, and particles from volcanic eruptions.

Equations of the Mie type have been derived for spheroids and infinite cylinders, forms not very similar to those of irregular atmospheric particles. It seems best to approach this problem experimentally. Chylek et al. (1976) have measured the phase functions of irregular particles of several pure substances as well as a sample of spherical particles. The latter was found to correspond closely to Mie computations. In general, the phase functions of the irregular particles fell below that for spheres in the backward hemisphere. These investigators have suggested a modification to the Mie phase function that provides a better fit to the experimental results.

Chylek (1977) has shown that the extinction cross section of large nonspherical randomly oriented particles is always larger than that of spherical particles of the same volume. He suggests that this is also probably true when the size parameters are not so very large.

## Problems

1.
On rare occasions the moon appears blue. Assuming that this is due to selective scattering by spheres of index of refraction 1.33, what is the radius of the spheres?

2.
It is desired to form an artificial fog that will minimize the transmissivity for a given total volume of the material of which the fog drops are formed. What is the optimal radius of the drops?

3.
A haze filter is often used on cameras to increase the contrast of distant objects. The filter absorbs ultraviolet and blue light while passing the rest of the spectrum. Under what conditions and why does the haze filter work as intended?

4.
Most methods that have been suggested for the dissipation of fog call for the removal of the fog drops by evaporation or precipitation. The visibility in fog can be increased by subdividing the water drops into much smaller droplets without removing any of the fog water. How small must the droplets be to increase the visibility by a factor of 10? The initial fog has a liquid water content of 0.2 $g/m^3$ of air and the drop radius is 10 $\mu m$. The visibility is given by $3.91/k_s$, where $k_s$ is the scattering coefficient.

5.
Why is it that a perfectly reflecting white target, seen against the horizon sky, is never as bright as the sky even when the target is near the observer?

# 3

# Solar Radiation and Its Disposition in the Atmosphere

## 3.1 Introduction

Absorbed solar radiation is the heat source of the atmosphere and the underlying surface. In comparison, the heat flux from radiogenic effects in the crust is four orders of magnitude smaller than the absorbed solar radiation. The distribution of solar radiation over the earth is dependent in the first instance on the geometry of the earth, the tilt of the earth's axis to the plane of the ecliptic, and the annual variation of the distance between the sun and the earth. These astronomical factors are well known and may be computed to any precision needed in meteorology. There is evidence of variable solar emission, particularly in the short-wavelength portion of the spectrum, which does not penetrate very far into the atmosphere. Measurements of the total incident solar radiation or solar constant made only in this century have shown a variability of not more than 1 or 2%, and much of this is probably due to errors of measurement. A more complete discussion of this question is presented in section 3.7.

As the solar beam enters the earth's atmosphere, it is depleted by absorption and scattering. The absorbed radiation is added directly to the atmosphere and contributes to its heat budget. Some of the scattered radiation is lost to space, and the remainder continues to penetrate the atmosphere, where it is subject to further scattering and absorption. In a cloudless atmosphere the greater part of the solar radiation reaches the surface, where most of it is absorbed, but a small fraction is reflected back into the atmosphere. Clouds exert a major influence and may backscatter up to 80% of the radiation incident upon them. The radiation that is absorbed at the surface is utilized in part to evaporate water into the atmosphere. Over a year the remainder of the surface-absorbed radiation is transferred to the atmosphere by thermal radiation and turbent transfer.

## 3.2 The Sun

Only a brief account of the relevant features of the sun will be given here. (For a much more complete discussion, see Kuiper, 1953.) The source of the radiant power of the sun is nuclear fusion deep within it. The visible solar disk is the photosphere, which is the immediate source of most solar radiation. The photosphere exhibits limb darkening (less radiation near the edge of the disk) and is covered with small rapidly changing cells, of the order of 1,000 km diameter, called granules, with lifetimes of only a few minutes. Above the photosphere is the solar "atmosphere," consisting of the chromosphere, 10,000–20,000 km deep, surmounted by the corona, which extends outward to several solar radii. The lower part of the chromosphere has a relatively low temperature of about 4,500°K. The temperature of the corona is on the order of $(10^6)$°K.

Sunspots are the best-known variable feature of the sun, having been recorded for more than 200 years. The commonly used Wolf or Zurich sunspot number is the sum of the number of individual sunspots and 10 times the number of sunspot groups. The sunspot number varies nearly cyclically, with a mean period of about 11.2 years. The Wolf sunspot number ranges from near 0 at minimum to more than 100 at maximum. There is evidence of a double sunspot cycle of about 22 years, and even longer periodicities have been found. Other evidence of variable solar activity is solar flares, originating near sunspots, solar prominences, which are photospheric eruptions, and coronal disturbances. Some of these are known to result in increased ultraviolet and x radiation, which, in turn, modify the ionosphere. The sun has an apparent period of rotation of about 26 days at its equator, increasing to about 30 days at 60° solar latitude.

## 3.3 Solar Spectrum

The solar spectrum extends from x rays to metric radio waves, but nearly all of the radiant power lies between 0.3 and 2.5 $\mu$m wavelength. In this region the emission is in the form of a continuum emitted by the solar photosphere. Line absorption in the relatively cool lower chromosphere results in the well-known Fraunhofer lines superimposed on the continuum. As the wavelength falls below about 0.3 $\mu$m, the absorption lines become more numerous and blacker, while the continuum decreases. Further into the ultraviolet, emission lines increase and the Fraunhofer lines disappear below 0.15 $\mu$m. At wavelengths below 0.14 $\mu$m the chromosphere and coronal emission deminate the emission. In addition to electromagnetic radiation, the sun emits plasma and associated magnetic fields, known as the solar wind, and also solar energetic particles from solar flares.

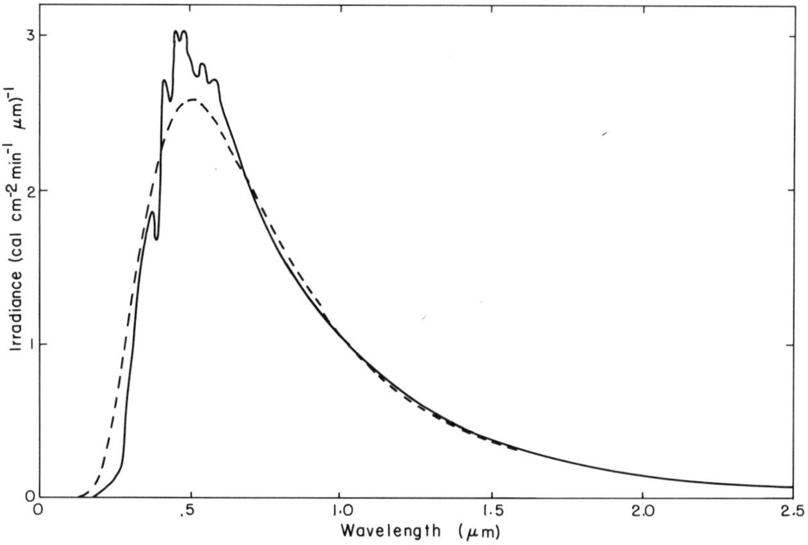


Figure 3.1 Solar spectral irradiance incident on the atmosphere. The dashed curve is for a 5,785°K blackbody for comparison. The spectral irradiance curve was drawn from data in table 16-1c of the *Handbook of Geophysics and Space Environments*, Air Force Cambridge Research Laboratories, Office of Aerospace Research, U.S. Air Force, 1965.

It would not be expected that the solar emission would correspond to that of a blackbody. In fact it is perhaps surprising that the solar spectrum resembles that of a blackbody as closely as it does. Figure 3.1 shows the solar spectrum outside the atmosphere at mean solar distance. The temperature of a blackbody of diameter equal to that of the sun and located at the mean solar distance that will yield the same total irradiance is about 5,785°K. Figure 3.1 shows that relative to the 5,785°K blackbody, the sun emits less ultraviolet but more visible radiation. Numerical values of the solar power spectrum outside the atmosphere are give in table 3.1.

Another blackbody temperature of the sun may be obtained from Wien's displacement law and is about 6,100°K. Equivalent blackbody temperatures are often given for specific spectral regions. Inspection of figure 3.1 shows that an equivalent temperature in the near ultraviolet would be less than 5,785°K. In portions of the radio wave spectrum the equivalent temperature is about $10^{6}$°K.

## 3.4 The Solar Constant

The solar constant is the solar power incident on a unit area normal to the solar beam outside the atmosphere at the mean distance between the sun

Table 3.1
Cumulative solar spectral radiance outside the atmosphere[a]

| Wavelength ($\mu$m) | Cumulative radiance (%) | Wavelength ($\mu$m) | Cumulative radiance (%) |
|---|---|---|---|
| 0.225 | 0.065 | 1.30$\mu$m | 83.33 |
| 0.25 | 0.192 | 1.40 | 85.94 |
| 0.27 | 0.413 | 1.50 | 88.05 |
| 0.29 | 0.840 | 1.60 | 89.77 |
| 0.30 | 1.269 | 1.70 | 91.18 |
| 0.35 | 4.68 | 1.80 | 92.35 |
| 0.40 | 9.15 | 1.90 | 93.32 |
| 0.45 | 16.0 | 2.00 | 94.14 |
| 0.50 | 23.5 | 2.20 | 95.42 |
| 0.55 | 30.6 | 2.40 | 96.35 |
| 0.60 | 37.4 | 2.60 | 97.08 |
| 0.65 | 43.6 | 2.80 | 97.62 |
| 0.70 | 49.1 | 3.00 | 98.03 |
| 0.75 | 54.0 | 3.50 | 98.723 |
| 0.80 | 58.3 | 4.00 | 99.133 |
| 0.85 | 62.1 | 5.00 | 99.542 |
| 0.90 | 65.5 | 6.00 | 99.725 |
| 0.95 | 68.6 | 7.0 | 99.822 |
| 1.00 | 71.3 | 11.0 | 99.948 |
| 1.10 | 76.1 | 30.0 | >99.999 |
| 1.20 | 80.07 | | |

a. Data are abstracted from the *Handbook of Geophysics and Space Environments*, Air Force Cambridge Research Laboratories, Office of Aerospace Research, U.S. Air Force, 1965.

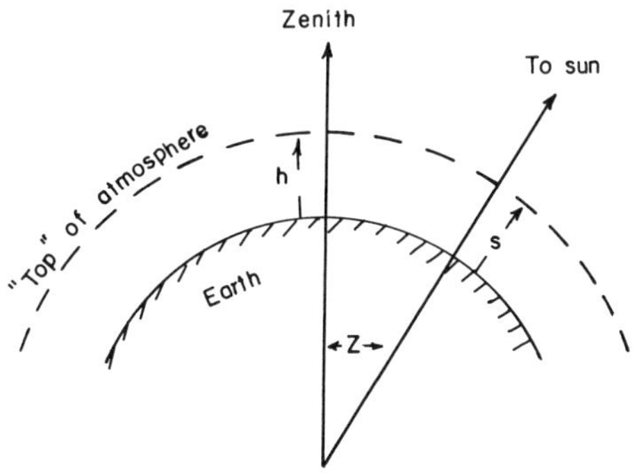


Figure 3.2 Geometry of the Smithsonian method for extrapolating the solar irradiance to the top of the atmosphere.

and the earth. The Astrophysical Observatory of the Smithsonian Institution under Langley and Abbot developed methods and instrumentation for the measurement of the solar constant and carried out an unbroken series of observations over the first half of the twentieth century as described in volumes 1–7 of the *Annals of the Astrophysical Observatory of the Smithsonian Institution.* Although the solar constant is now measured from spacecraft, a review of the Smithsonian work is warranted by the unique length of the record, the many by-products of the work, and its historical significance.

Of necessity, the Smithsonian observations were made within the atmosphere, and it was necessary to extrapolate to the outer limits of the atmosphere by the method illustrated in figure 3.2. At a given wavelength the transmissivity of the atmosphere is represented by the exponential law

$$E_\lambda/E_{0\lambda} = \exp\left[-\left(\int_0^s k_\lambda\, ds\right)\right] \qquad (3.1)$$

where $E_\lambda$ and $E_{0\lambda}$ are the parallel beam irradiances at the surface of the earth and outside the atmosphere, respectively, and $k_\lambda$ is the extinction coefficient along the slant path from 0 to $s$ through the atmosphere. As long as the zenith angle of the sun, $Z$, does not exceed 60–70°, $ds - dh \sec Z$, where $h$ is measured along the local vertical. Then (3.1) can be written

$$E_\lambda/E_{0\lambda} = \exp\left[-\left(\sec Z \int_0^h k_\lambda\, dh\right)\right], \qquad (3.2)$$

or

$$\ln E_{0\lambda} = \ln E_\lambda + \sec Z \int_0^h k_\lambda\, dh.$$

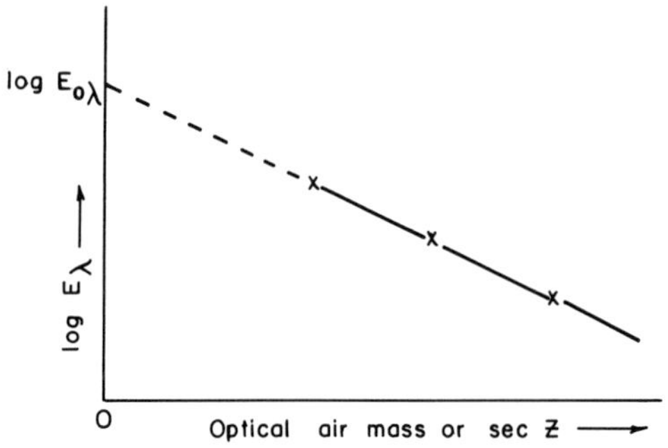


Figure 3.3 Illustrating the extrapolation of ground-based measurements of irradiance to the top of the atmosphere.

The transmissivity of the atmosphere with the sun in the zenith is $\exp[-\int_0^h k_\lambda\, dh]$. The secant of the zenith angle is a measure of the length of the path through the atmosphere in terms of the vertical path and is usually called the optical air mass. If observations of $E_\lambda$ are made at a number of zenith angles, a plot of $\ln E_\lambda$ against $\sec Z$ determines a straight line, as illustrated in figure 3.3. The intercept on the zero air mass ordinate then yields $E_{0\lambda}$. This procedure assumes that the atmosphere is horizontally homogeneous and that its optical properties do not change during the several hours required to yield a reasonable range of $Z$. If these conditions are not met, the data will not fall on a straight line and the observations are discarded. This procedure is the basis of the Smithsonian determinations of the solar constant.

## 3.5 The Smithsonian Long Method

This method is based on the procedure outlined earlier. The principal instruments are a spectrobolometer and a normal incidence pyrheliometer. The spectrobolometer is a prism spectrometer employing a bolometer to measure the relative radiant power. A recording galvanometer and an automatic drive yield a record of the relative irradiance spectrum, which is known as a spectrobologram. The general appearance of a spectrobologram is illustrated in figure 3.6. A series of spectrobolograms and simultaneous pyrheliometer readings is taken at different zenith angles on a clear day. The spectrobolometer is a rather complex instrument, so no attempt was made to calibrate it in absolute units. Because of the glass optics, it is not equally transparent to all wavelengths, and it must be calibrated so that the readings can be corrected to the true relative irradiance at all wavelengths. Further, it is useful only between about 0.346 and 2.442 $\mu$m, and small corrections must be added for the solar radiation outside these limits. The calibration in radiation units is made by equating the pyrheliometer

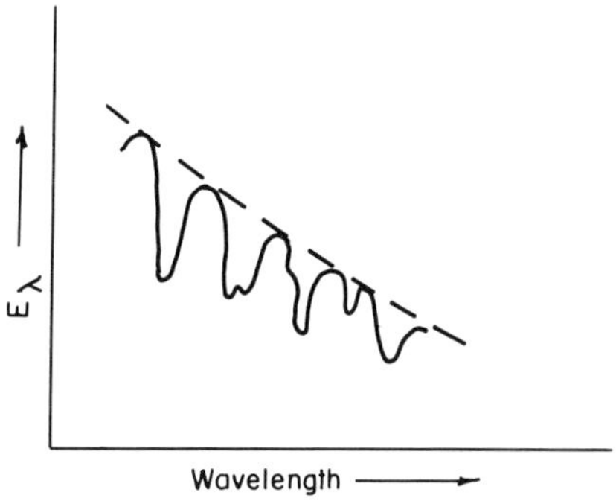


Figure 3.4 A schematic representation of a smooth curve drawn tangent to the regions between absorption bands.

reading of total irradiance to the area under the spectrobologram plus the end corrections.

The spectrobolograms were read at 44 selected wavelengths, and these readings were extrapolated to the top of the atmosphere by the procedure illustrated in figure 3.3. The 44 wavelengths are not sufficient to resolve the absorption bands shown in figure 3.6. The Smithsonian procedure was to draw a smooth curve tangent to the regions between absorption bands, as shown schematically in figure 3.4. In the region of absorption bands, values were read from the dashed curve and extrapolated to the top of the atmosphere. Some slight smoothing of the resultant curve outside the atmosphere was done to minimize errors in drawing the tangent curves.

It should be noted that the method of calibrating the area under the spectrobologram against the pyrheliometer requires end corrections for the radiation measured by the pyrheliometer that fall outside the limits of the spectrobolometer (0.346–2.442 $\mu$m). These corrections are smaller than the similar corrections applied outside the atmosphere and depend on the optical air mass.

Because of the intense absorption by ozone and oxygen, the surface solar spectral irradiance curve cuts off sharply at about 0.29 $\mu$m. (The cutoff shifts toward longer wavelengths as the solar zenith angle increases.) Further, the optical properties of the prism effectively limit the spectrum at 0.346 $\mu$m. Thus it was not possible to obtain values of the ultraviolet radiation outside the atmosphere by extrapolation. The Smithsonian procedure was to fit a 6000°K blackbody curve to the blue end of the measured spectrum and to use this curve, reduced by Fraunhofer line absorption, to represent the unmeasured solar ultraviolet spectrum. No direct information on the number of Fraunhofer lines in this region existed, and it was necessary to estimate this by extrapolation from the longer wavelength regions. The trend to increasing numbers of Fraunhofer lines with decreasing wavelengths was so marked that the procedure used by Abbot et al. (1932) effectively terminated the solar spectrum at 0.29 $\mu$m. Thus the Smithsonian ultraviolet correction is a correction only for the spectral range from 0.29 to 0.346 $\mu$m. The resultant solar constant therefore does not include the radiation absorbed by ozone and oxygen in the ultraviolet. It is now known that about 0.75% of the quiet sun emission is in the wavelength region shorter than 0.29 $\mu$m.

A similar infrared correction for the unmeasured radiation beyond 2.442 $\mu$m is applied, and this procedure is less uncertain than the ultraviolet correction. The solar constant is determined from the area under the corrected solar irradiance curve outside the atmosphere adjusted to the mean solar distance by means of the inverse square law.

## 3.6 The Short Method

The so-called short method of determining the solar constant is essentially an empirical approach based on many years of long-method observations. The loss of radiation as the solar beam passes through the atmosphere is due to scattering and to absorption, primarily by water vapor. The scattering was characterized by a pyranometer reading that measures the scattered light in the vicinity of the sun (forward scattering) and the absorption by the depression at the center of a selected water vapor absorption band. These were combined empirically to yield the transmissivity at each of the 44 wavelengths, that is, the slope of the straight line in figure 3.3. This, together with a single spectrobologram and pyrheliometer reading, makes it possible to extrapolate to the outer limits of the atmosphere. The advantages of the short method are that only a single observation time is required, thus permitting measurements to be made on days not optically stable enough for the long method, several observations can be made in one day, and the reduction procedures may be carried out more rapidly. For these reasons most of the later determinations of the solar constant by the Smithsonian were made by the short method.

## 3.7 Numerical Values of the Solar Constant

As a result of many years of observations at a number of mountain observatories in various parts of the world, the Astrophysical Observatory of the Smithsonian Insitution announced in 1952 that the preferred value of the solar constant is 1.94 cal cm$^{-2}$ min$^{-1}$.

The principal criticism of the Smithsonian procedure has been directed at the ultraviolet correction. The Smithsonians contended that their procedure for calibrating the spectrobologram against the pyrheliometer compensates for any errors in the ultraviolet (and infrared) corrections. They argue that if the ultraviolet correction is too small, the scale of the spectrobologram will be slightly too large, thus minimizing any error in the solar constant. There is no question that some compensation results, but it may be shown that it is complete only for certain conditions that are not generally met.

Renewed attention was directed to the solar constant when the U.S. Naval Research Laboratory obtained solar ultraviolet spectra from V-2 rockets that ascended above the ozone layer. The ultraviolet spectrum in figure 3.5 is based on the work of Furukawa et al. (1967). Also shown in figure 3.5 are three blackbody curves and the result of the Smithsonian extrapolation.

Note that the ultraviolet spectrum of figure 3.5 is too broadband to show the many spectral lines in this region. The underestimate of the Smithsonian ultraviolet correction is clearly evident.

A number of investigators recomputed the solar constant on the basis of the new information on the ultraviolet solar spectrum and, in some cases, new measurements in the near ultraviolet and the visible light regions. One of the more complete reexaminations was that of F. S. Johnson (1954), who gave a solar constant of $2.00 \pm 0.04$ cal cm$^{-2}$ min$^{-1}$. Others found solar constants ranging from 1.95 to 2.05 cal cm$^{-2}$ min$^{-1}$.

The uncertainties in the value of the solar constant result from the extrapolation to the outer limits of the atmosphere, calibration of the instruments in absolute units of power, and instrumental errors. Extrapolation errors have been minimized by carrying the instruments aloft on airplanes (10–12 km), free balloons (28–36 km), rockets (100–250 km), and, most recently, spacecraft. Differences in the absolute scale of pyrheliometry of 1–2% persisted until the recent development of absolute cavity radiometers, in which the solar power is equated to electrical power with an accuracy of better than 0.5%. This has made it possible to bring all of the more recent measurements to a common absolute scale, as shown by Froehlich (1977) and, slightly modified, Willson and Hickey (1977).

The results of this comparison and the most recent value of the solar constant from an absolute cavity radiometer on the *Solar Maximum Mission* (*SMM*) spacecraft launched in 1980 (Willson et al., 1981) are given in table 3.2. The uncertainty in the International System of Units of the absolute cavity radiometer is less than 0.5%, and is probably less than 0.1%. The relative precision of the radiometer is much smaller than 0.1%. The solar constant measured by the *SMM* spacecraft and given in table 3.2, is the weighted mean of observations over 153 days. The last item in table 3.2 is only of historical interest, and the scale of radiometry has not been adjusted to the absolute scale of the cavity radiometer. For the purpose of this book, the solar constant may be taken as 1,370 W m$^{-2}$.

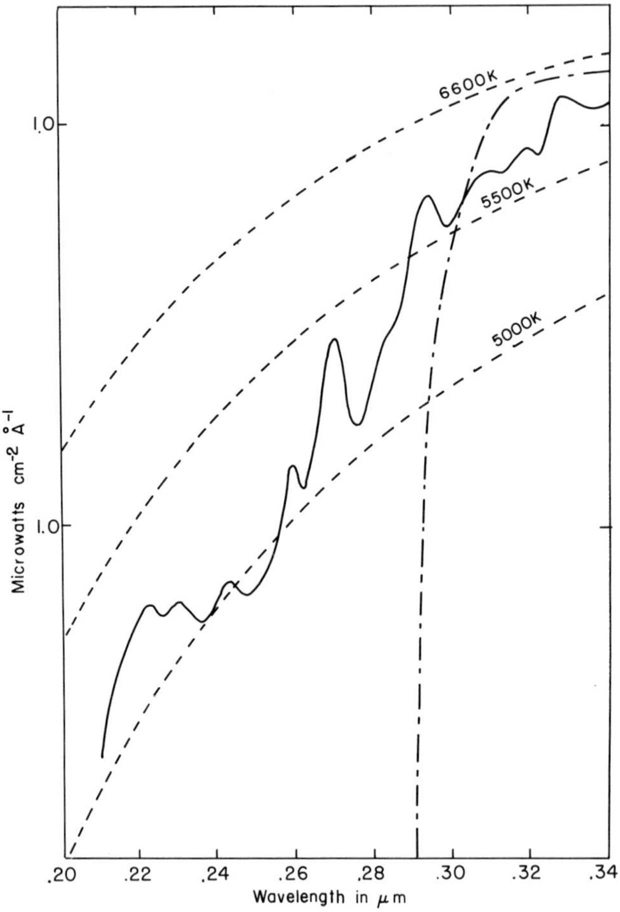


Figure 3.5 The solid curve is an ultraviolet spectrum of the sun. The dashed curves are for blackbodies at the indicated temperatures. The dash-dot curve is the Smithsonian ultraviolet correction.

Table 3.2
Comparison of the solar constants obtained from different observing platforms

| Platform | Solar constant ($W\ m^{-2}$)[a] |
|---|---|
| Aircraft | 1,372.5 |
| Balloons | 1,369.2 |
| Rockets | 1,370.8 |
| *Mariner* spacecraft | 1,366.8 |
| *Nimbus 7* spacecraft | 1,376.0 |
| *SMM* spacecraft | 1,368.3 |
| Surface of the earth (Smithsonian) | (1,354) |

a. Multiply radiant power in cal $cm^{-2}$ $min^{-1}$ by 697.8 to convert to $W\ m^{-2}$.

## 3.8 Variability of the Smithsonian Solar Constant

Solar activity as monitored by sunspot numbers and many other features follows an 11-year cycle, and it has been speculated that the solar constant may vary in the same way but with an unknown amplitude. The only time series of the solar constant that exceeds the 11-year period is that obtained by the Smithsonian workers. In fact, they claimed years ago that there was a variation in the solar constant with an amplitude on the order of 1%. Since they believed that their monthly mean values were repeatable to 0.1%, they considered the 1.0% variations real. Independent analyses of the Smithsonian data led to the conclusion that the uncertainties of measurement were about the same as the claimed variability. These analyses may be interpreted as suggesting that there has been no variation exceeding 1% in the 0.346–2.442 $\mu$m spectral region during the first half of the twentieth century.

## 3.9 Indirect Evidence of Solar Variability

Observations made from spacecraft have shown that the solar flux in the extreme ultraviolet ($< 0.1025\ \mu$m) varies by a factor of two or more in time, and there is some evidence that this variation follows the solar cycle. In the x-ray region there is an even larger temporal variation. Variation is also observed in the dekametric solar radiation. The far-ultraviolet and x-ray fluxes represent only a very small fraction of the solar constant, but they are absorbed in the high atmosphere, where they provide an important energy source.

Over the years hundreds of correlations of solar activity with tropospheric weather parameters have been explored, only a very few of which have stood the tests of time and statistical significance. In the absence of any convincing physical mechanism linking solar variability with the troposphere, most meteorologists have considered these few positive results statistical accidents. However, it is dangerous to ignore significant statistical relations because of the lack of a physical model.

The 11-year solar cycle that is so clear in the records of sunspot numbers and magnetic indices has generally been accepted as a basic and continuing feature of the sun. Theoretical models of the sun have as yet failed to explain the solar cycle. The only theoretical result of pertinence here is that the solar output was apparently only 0.7–0.8 of the current value shortly after the sun began to burn hydrogen in the main sequence. This corresponds to an average change of about a millionth of 1% per century.

Eddy (1976) has reexamined the evidence for a near disappearance of the solar cycle from about 1645 to 1715 first noted by Maunder. Eddy found that there were very few reports of sunspots in this period as compared with the periods before and after. He also found that reports of four total solar eclipses in this period described coronas of very limited extent with no streamers or structure. He also found that very few auroras occured during this period. Eddy also determined that the $^{14}C$ content of dated tree cores showed a maximum in the 1650–1790 period. The relevance of this finding is that $^{14}C$ is formed by cosmic radiation, which is modulated by solar activity to yield a minimum at maximum solar activity, and conversely. Eddy has called this period of low solar activity the Maunder minimum. He has also found another similar minimum which he christened the Spoerer minimum, from the midfifteenth to the early sixteenth centuries, and he believes there are still others. Eddy points out that the Maunder minimum corresponds in time to one of the colder periods of the Little Ice Age. Interesting as this is, the historical records on which it is based are of doubtful quality, and a relation between solar variability and the solar constant has not yet been firmly established.

## 3.10 Effects of Orbital Changes

The orbit of the earth about the sun is subject to changes in time due to changes in eccentricity, with a period of about $10^5$ years, changes in obliquity with a period of about $4 \times 10^4$ years, and changes in the longitude of the perihelion with a period of about $2 \times 10^4$ years. Without any change in solar output, these orbital variations lead to changes in the distribution of the solar radiation on the earth in space and time, affecting seasonal variations and meridional profiles. The changes in solar radiation due to these orbital variations have been calculated by Milankovitch (1930), Vernekar (1972), and others. The concomitant effects on climate have been sought by the use of numerical climate models. The rather crude models now available predict that the climatic changes resulting from the orbital perturbations would be too small to explain the pleistocene glaciations. This conclusion may be changed as more complete numerical models of climate are developed.

Hays et al. (1976) have used an approach that does not require that there be knowledge of the climatic response to changes in the distribution of insolation but only that the relation be linear. Their procedure is based on measurements made on two deep-sea sediment cores (43.5S–80E and 46S–90E) that were spliced together to provide a long-period record. The cores were dated and analyzed for $^{18}O$ and two species of radiolarians. Changes in $^{18}O$ are due principally to the waxing and waning of the Northern Hemisphere ice sheet; one of the radiolarians is sensitive to the sea surface temperature at the core site and the other to the temperature and salinity structure of Antarctic water. The procedure was to apply statistical spectral analysis to these three climate-related time series and to stretch or shrink the time scale to yield the strongest peaks. All of the climatic frequency peaks were found to correspond reasonably well to the three orbital frequencies listed earlier. An important and unexpected result was that the dominant peak occurred at the $10^5$-year period, corresponding to changes in the eccentricity of the orbit. The computed changes in insolation due to this effect are the smallest of the three, not exceeding 0.1%. Hayes et al. (1976) suggested that there was a nonlinear relation, whereas their analysis assumed linearity. This may be a result of strong feedback between insolation and the surface albedo resulting from changes in the snow and ice coverage.

This and other studies have convinced many that orbital changes are a cause of climatic changes but that they are responsible for only a fraction of the total change. The dominance of the $10^5$-year cycle is still unexplained. The Hays et al. approach is promising, but better dating of the deep-sea cores is needed. Orbital variations cannot explain climatic changes on time scales of the order of 100 years or those with time scales in the millions or billions of years; continental drift and changes in sea level must have contributed to the latter.

## 3.11 Astronomical Factors

The solar constant is defined for the average distance of the earth from the sun. The eccentricity of the earth's orbit[1] is 0.0167, the maximum earth-sun distance (aphelion) is $1.521 \times 10^8$ km, the minimum (perihelion) is $1.471 \times 10^8$ km, and the mean distance is $1.4960 \times 10^8$ km, or one astronomical unit. By the application of the inverse square law, the factors by which the solar constant must be multiplied to give the irradiance

1. 0.0167 is not the eccentricity used in analytic geometry, which is 0.254, but is the maximum deviation of the orbit from a circular orbit of mean radius expressed as a fraction of the mean radius.

Table 3.3
Factors to convert solar constant to solar irradiance outside the atmosphere

| Date | Factor | Date | Factor |
|---|---|---|---|
| January 1 | 1.0343 | July 1 | 0.9675 |
| February 1 | 1.0306 | August 1 | 0.9704 |
| March 1 | 1.0199 | September 1 | 0.9864 |
| April 1 | 1.0030 | October 1 | 0.9973 |
| May 1 | 0.9863 | November 1 | 1.0148 |
| June 1 | 0.9731 | December 1 | 1.0282 |

outside the atmosphere at any date may be readily computed. For convenience, these factors are given in table 3.3 for perihelion on January 4.

## 3.12 Solar Zenith Angle

The zenith angle of the sun determines the relative length of the path through the atmosphere and also the irradiance on a horizontal area. Thus $E_h = E_s \cos Z$, where $E_h$ is the irradiance on a horizontal surface and $E_s$ is the irradiance on a surface normal to the solar beam. From spherical trigonometry it is readily found that

$$\cos Z = \sin\phi \sin\delta + \cos\phi \cos\delta \cos h, \tag{3.3}$$

where $\phi$ is the latitude, $\delta$ is the solar declination angle, and h is the hour angle, measured from local solar noon (1 day = 24 hr = 360°). The solar declination is a function of the time of year and also varies slowly from year to year due to astronomical factors. Table 3.4 gives the declination as a function of date for 1983.

It may be noted that the angular diameter of the sun as seen from the earth is 32′. Sunrise or sunset occur when the upper limb of the sun is at the horizon. By allowing for an average atmospheric refraction of 34′, sunrise or sunset occurs when the center of the solar disk is $34 + (32/2) = 50'$ below the geometric horizon. Thus radiation from the direct solar beam is zero at a zenith angle of 90°50′.

## 3.13 Optical Air Mass

The optical air mass, or the slant path through the atmosphere relative to the normal incidence path, is given closely by sec $Z$ for zenith angles smaller than about 60°. At larger zenith angles, the curvature of the earth and the average atmospheric refraction must be taken into account. Table 3.5 gives the optical air mass as a function of zenith angle taken from the

Table 3.4
Solar declination angles at 1,200 GMT for 1983[a]

| | | | | | |
|---|---|---|---|---|---|
| January | 1 | −23°4.2′ | July | 1 | +23°9.9′ |
| | 15 | −21°16.5′ | | 15 | 21°39.5′ |
| February | 1 | −17°20.2′ | August | 1 | 18°13.9′ |
| | 15 | −12°57.8′ | | 15 | 14°19.0′ |
| March | 1 | −7°54.3′ | September | 1 | 8°35.4′ |
| | 15 | −2°28.2′ | | 15 | 3°21.3′ |
| April | 1 | +4°12.8′ | October | 1 | −2°51.4′ |
| | 15 | 9°27.8′ | | 15 | −8°12.1′ |
| May | 1 | 14°49.1′ | November | 1 | −14°9.4′ |
| | 15 | 18°39.9′ | | 15 | −18°16.0′ |
| June | 1 | 21°56.2′ | December | 1 | −21°40.1′ |
| | 15 | 23°16.3′ | | 15 | −23°13.1′ |

a. Abstracted from the *Nautical Almanac* for 1983, U.S. Naval Observatory. Washington, D.C.: U.S. Government Printing Office.

Table 3.5
Optical air mass as a function of zenith angle[a]

| $Z$ (degrees) | Air mass | sec $Z$ | $Z$ (degrees) | Air mass | sec $Z$ |
|---|---|---|---|---|---|
| 60 | 2.00 | 2.000 | 80 | 5.60 | 5.759 |
| 62 | 2.12 | 2.130 | 81 | 6.18 | 6.393 |
| 64 | 2.27 | 2.281 | 82 | 6.88 | 7.185 |
| 66 | 2.45 | 2.459 | 83 | 7.77 | 8.205 |
| 68 | 2.65 | 2.669 | 84 | 8.90 | 9.567 |
| 70 | 2.90 | 2.924 | 85 | 10.39 | 11.47 |
| 72 | 3.21 | 3.236 | 86 | 12.44 | 14.33 |
| 74 | 3.59 | 3.628 | 87 | 15.36 | 19.11 |
| 76 | 4.07 | 4.134 | 88 | 19.79 | 28.65 |
| 78 | 4.72 | 4.810 | 89 | 26.96 | 57.31 |

a. After the *Smithsonian Meteorological Tables* (1951).

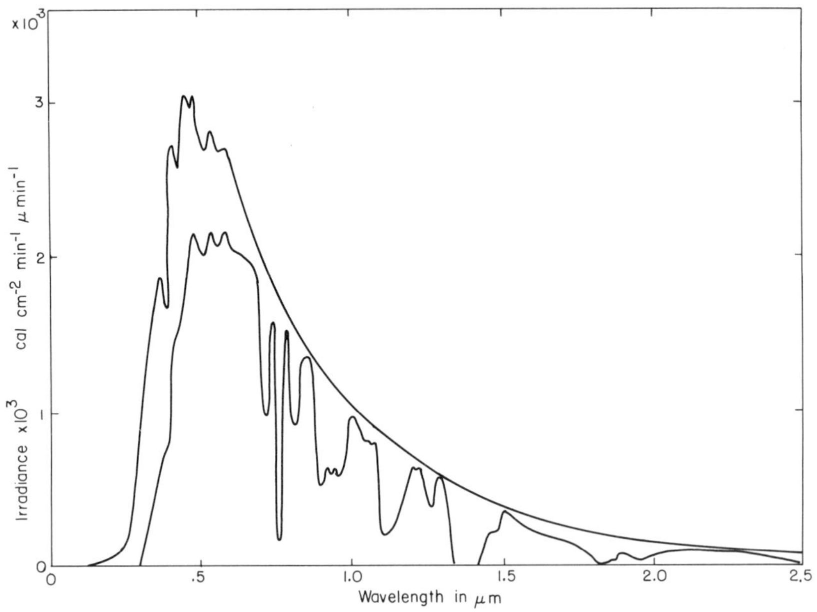


Figure 3.6 Solar spectral irradiance outside the atmosphere and at the surface of the earth. The optical air mass is 2.0, the precipitable water is 1 cm, and the atmosphere is free of aerosol. Based on Smithsonian data.

*Smithsonian Meteorological Tables* (1951). The secant of the zenith angle is included for comparison.

## 3.14 Transmissivity of the Atmosphere

In a cloud-free atmosphere the solar beam is attenuated by scattering and gaseous and aerosol absorption. The effects of this depletion are depicted in figure 3.6. The upper curve is the spectral irradiance at the top of the atmosphere, copied from figure 3.1, while the lower curve is the spectral irradiance at sea level for 1 cm of precipitable water and optical air mass 2. As already noted, the sharp cutoff near 0.3 $\mu$m is due to absorption by ozone and oxygen. Scattering by the air molecules reduces the incident radiation mostly in the ultraviolet and visible regions. The deep depressions in the infrared are due to gaseous absorption by water vapor, carbon dioxide, oxygen, and ozone. Attenuation due to the atmospheric aerosol has been omitted since it is so variable. The aerosol causes additional depletions due to scattering and absorption. Note that figure 3.6 refers only to the irradiance of the direct solar beam and does not include skylight. The latter is richer in the blue and partially compensates for the deficiency in blue of the direct solar beam.

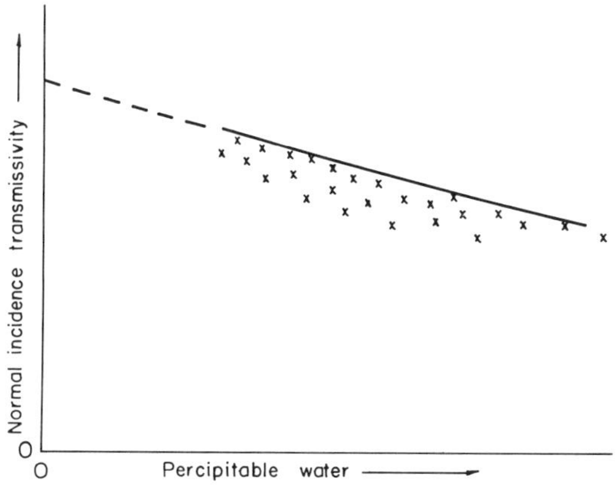


Figure 3.7 Schematic plot of the normal incidence transmissivity at a selected wavelength as a function of the precipitable water. The solid line is the upper envelope of the observations represented by crosses. The extrapolation of this line to zero precipitable water (the dashed line) gives the transmissivity due to water vapor scattering at the given wavelength.

A by-product of each long-method determination of the solar constant is the transmissivity of the atmosphere at each of the 44 wavelengths; this is simply the slope of the straight line of figure 3.3. As already noted, these transmissivities do not include gaseous absorption, but are based on the tangent curve shown in figure 3.4.

## 3.15 Transmissivity Due to Molecular Scattering

When a large number of the observed transmissivities at one of the 44 wavelengths is plotted against the precipitable water, the result is as shown schematically in figure 3.7. The solid line is drawn as the upper envelope of the experimental points and corresponds to the minimum aerosol concentration. Extrapolation of the line to zero precipitable water (dashed line) gives, approximately, the transmissivity due to molecular scattering, called "dry air scattering," by the Smithsonian workers. The results of these extrapolations for the summit of Mount Wilson are given in table 3.6, together with those computed from Rayleigh molecular scattering. The agreement is generally good except in the infrared, where the residual aerosol has a proportionally larger effect. Both the Smithsonian and subsequent investigators have preferred to use the computed transmissivities. For an atmosphere of uniform composition the number of molecules in a vertical atmospheric path is proportional to the pressure at the base of the column. Water vapor is a variable component, but little error will result from the use of the total pressure. Thus

Table 3.6
Computed and observed transmissivities due to molecular scattering at Mount Wilson

| Wavelength ($\mu$m) | 0.370 | 0.400 | 0.430 | 0.460 | 0.500 | 0.750 | 1.00 | 1.50 |
|---|---|---|---|---|---|---|---|---|
| Observed | 0.683 | 0.757 | 0.808 | 0.851 | 0.885 | 0.977 | 0.987 | 0.990 |
| Computed | 0.680 | 0.755 | 0.808 | 0.850 | 0.890 | 0.977 | 0.993 | 0.999 |

$$a_a(m,p,\lambda) = a_a(1,p_0,\lambda)^{mp/p_0}, \tag{3.4}$$

where $a_a(m,p,\lambda)$ is the dry air transmissivity at wavelength $\lambda$, total pressure $p$ and optical air mass $m$, and $a_a(1,p_0,\lambda)$ is the transmissivity at unit air mass and standard surface pressure $p_0$.[2]

## 3.16 Water Vapor Scattering

The slope of the line in figure 3.7 shows that there is an additional process attenuating the direct beam that is a function of the water vapor path length. Since absorption has been eliminated, this must result from scattering. For this reason it was called "water vapor scattering" by the Smithsonian investigators. This name led some to believe that scattering by the water vapor molecules was meant, a contention that is easily proved wrong. The Smithsonian workers did not explore the source of this scattering, but it is now clear that it is scattering by aerosol particles. Most of these are hygroscopic and wax and wane with the relative humidity, thus accounting for the variation with precipitable water. The upper tropospheric and stratospheric aerosols are less variable than the denser lower-tropospheric aerosol, but considerable variation of the water vapor scattering is to be expected. The mean transmissivity due to water vapor scattering can be obtained by fitting a least-squares line to the data points of figure 3.7.

Numerical values of the transmissivities due to dry air scattering and to water vapor scattering are given in table 3.7. Those for dry air scattering are for a vertical column extending from standard sea level pressure to the outer limit of the atmosphere. Those for water vapor scattering are for a vertical path containing 1 cm of precipitable water. They are representative only of the mean aerosol above the mountaintop observatories. (For example, for Mount Wilson $p = 831$ mbar, about 1,650 m above sea level.) They are included here for historical interest and such information as they may provide on the optical properties of the aerosol. Note that $a'_w$ varies

2. Confusion may be avoided if it is remembered that $p/p_0$ is the relative path length in the vertical and that $m$ is the geometric factor that converts a vertical to a slant path.

Table 3.7
Transmissivities due to dry air scattering and to water vapor scattering as a function of wavelength[a]

| $\lambda$ ($\mu$m) | $a_a$ | $a'_w$ | $\lambda$ ($\mu$m) | $a_a$ | $a'_w$ |
|---|---|---|---|---|---|
| 0.20 | 0.0026 | — | 0.90 | 0.987 | 0.986 |
| 0.25 | 0.087 | — | 1.00 | 0.992 | 0.988 |
| 0.30 | 0.307 | — | 1.20 | 0.996 | 0.988 |
| 0.35 | 0.543 | 0.926 | 1.40 | 0.998 | 0.988 |
| 0.40 | 0.704 | 0.950 | 1.60 | 0.999 | 0.988 |
| 0.45 | 0.805 | 0.961 | 1.80 | 1.000 | 0.987 |
| 0.50 | 0.869 | 0.968 | 2.00 | 1.000 | 0.986 |
| 0.60 | 0.936 | 0.976 | 2.20 | 1.000 | 0.984 |
| 0.70 | 0.965 | 0.981 | 2.40 | 1.000 | 0.982 |
| 0.80 | 0.980 | 0.985 | | | |

a. The values of $a'_w$ are from measurements that did not extend to the ultraviolet.

with wavelength in the same sense as $a_a$, but less rapidly. This shows that the aerosol particles have radii of a few tenths of a micrometer. The reversal of the trend in $a'_w$ in the infrared, if correct, indicates the presence of a few much larger particles of radius of a few micrometers. Values of $a_a$ are taken from Hoyt (1977), while those of $a'_w$ are from the *Smithsonian Meteorological Tables* (1939).

For later reference, table 3.8 gives the transmissivity due to molecular scattering, integrated over the entire solar spectrum for various optical air masses.

## 3.17 Gaseous Absorption

The principal absorbing gases in the atmosphere are $H_2O$, $CO_2$, $O_3$, CO, $N_2O$, $CH_4$, $NO_2$, O, and $N_2$ and their isotopic modifications. Absorption in the ultraviolet is due to $O_3$, $O_2$, O, and $N_2$ and, with the exception of $O_3$, occurs predominantly in the high atmosphere above about 100 km. The maximum absorption by ozone ($O_3$) occurs around 40–50 km. Ionization due to the extreme ultraviolet and x radiation forms and maintains the ionosphere. Discussions of absorption, ionization, and dissociation by high-energy radiations are more appropriate to books on the physics of the high atmosphere. It will suffice to say that essentially all of the radiation of wavelength shorter than about 0.3 $\mu$m is absorbed above 40–50 km. In total this amounts to less than 1% of the solar constant. This does not mean that the total absorption by ozone is less than 1% because ozone absorbs also at longer wavelengths.

Table 3.8
Dry air scattering averaged over the solar spectrum

| Optical air mass | Transmissivity | Optical air mass | Transmissivity |
|---|---|---|---|
| 0.1 | 0.987 | 4 | 0.744 |
| 0.3 | 0.963 | 5 | 0.711 |
| 1.0 | 0.899 | 6 | 0.686 |
| 2 | 0.833 | 7 | 0.668 |
| 3 | 0.783 | | |

In the lower atmosphere (below 40–50 km) the absorption of the solar radiation is due primarily to $H_2O$, $CO_2$, $O_2$ and $O_3$. The important absorption bands of water vapor are centered at 0.72, 0.81, 0.94, 1.10, 1.38, 1.87, 2.7, 3.2, and 6.3 $\mu$m. The carbon dioxide bands of interest are centered at 1.4, 1.6, 2.0, 2.7, 4.3, 4.8, and 5.2 $\mu$m. Oxygen bands are found at 0.69, 0.76, and 1.25 $\mu$m. Nitrogen dioxide absorbs from 0.30 to 0.55 $\mu$m and may be of importance in polluted air according to Shaw (1976).

Water vapor is the most important absorber of solar radiation, but the neglect of the other absorbers in some instances has led to considerable confusion. The original measurements of the absorption by water vapor under atmospheric conditions were made by Fowle (1915). He made measurements over horizontal paths at sea level using a Nernst glower source and a low-resolution spectrometer. Optical and meteorological factors limited the maximum path length to 0.5 cm precipitable water. The data were extended by selecting a day when the precipitable water above Mount Wilson was less than 0.5 cm and then taking observations at increasing solar zenith angles, thereby extending the path length in a known manner. Fowle presented his data in the form of separate absorptivity curves for six water vapor bands covering the spectral region from 0.70 to 2.10 $\mu$m. Comparisons with more recent data for the same bands show that Fowle's results were remarkably good. Fowle's data included only minor parts of the absorption due to $CO_2$ and $O_2$ because his basic sea level data were taken with large water vapor concentrations and consequently relatively short $CO_2$ and $O_2$ paths. Equally important, Fowle's measurements failed to include the water vapor bands centered at 2.7, 3.2, and 6.3 $\mu$m. It has been widely assumed that Fowle's data represented essentially all of the atmospheric absorption and that they were used in all of the subsequent Smithsonian work. An examination of the total absorption reported in many papers by the Smithsonian staff shows values significantly larger than those derived from Fowle's results, but no reference to the method actually used is made. It is my belief that the total absorption was derived from the accumulated results of their solar constant observations, which were described in sections 3.4–3.6. By starting with a given spectrobologram, it is possible to construct the spectrum that would be received without absorption by using the spectral transmissivities for dry air and water vapor scattering. The area between this synthetic spectrum and the actual spectrobologram represents the loss due to absorption. This will include absorption due to all of the atmospheric gases and the aerosol. Unfortunately no summary of this approach and the results appears to exist in the literature.

The most complete reexamination of the gaseous absorption of solar radiation appears to be that of Yamamoto (1962). His computations are based on Howard et al. (1955), laboratory data on the absorption of water vapor and carbon dioxide, supplemented by the data of others when required. The effects of the total and partial pressures on the absorption were taken into account. The mixing ratios of carbon dioxide and oxygen are independent of height, while that of water vapor tends to decrease exponentially with height. For this reason, Yamamoto had to prescribe the vertical distribution of water vapor, and he chose to use those selected by London for the latter's studies of the atmospheric heat balance. It is probable that Yamamoto's results are adequate for other vertical distributions of water vapor that fall within the usual range found in the atmosphere.

Yamamato presents results only for optical air mass 1, but it is easy to extend them to longer paths. The first three columns of table 3.9 were read from the graph given in his figure 1. The ratio of the third to the second columns is the transmissivity due only to carbon dioxide and molecular oxygen for air mass 1. The relative constancy of the fourth column shows that the absorption due to $CO_2$ and $O_2$ is nearly independent of that due to $H_2O$ (small overlap). This makes it possible to extend the table to larger optical air masses. This is a double-entry table because the absorption due to water vapor depends on the product of the air mass and the precipitable water in the vertical, while the absorption due to $CO_2$ and $O_2$ depends only on the air mass, $(p/p_0) \sec Z$. As an example, suppose the optical air mass $m$ is 3 and the precipitable water $w$ in the vertical is 1.0 cm. Then $wm = 3.0$, and the transmissivity along the slant path is 0.833.

## 3.18 Absorption by Ozone

The principal absorption bands of ozone in the solar spectrum are the Hartley band, 0.2–0.3 $\mu$m, the Huggins band, 0.31–0.34 $\mu$m, and the Chappuis bands, 0.4–0.8 $\mu$m. The strong ultraviolet bands of oxygen remove nearly all of the solar radiation shorter than about 0.2 $\mu$m at heights below about 50 km. The strong Hartley band of ozone is responsible for the sharp cutoff of the solar spectrum at 0.29–0.30 $\mu$m. The weaker Huggins band and the even weaker Chappuis bands are of con-

Table 3.9
Transmissivities due to the absorption of solar radiation[a]

| Transmissivity for $m = 1$ | | | | Transmissivity due to $H_2O + CO_2 + O_2$ | | | |
|---|---|---|---|---|---|---|---|
| Slant path $wm$ (cm) | $H_2O$ alone | $H_2O$ + $CO_2$ + $O_2$ | $CO_2$ + $O_1$ | $m = 2$ | $m = 3$ | $m = 5$ | $m = 7$ |
| 0.01 | 0.983 | 0.973 | 0.990 | 0.963 | 0.954 | 0.935 | 0.916 |
| 0.02 | 0.976 | 0.966 | 0.990 | 0.957 | 0.947 | 0.928 | 0.910 |
| 0.03 | 0.972 | 0.962 | 0.990 | 0.953 | 0.943 | 0.924 | 0.906 |
| 0.05 | 0.965 | 0.955 | 0.990 | 0.946 | 0.936 | 0.918 | 0.899 |
| 0.07 | 0.960 | 0.950 | 0.990 | 0.941 | 0.931 | 0.913 | 0.895 |
| 0.10 | 0.954 | 0.944 | 0.990 | 0.935 | 0.925 | 0.907 | 0.889 |
| 0.20 | 0.942 | 0.932 | 0.989 | 0.921 | 0.911 | 0.891 | 0.872 |
| 0.30 | 0.932 | 0.922 | 0.989 | 0.912 | 0.901 | 0.882 | 0.863 |
| 0.50 | 0.920 | 0.910 | 0.989 | 0.900 | 0.890 | 0.870 | 0.851 |
| 0.70 | 0.911 | 0.900 | 0.988 | 0.889 | 0.878 | 0.857 | 0.837 |
| 1.0 | 0.901 | 0.890 | 0.988 | 0.880 | 0.869 | 0.848 | 0.828 |
| 2.0 | 0.879 | 0.868 | 0.987 | 0.856 | 0.845 | 0.824 | 0.802 |
| 3.0 | 0.866 | 0.855 | 0.987 | 0.844 | 0.833 | 0.811 | 0.790 |
| 5.0 | 0.847 | 0.837 | 0.988 | 0.827 | 0.817 | 0.797 | 0.778 |
| 7.0 | 0.834 | 0.824 | 0.988 | 0.814 | 0.804 | 0.785 | 0.766 |
| 10.0 | 0.821 | 0.811 | 0.988 | 0.801 | 0.791 | 0.773 | 0.754 |

a. After Yamamoto (1962).

siderable importance because of the much greater solar power at these longer wavelengths.

Integration over the solar spectrum of the atmospheric model of Elterman (1968), which contains 0.35 cm STp (standard temperature and pressure) of ozone, leads to the transmissivity due to ozone in table 3.10. At air mass 1 essentially all of the radiation of wavelength less than 0.3 $\mu$m has been absorbed; thereafter the increasing absorption is due to the weaker Huggins and Chappuis bands.

## 3.19 Absorption by the Aerosol

Knowledge of the optical properties of the atmospheric aerosol is incomplete, particularly of the imaginary part of the index of refraction or the absorptivity (see section 2.13). The transmissivity of the aerosol may be found as the ratio of the observed direct solar beam irradiance to that

Table 3.10
Transmissivities due to ozone absorption, integrated over the solar spectrum for 0.35 cm (STp) of ozone

| Optical air mass | Transmissivity | Optical air mass | Transmissivity |
|---|---|---|---|
| 1 | 0.971 | 4 | 0.941 |
| 2 | 0.960 | 5 | 0.933 |
| 3 | 0.950 | 7 | 0.917 |

computed for an aerosol-free atmosphere. A near-simultaneous measurement of the precipitable water is necessary. Normal incidence pyrheliometers have an angle of view greater than that subtended by the sun and therefore accept a portion of the solar aureole as transmitted radiation. The aureole is the bright region surrounding the sun that results from the forescattering, particularly from the aerosol. Sutherland et al. (1975) state that the error due to the aureole is negligible for visible light, but not for the ultraviolet, if the acceptance angle of the pyrheliometer is no more than a few degrees.

The aerosol transmissivity may be determined for the entire solar spectrum, in which case aerosol absorption is included, or in a narrow spectral interval (typically at about 0.5 $\mu$m) where absorption is small. Some observations of the aerosol transmissivity are given in table 3.11. No conclusions regarding secular changes in the aerosol should be drawn from this table.

Ångstrom (1929 a, b) observed the variation of the aerosol transmissivity in the visible spectrum and expressed his results by

$$k_e = \beta\lambda^{-a}, \tag{3.5}$$

where $k_e$ is the aerosol extinction coefficient and $a$ was found to be about 1.3. The Ångstrom turbidity coefficient $\beta$ is usually measured at $\lambda = 0.50$–$0.55$ $\mu$m, where gaseous absorption is minimal. A somewhat similar turbidity coefficient has been defined by Volz as the decadic aerosol extinction coefficient at a wavelength of 0.50 $\mu$m. The Linke and Boda (1922) turbidity factor includes gaseous absorption with the aerosol depletion and is of limited utility.

The fraction of the extinction by the aerosol that is due to absorption is still poorly known. There has been increasing evidence that the aerosol absorption is often as large as the gaseous absorption. Robinson (1962) used clear sky observations of the direct beam radiation and the skylight to deduce the aerosol scattering and absorption. The total aerosol loss was computed as outlined earlier. The difference between the observed skylight and that computed for molecular scattering was taken as the aerosol

Table 3.11
Aerosol Transmissivities for unit air mass

| Location | Transmissivity | Comments |
|---|---|---|
| Washington, D.C. | 0.90 | These are the ratios of a normal incidence pyrheliometer, which accepts the total solar spectrum, to that computed for aerosol-free air. Taken mostly in 1920–1930 from various sources. |
| Lincoln, Nebraska | 0.96 | |
| Davos, Switzerland | 0.97 | |
| Frankfurt, W. Germany | 0.74 | |
| Potsdam, E. Germany | 0.87 | |
| Open country, Argentina | 0.96 | |
| Oceans | 0.86–0.94 | |
| Western United States | 0.80–0.83 | Annual means 1961–1966 as estimated from figure 2 of Flowers et al. (1969). Data are taken at 0.5 $\mu$m wavelength. |
| Great Plains, United States | 0.83–0.87 | |
| Midwest, United States | 0.73–0.80 | |
| Eastern United States | 0.65–0.80 | |
| Kew, London | 0.80 | Deduced from data given by Robinson (1962), total solar spectrum. |
| Halley Bay, Antartica | 0.95 | |
| Vienna, Austria | 0.75 | |
| Lerwick, Scotland | 0.83 | |
| Pretoria, Union of South Africa | 0.87 | |

downscatter. This was coverted to total scatter on the basis of some measurements made by Waldram (1945) summarized in table 3.12. Robinson considered these ratios only approximate, but recent Mie computations for several aerosol size distributions by Joseph and Wolfson (1975) bracket Waldram's results. The difference between the total aerosol depletion and that due to scattering was attributed to absorption. Robinson found that the fraction of the total aerosol extinction due to absorption was two-thirds in London and Pretoria, one-half in Lerwick, Scotland, and Vienna, and one-sixth in Halley Bay, Antarctica. These results cannot be considered definitive because the aerosol absorption is the final remainder and thus incorporates all of the uncertainties.

Joseph and Wolfson (1975), following Robinson's procedure, used carefully selected data from a station in the coastal plain and one in the Judean Hills, both in Israel. On the average, they found that one-half of the aerosol depletion was due to absorption, with a range of one-fifth to two-thirds. Eleven out of the 26 selected days were khamsin, the hot, dry desert wind of the region. This is probably representative of the desert aerosol, which often extends over large areas downwind of major deserts. They found ratios of aerosol absorption to backscatter of about 9 : 15 at air mass 1.2, the lower ratios being on khamsin days, when the aerosol is dry.

Table 3.12
Partition of aerosol scattering into down and up parts[a]

| Optical air mass | 1.0 | 1.1 | 1.2 | 1.4 | 1.7 | 2.0 | 2.5 | 3.3 | 5.0 |
|---|---|---|---|---|---|---|---|---|---|
| Ratio: down/up | 12 | 10 | 8 | 6 | 5 | 3.5 | 2.5 | 2.0 | 1.5 |

a. After Waldram (1945)

## 3.20 Effects of the Aerosol on the Planetary Albedo

It has been conjectured that a presumed increase in the aerosol due to man's activities raises the planetary albedo by increasing the backscatter from the atmosphere. As noted in section 1.18 the evidence for a significant global increase in the aerosol is equivocal. Further, it has been pointed out that an aerosol that absorbs as well as scatters may either increase or decrease the planetary albedo depending on the ratio of the aerosol backscatter to the absorption and the reflectivity of the ground. For typical surface albedos of 0.1–0.2, most workers agree that an aerosol in the lower atmosphere will slightly increase the planetary albedo but that this result is quite sensitive to the imprecisely known ratio of backscatter to absorption. Absorption by the tropospheric aerosol increases the heating of the air and reduces the radiation reaching the surface, thereby increasing the stability of the lower atmosphere, even if the planatery albedo is unchanged.

A stratospheric aerosol will always decrease the solar radiation reaching the troposphere and the surface. However, whether the aerosol layer will increase or decrease the planetary albedo again depends on the ratio of the aerosol backscatter to absorption. The net result of an increase in the stratospheric aerosol is a cooling of the troposphere and a heating of the stratosphere. This is the basis for suggestions that past climatic changes may have resulted from marked changes in volcanism.

## 3.21 Effects of Clouds on Solar Radiation

It is evident from the most casual observations that clouds grossly modify the course of solar radiation through the atmosphere. The water drops or ice particles of which the clouds are composed strongly scatter the radiation, absorb some of it, and markedly reduce the radiation that reaches the surface in most cases. In a gross sense a cloud can be considered to be a corpus, and we speak of its albedo, absorptivity, and transmissivity. These terms are all relative to the radiation incident on the cloud, usually on its top surface. The wide variation in cloud forms, dimensions, and opacity suggests a similarly broad range of albedo, absorptivity, and transmissivity. The standard meteorological classification of cloud forms is essen-

Table 3.13
Some measured albedos of various cloud forms

| Cloud form | Albedo (%) |
|---|---|
| Stratus (500 ft thick) | 5–85 (average 30) |
| Stratus (500–1,000 ft thick) | 9–77 (average 60) |
| Stratus (1,000–2,000 ft thick) | 50–73 (average 70) |
| Stratocumulus | 35–80 |
| Subtropical cumulus | 38–45 |
| Nimbostratus | 64–70 |
| Cirrostratus | 44–59 |
| Cirrus | 15–20 |

tially morphological, and it cannot be expected that there will be any unique relation between cloud form and albedo or the other optical properties. Some examples of measured cloud albedos, taken from a variety of sources, are given in table 3.13 and confirm that cloud type is not an adequate measure of the albedo.

The first three lines of table 3.13 were taken from the work of Neiburger (1949) and suggest that cloud thickness is a useful parameter. These data on California stratus are plotted in figure 3.8, together with observations of Cheltzov (1952). The curves were extrapolated to the surface albedo of the sites in Archangel (USSR) and of the ocean. Each point in the Cheltzov data is an average of about 40 observations. The very considerable scatter of the points shows that other properties, such as the microphysics of the clouds and the solar zenith angle, are important.

## 3.22 Absorptivity of Clouds

In principle, the absorptivity of a cloud layer may be determined from measurements of the up and down hemispheric fluxes above and below the cloud. The radiation absorbed in the cloud is then the difference between the net fluxes (down minus up) at the top and at the bottom of the cloud. This difference of the differences is sensitive to observational errors. It is also seldom possible to make the measurements above and below the cloud simultaneously and in the same horizontal position. Results of such observations are summarized in table 3.14.

The expected uncertainties are in evidence, particularly in the case of the California stratus, where the range of the observations is large and encompasses physically meaningless negative values. The absorptivities of deeper clouds and cloud systems seem a little more consistent, but this may

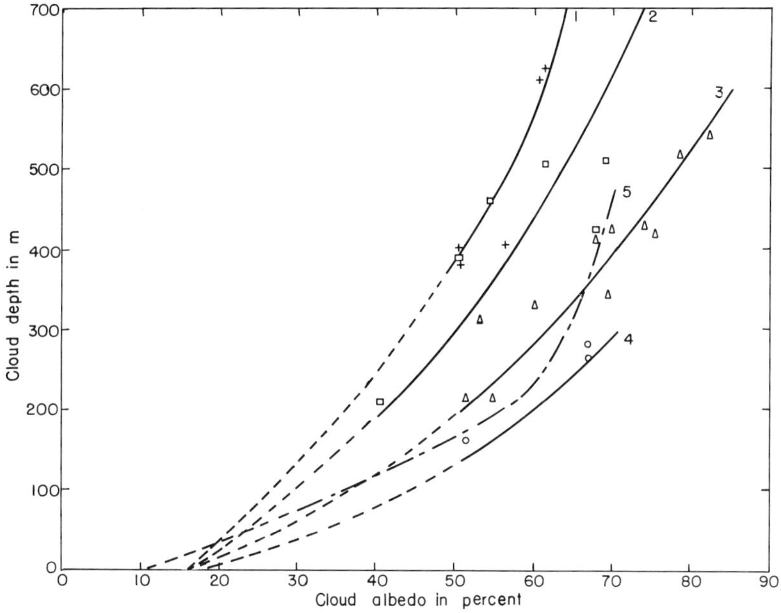


Figure 3.8 Observations of cloud albedos versus cloud depth for several cloud types. Key: curve 1 (+), stratocumulus with breaks; curve 2 (□), stratocumulus and cumulus; curve 3 (△), stratocumulus; curve 4 (○), altocumulus; curve 5, California stratus. After Cheltzov (1952) and Neiburger (1949). Figure 3.8 and table 3.13 are based on the same data.

be due to the small number of observations. It must be remembered that the measured cloud absorptivities include the gaseous absorption. Thus the additional absorption due to the cloud as compared with that in a cloud-free atmosphere is smaller than the above figures would indicate for the deep cloud systems that include much of the atmospheric water vapor.

### 3.23 Average Effect of Clouds

Another simpler means for determining the effect of clouds on the down-coming solar radiation is to compare the radiation received at the surface under overcast skies with that received at the same station at the same time of year with clear skies. Results of such a study by Haurwitz (1948) for Blue Hill, Massachusetts, are shown in table 3.15 in terms of the ratio of the radiation reaching the surface pyranometer to that on a comparable clear day. This cloud transmissivity evidently includes the effects of different amounts of precipitable water and aerosol attenuation on clear versus cloudy days. On the average, the precipitable water is larger on cloudy than on clear days. The results are as consistent, as can be expected with those

Table 3.14
Measured cloud absorptivities[a]

| Cloud description | Absorptivity |
|---|---|
| California Stratus | |
| <500 ft thick | 0.03 (−0.11–0.21) |
| 500–1,000 ft thick | 0.07 (−0.18–0.32) |
| 1,000–2,000 ft thick | 0.06 (−0.15–0.24) |
| Nimbostratus | 0.21 (0.16–0.25) |
| Stratocumulus | 0.22 (0.13–0.29) |
| Stratocumulus-subtropical | 0.12–0.36 |
| Cumulus-stratocumulus | 0.22 (0.13–0.28) |
| Cloud (?) 360 m thick | 0.035 |
| Cloud (?) 530 m thick | 0.072 |
| Deep cloud, 22,000 ft thick | 0.17 ± 0.07 |
| Deep cloud, 18,000 ft thick | 0.27 ± 0.03 |
| Deep cloud, 23,000 ft thick | 0.14 ± 0.06 |
| Cirrus and cirrostratus | 0.13–0.14 |

a. Based on the observations of Neiburger (1949), Griggs and Margraff (1967), Robinson (1958), Reynolds et al. (1975), Cheltzov (1952), and Fritz and MacDonald (1951).

presented in table 3.13. The variation with optical air mass is small and of both signs.

A more common method of representing the effect of clouds on the solar radiation reaching the surface is the establishment of a linear regression between the cloud amount and the fraction of clear sky radiation received. A typical example is

$$Q/Q_0 = 0.29 + 0.71(1 - C), \tag{3.6}$$

where $Q$ and $Q_0$ are the radiation received with cloud amount $C$, in tenths, and that received on cloudless days. The quantities $Q$, $Q_0$, and $C$ are average values over a period of the order of a month. Such expressions are not useful for shorter periods because $C$ is the total cloud amount without regard to cloud type. Different values of the empirical constants are found at different stations and seasons presumably because of the differing mix of cloud types incorporated without distinction in $C$. It follows that expressions of this type are of limited value and then only at the stations at which the regression was established.

A somewhat more satisfactory empirical expression is

$$Q/Q_0 = a + bS, \tag{3.7}$$

Table 3.15
Ratio of insolation with overcast to clear sky (in %)[a]

| Air mass | Ci | Cs | Ac | As | Sc | St | Ns | Fog |
|---|---|---|---|---|---|---|---|---|
| 1.1 | 85 | 84 | 52 | 41 | 35 | 25 | 15 | 17 |
| 1.5 | 84 | 81 | 51 | 41 | 34 | 25 | 17 | 17 |
| 2.0 | 84 | 78 | 50 | 41 | 34 | 25 | 19 | 17 |
| 2.5 | 83 | 74 | 49 | 41 | 33 | 25 | 21 | 18 |
| 3.0 | 82 | 71 | 47 | 41 | 32 | 24 | 25 | 18 |
| 3.5 | 81 | 68 | 46 | 41 | 31 | 24 | | 18 |
| 4.0 | 80 | 65 | 45 | 41 | 31 | | | 18 |
| 4.5 | | | | | 30 | | | 19 |
| 5.0 | | | | | 29 | | | 19 |

a. After Haurwitz (1948).

where $S$ is the ratio of the hours of sunshine, as measured by a sunshine recorder, to the maximum possible hours of sunshine and $a$ and $b$ are empirical constants. On the basis of linear regressions on the monthly means from 11 selected stations in the United States, Fritz and MacDonald (1949) gave $a = 0.35$ and $b = 0.61$. Other values of $a$ and $b$ have been reported in the literature—for example, 0.22 and 0.78. The principal advantages of (3.7) over (3.6) are that $S$ is recorded continuously, while $C$ is a spot observation, and that it is not affected by thin cirrus, while it carries full weight in $C$.

Scattered and broken cumulus clouds are very common over subtropical ocean areas. Reflection from the sides of such clouds often leads to larger than clear sky radiances at the surface at a given moment. Thus Kaiser and Hill (1976) found that the irradiance increased on the approach of a cloud until the cloud obscured the sun. In the mean and with a cloud cover less than six-tenths, the average irradiation exceeds that which would be predicted by expressions like (3.6) and (3.7). Davies (1978) has carried out a very complete analysis of the transfer of solar irradiance in three-dimensional cuboidal clouds that may be consulted by interested readers.

## 3.24 Computed Cloud Albedos

In all except optically thin clouds, multiple scattering dominates the course of solar radiation in clouds.[3] For this reason a successful optical cloud

3. A cloud through which the direction to the sun can be observed as a bright area may be treated as a single-scattering problem; for all other clouds multiple scattering must be used.

model must be based on one of the methods outlined in sections 2.16–2.18 for the numerical treatment of radiative transfer by multiple scattering. This section is devoted to a presentation of some of the results of the calculations of cloud albedo by one or another of these methods.

Typical cloud drops have radii that are large compared with the wavelength in the more important parts of the solar spectrum, and hence their scattering efficiency factor is essentially independent of both the wavelength and the drop radius. This greatly simplified the computations. It also suggests that the important parameter is the optical depth, the product of the distance (depth of cloud) and the volume-scattering coefficient.

Absorption occurs both in the drops (or ice crystals) and the gaseous absorbers, principally water vapor. Almost all of this absorption occurs at wavelengths greater than 0.7 $\mu$m, which is also the approximate upper limit of visible light. If absorption is neglected, one obtains the visual cloud albedo, which will be somewhat larger than the albedo averaged over the entire solar spectrum when absorption is taken into account.

The first attempts to compute the albedos of terrestrial clouds were based on what is now called the "two-stream model." This method was introduced by Schuster (1905) for an astrophysical problem and was applied by Mecke (1921) and Hewson (1943) to water clouds. It will suffice to present some of Hewson's results.

The basic differential equation for thc two-stream model is equation (2.14) The solution for the cloud albedo, assuming that the fractional backscatter and the scattering coefficient are invariant with height in the cloud and that the surface albedo is zero, is

$$\text{cloud albedo} = Bt/(1 + Bt), \tag{3.8}$$

where $t$ is the optical depth and $B$ is the fraction of the scattered radiation that is in the backward hemisphere. It will be remembered that the two-stream model assumes that all of the radiation is isotropic, so that it is usually not possible to include the effect of the solar zenith angle. Hewson included absorption in the drops, but not the gaseous absorption. It is considered here that the albedos he obtained are essentially for the visible spectrum and for a solar zenith angle of 67°, the latter based on arguments given in the reference. The results are plotted in figure 3.9, together with others to be considered.

The next significant advance was due to Fritz (1954), who divided a cloud into a series of horizontal layers each thin enough to make a second collision of a photon with a drop unlikely. Starting with an incident solar beam at an arbitrary zenith angle, he distributed the radiation scattered in the top layer according to the phase function computed by Wiener (1900) from geometric optics. In the second layer he treated the remainder of the incident solar beam in the same way as in the top layer, and in addition,

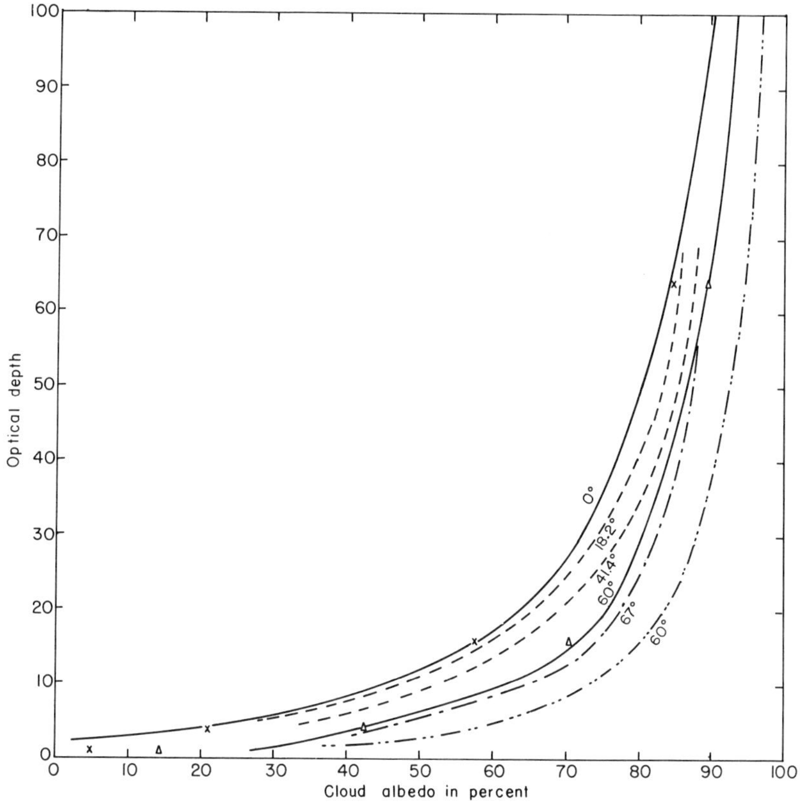


Figure 3.9 Computed cloud albedos as a function of the optical depth. (The optical depth is the product of cloud depth and the scattering coefficient or the extinction coefficient.)

| Key | Solar zenith angle (degrees) | Reference |
|---|---|---|
| ——— | 0 | |
| | 60 | Danielsen et al. (1969) |
| – – – | 18.2 | Twomey et al. (1966) |
| | 41.4 | Twomey et al. (1967) |
| –·–· | 67.0 | Hewson (1943) |
| –·–·– | 60 | Fritz (1954) |
| × | 0 | Liou (1973a) |
| △ | 60 | |

computed the second scattering of that scattered in the first layer. This procedure rapidly becomes ponderous, but Fritz carried it through on a desk calculator. At each level he subtracted the isotropic part of the radiation that he could follow analytically. He also assumed that the edge-diffracted radiation could be treated as a part of the direct solar beam. Fritz was evidently the first to show quantitatively how the cloud albedo varies with the solar zenith angle. The albedos obtained by Fritz for a zenith angle of 60° are plotted in figure 3.9, and it appears that his technique led to albedos that are somewhat too large.

More adequate means of handling radiative transfer in a multiple scattering medium are discussed in sections 2.15–2.17. These procedures have been applied to clouds by several investigators. The results of three of the studies are plotted in figure 3.9. The general agreement among them is apparent; the curves of Danielson et al. (1969) and Liou (1973a) are so nearly coincident that only the data points for the latter are plotted. The curves of Twomey et al. (1966, 1967) are intermediate between these of Danielson et al., as would be expected from the solar zenith angles. The close correspondence of these three sets of curves in spite of a number of different assumptions regarding the cloud drop size and concentration, the phase function, and the single-scattering albedo is unexpected. It is also of interest that Danielson et al. used the Monte Carlo method, Twomey et al. a matrix form of the doubling method, and Liou the discrete ordinate method.

It may be concluded that the cloud albedo may be obtained from figure 3.9 if the absorption is small, when the solar zenith angle and the optical depth are given. Unfortunately the optical depth depends on the size distribution and concentration of the cloud drops, quantities that are not usually available. Equation (2.24) gives a relation between the visual range and the scattering coefficient. Thus if the visual range is known or estimated and the cloud depth is known, the optical depth is specified. The estimated visual ranges in cloud at the summit of Mount Washington, New Hampshire, ranged from about 30 to 100 m, with a mode at about 50.

It must be remembered that the computed cloud albedos are for homogeneous clouds of indefinite horizontal extent with smooth horizontal tops and bases. Real clouds often contain billows, turrets, and other inhomogeneities. The effects of these on the cloud albedo are not known, although one would expect that an irregular top surface would reduce the dependence of the albedo on the solar zenith angle. It is common to assume that the albedo of a broken or scattered cloud field is the linear product of the fractional cloud amount and the albedo of a continuous cloud cover. This may be a satisfactory approximation for relatively thin layer clouds, but is almost certainly invalid for scattered and broken convective clouds because of the radiation scattered from their sides.

Table 3.16
Properties of four model clouds selected by Twomey[a]

| Name | Droplet concentration ($cm^{-3}$) | Liquid water ($g\,m^{-3}$) | Temperature (°C) |
|---|---|---|---|
| Maritime | 25 | 0.33 | 10 |
| Moderately continental | 200 | 0.33 | 10 |
| Strongly continental | 1,000 | 0.33 | 10 |
| Ice cloud | 25 | 0.33 | −20 |

a. Twomey (1976b).

## 3.25 Computed Cloud Absorptivities

The cloud absorptivity may be computed in much the same way as the albedo, but with the further complications that the absorption coefficients of the gases and the water (or ice) are highly dependent on wavelength and that an integration over the spectrum is required to get the absorptivity. Some investigators have included only the absorption by the cloud particles, usually by specifying a single-scattering albedo[4] of around 0.999. The gaseous absorption is not negligible and may exceed that in the droplets at warm temperatures. Since the water vapor concentration in the clouds is dependent on temperature, a very large number of computations would be required to encompass the properties of natural clouds.

Twomey (1976b) has computed the absorptivity of a few model clouds using the doubling method and detailed data on the absorption coefficients of water, ice, and water vapor. He selected four cloud models, as listed in table 3.16 In each case a Gaussian drop size distribution was used with a relative dispersion of 0.2 (standard deviation divided by mean radius). Twomey notes that any other reasonable distribution of dispersion 0.2 or larger will lead to nearly the same result. Twomey's results are shown in figure 3.10 and 3.11.

For a given liquid water content, the cloud with the largest drops (maritime) gives the greatest absorptivity. Deep clouds of the maritime type have absorptivities of about 0.2, even larger absorptivities are to be expected for liquid water contents greater than that assumed by Twomey. These results agree as well as can be expected with the values given in table 3.14. Some more recent observations over the subtropical ocean have given absorptivities of more than 0.5. It seems impossible to explain such large

4. The single-scattering albedo is the ratio of the absorption coefficient to the extinction coefficient or the corresponding efficiency factors.

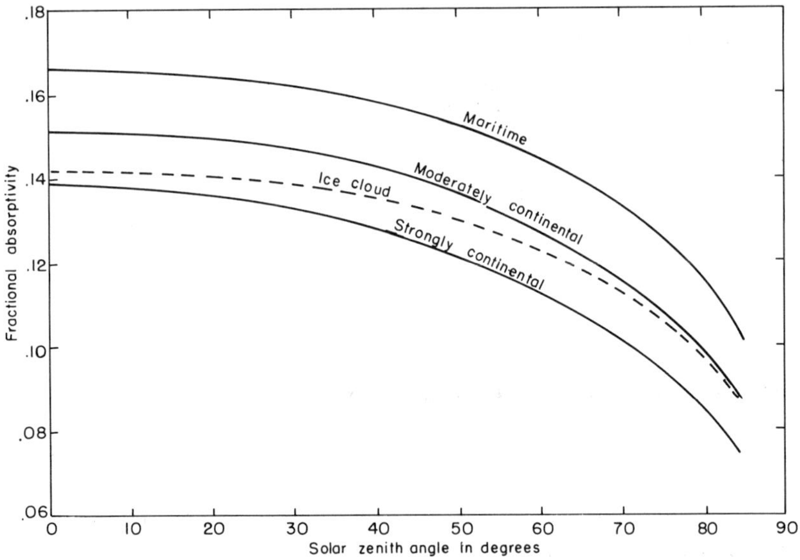


Figure 3.10 Computed absorptivities of clouds as a function of the solar zenith angle and the cloud microphysics. After Twomey (1976b).

values by a model consisting of a continuous cloud layer with a horizontal base and top. Perhaps the discrepancy is related to marked differences between this idealized cloud and the natural cloud field.

Twomey does not discuss his ice cloud model in detail, but it is likely that the ice particles were considered to be spherical. In natural ice clouds, the particles are ice crystals of a variety of habits, few of which approximate spheres (see section 7.12). In cirrus clouds the crystals are often hexagonal columns. Liou (1973b) and Jacobowitz (1971) found that the phase functions of long cylinders and hexagonal columns differ significantly from those for an "equivalent sphere." Calculations made by Liou and Jacobowitz cannot yet be considered definitive because the microphysics of cirrus clouds are poorly known. Unanswered questions include the extent to which crystals are oriented in cirrus clouds and the prevalence of aggregates of columnar and bullet crystals.

## 3.26 The Surface Albedo over Land

A significant fraction of the solar radiation that reaches the land surface is reflected back into the atmosphere. The surface albedo depends greatly on the nature of the surface, the vegetative cover, and the snow cover. Table 3.17 is a general overview of surface albedos taken from various sources. The ranges in the table do not represent extremes, but only the range

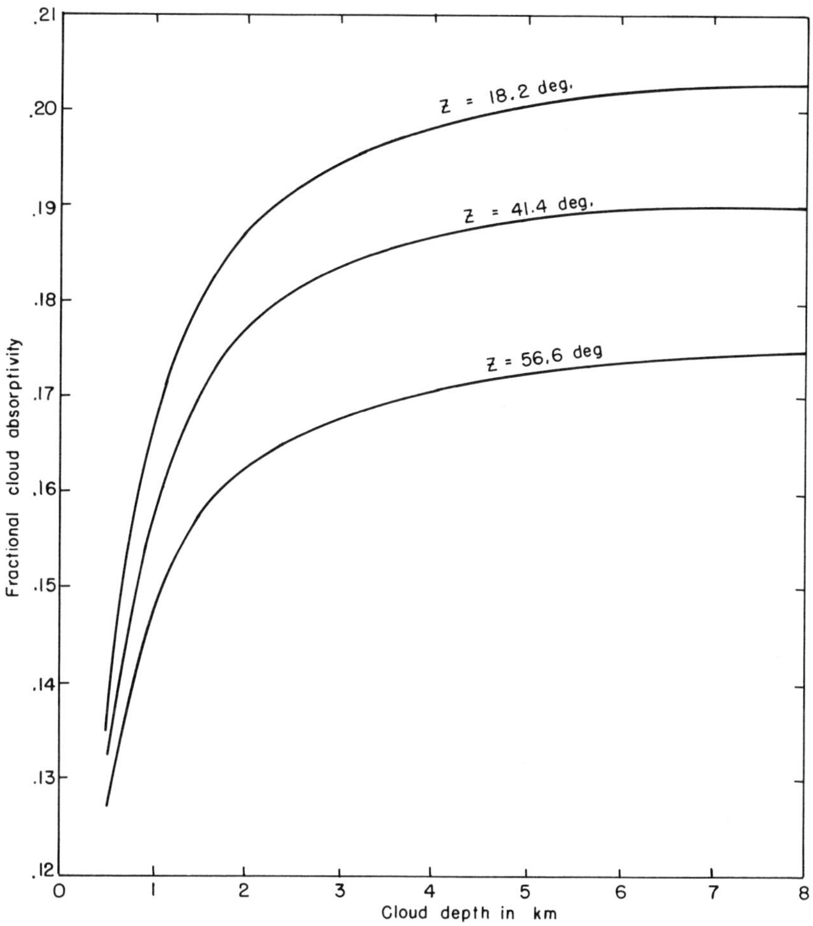


Figure 3.11 Computed absorptivities of maritime clouds as a function of cloud depth for three solar zenith angles. After Twomey (1976b).

Table 3.17
Surface albedos[a]

| Surface | Albedo (%) | Surface | Albedo (%) |
|---|---|---|---|
| Bare ground | 10–12 | General woodlands | 13–19 |
| Sand and desert | 18–28 | Coniferous woods | 9–14 |
| Grasslands | 16–20 | Deciduous woods in leaf | 15–19 |
| Swamps and marshes | 8–12 | Fresh snow | 70–95 |
| Green crops | 15–25 | Old snow | 45–70 |
| Prairie | 14–16 | Cities and towns | 14–18 |

a. Averaged over the solar spectrum.

Table 3.18
Surface albedos of North America versus latitude[a]

| Latitudinal zone (°N) | Winter mean snow cover (%) | Summer (%) | Weighted annual mean (%) |
|---|---|---|---|
| 20–25 | 15.8 | 15.8 | 15.8 |
| 25–30 | 17.8 | 17.9 | 17.9 |
| 30–35 | 19.1 | 17.2 | 17.8 |
| 35–40 | 28.5 | 16.5 | 19.5 |
| 40–45 | 37.9 | 15.8 | 20.3 |
| 45–50 | 46.4 | 14.8 | 19.7 |
| 50–55 | 50.3 | 14.6 | 18.4 |
| 55–60 | 59.1 | 16.5 | 18.9 |
| 60–65 | 67.3 | 15.6 | 16.4 |
| 65–70 | 82.8 | 16.1 | 16.1 |

a. From Kung et al. (1964).

of average values due to variations in surface properties within each category.

Kung et al. (1964) give an account of the surface albedo of North America based on 210,000 albedo observations and maps of the vegetative cover and of the extent of snow cover. Their latitudinally averaged surface albedos for a winter with average snow cover and for summer are given in the second and third columns of table 3.18. The values in the last column, not given by Kung et al., were obtained by weighting the winter and summer albedos in proportion to the radiation received. This shows that the large winter albedos due to snow cover in high latitudes are of relatively little importance in the annual mean due to the small amount of radiation received there in winter.

Ground surfaces are not perfect Lambertian reflectors, but their albedos vary with the solar zenith angle. In addition, the albedos are dependent on wavelength, particularly green plants, which reflect most strongly in the near infrared. Figures 3.12 and 3.13, taken from the scanty literature on these subjects, illustrate the complex behavior of the surface albedo with solar zenith angle and wavelength. The rapid increase in albedo beginning at 0.7 $\mu$m is characteristic of green leaves; it is followed by a decline at about 15 $\mu$m and finally approaches zero in the thermal infrared. Snow is a surface cover of considerable importance because of its seasonal and shorter-term variability. It has a high albedo for solar radiation, but is nearly black in the thermal infrared. Dramatic changes in the radiative budget follow the freezing of water areas and a subsequent snow fall.

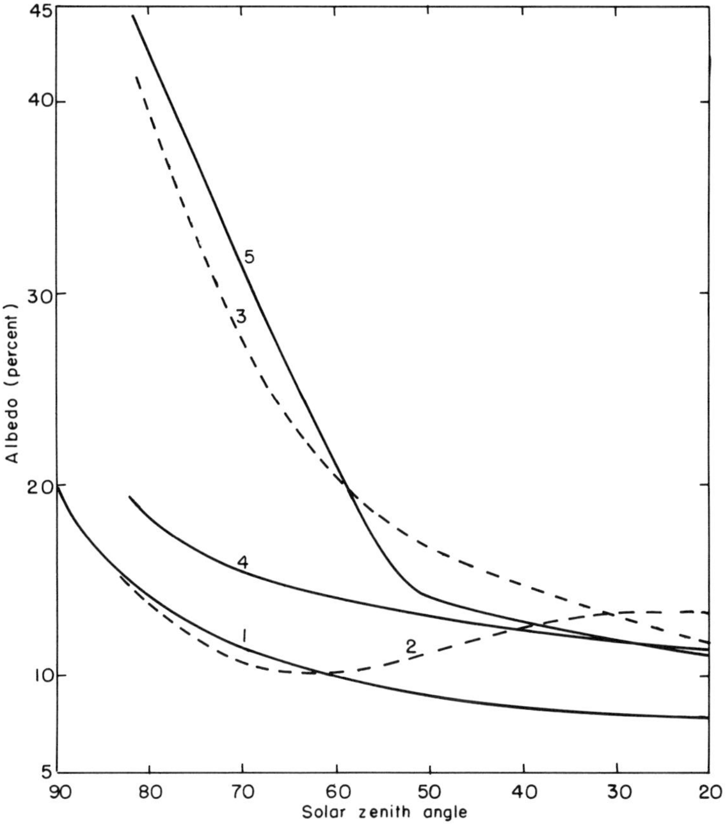


Figure 3.12 Albedos of some ground surfaces and their variation with the solar zenith angle: curve 1, pine forest; curve 2, bare ground (morning); curve 3, bare ground (afternoon); curve 4, irrigated potatoes (morning); curve 5, irrigated potatoes (afternoon). Curve 1 is after Stewart (1971); curves 2–5 are after Nkemdirin (1972).

## 3.27 Albedo of the Sea Surface

The albedo of the sea is more uniform than that of land areas, except in high latitudes, where there are seasonal changes from water to ice and snow. The reflectivity of a plane water surface for direct solar radiation may be computed from the Fresnel coefficients as a function of the angle of incidence. The results, which are given in table 9.2, show that the reflectivity is only 2% at normal incidence, but increases rapidly toward unity at glancing incidence.

The albedo of the real ocean surface depends on the angular distribution of the incident radiation, both diffuse and direct beam, and the roughness

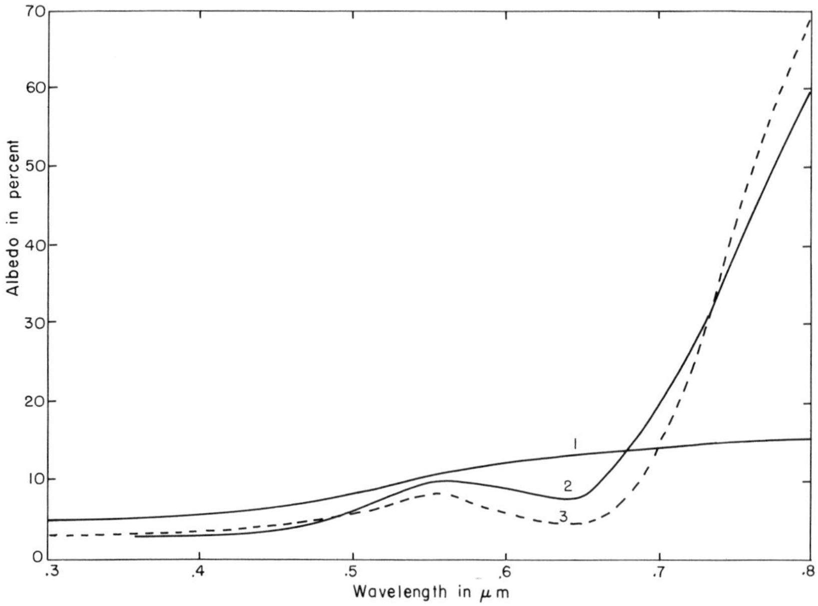


Figure 3.13 Albedos of three ground surfaces as a function of the wavelength, all at a solar zenith angle of 60°: curve 1, dry disked Yolo loam; curve 2, blue grass; curve 3, alfalfa. After Coulson and Reynolds (1971).

of the surface. The diffuse radiation depends primarily on the cloudiness, and its reflectivity is about 6%. Because of these complications, some investigators have chosen to give albedos for perfectly clear skies and/or a solid overcast, often at local noon. For many purposes such albedos are of limited utility.

In my opinion the most satisfactory sea surface albedos are those of Payne (1972), which are based on a large number of instrumental observations taken at the Buzzards Bay Entrance Light Station (41°24′N, 71°02′W). The data were automatically recorded during the period from 25 May to 28 September in 1970. Payne found that the albedos could be stratified by the altitude angle of the sun and the total transmissivity of the atmosphere, taken as the ratio of the measured hemispheric radiation to the radiation at the top of the atmosphere. This procedure was found to account for the effects of cloud cover as well as the atmospheric scattering and absorption. Figure 3.14 shows the albedo for clear skies (transmissivity 0.60–0.65) and a thin or broken cloud cover (transmissivity 0.35–0.40). The computed albedo for a plane water surface and a parallel incident beam is included for comparison. For heavily overcast skies (transmissivity $\leq 0.1$), the albedo was found to be $0.060 \pm 0.005$, independent of the solar elevation angle.

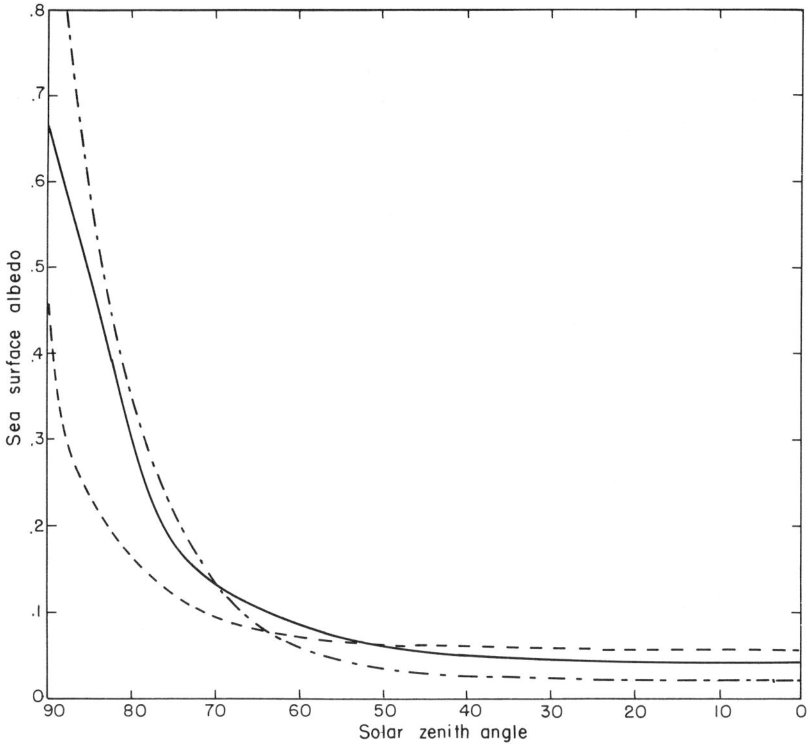


Figure 3.14 The albedo of the ocean surface as a function of the solar zenith angle and the vertical transmissivity of the atmosphere: ———, clear (transmissivity 0.60–0.65); – – –, thin overcast (transmissivity 0.35–0.40); –·–·–·, computed for a parallel beam and a smooth surface. After Payne (1972).

Using Budyko's (1963) values of the monthly mean surface irradiation, Payne (1972) has deduced monthly mean sea surface albedos for the North Atlantic Ocean as a function of latitude; his results, in abridged form, are given in table 3.19.

## 3.28 Zonal Average Surface Albedo

For studies of the radiation budget of the earth, there is often a need for the zonal average surface albedo. Several studies have yielded similar results, but important quantitative differences remain. These seem to result from various assumptions about the cloudiness, the albedo of the ocean surface, the albedo over the continents, and the snow and ice in high latitudes. In comparison with earlier results the Hummel and Reck (1979), albedos are lower from 30°S to 30°N and higher from 50°–90° in both hemispheres. Their average global albedo is found to be 15.4% which may be compared

Table 3.19
Mean albedos of the North Atlantic[a]

| Latitude (°N) | January | April | July | October |
|---|---|---|---|---|
| 80 | — | 0.14 | 0.08 | — |
| 70 | — | 0.10 | 0.07 | 0.25 |
| 60 | 0.28 | 0.07 | 0.06 | 0.10 |
| 50 | 0.11 | 0.07 | 0.06 | 0.08 |
| 40 | 0.10 | 0.07 | 0.06 | 0.07 |
| 30 | 0.09 | 0.06 | 0.06 | 0.06 |
| 20 | 0.07 | 0.06 | 0.06 | 0.06 |
| 10 | 0.07 | 0.06 | 0.06 | 0.06 |
| 0 | 0.06 | 0.06 | 0.06 | 0.06 |

a. After Payne (1972).

Table 3.20
Zonal surface albedos[a]

| Latitude (°N) | Winter, January–March (%) | Summer, July–September |
|---|---|---|
| 90 | 70 | 61 |
| 80 | 71 | 54 |
| 70 | 65 | 29 |
| 60 | 46 | 13 |
| 50 | 30 | 11 |
| 40 | 20 | 11 |
| 30 | 14 | 10 |
| 20 | 10 | 9 |
| 10 | 7 | 7 |
| 0 | 7 | 7 |

a. After Hummel and Reck (1979).

with earlier estimates of about 13.0%. No attempt is made here to resolve these differences because of the many complexities. The Hummel and Reck albedos are given in abridged form for the Northern Hemisphere in table 3.20.

## 3.29 Computations of Solar Radiation

It is often desired to compute the solar radiation received at the surface with clear skies. The same computation yields an estimate of the absorption of solar radiation in the atmosphere. An approximate method for making these computations will now be presented.

The irradiance on a horizontal surface area due to the direct solar beam in a cloud-free atmosphere is given by

$$E = S \cos Z \, [a_a(mp/p_0)] \, [a_w(wm)] \, [a_0(m)] \, [a_d(m)], \tag{3.9}$$

where $S$ is the solar constant, corrected for the solar distance, $Z$ is the zenith angle of the sun, and $a_a$, $a_w$, $a_0$, $a_d$ are, respectively, the transmissivities due to molecular scattering, gaseous absorption, ozone absorption, and the aerosol. The quantities in parentheses indicate the parameters on which each transmissivity depends, where $m$ is the optical air mass, $w$ the precipitable water, $p$ the pressure at the surface, and $p_0$ the standard sea level pressure (1013.3 mbar). Strictly, all of the transmissivities should be for a single wavelength and the product then integrated over the solar spectrum. Experience has shown that this onerous procedure is not necessary, particularly because of the uncertainty of the aerosol transmissivity. Instead, each partial transmissivity is weighted by the solar spectrum. The data in tables 3.8–3.11 are in this form.

In order to estimate the contribution of skylight to the surface irradiation, it is necessary to separate the total depletion of the direct solar beam into absorption and scattering. Gaseous absorption occurs primarily in the red and infrared (except for ozone absorption in the ultraviolet), while molecular scattering occurs predominantly at the shorter wavelengths (see figure 3.6). Thus molecular scattering and gaseous absorption are nearly independent. Radiation scattered from the solar beam is still subject to absorption; that scattered forward traverses nearly the same amount of absorber as the direct beam, while that scattered backward at high levels traverses little absorber, and that scattered up from near the surface traverses the absorbers twice. From these arguments, the location of most of the ozone in the stratosphere, and that of the aerosol in the lower atmosphere, it is reasonable to assume that ozone absorption occurs first, followed by gaseous absorption, molecular scattering, and the aerosol depletion. This is summarized in table 3.21. These are fractions of the irradiance at the top of the atmosphere on a horizontal surface. A further

Table 3.21
Scheme for separating absorption and scattering

| | |
|---|---|
| Fraction absorbed by ozone | $[1 - a_0(m)]$ |
| Fraction absorbed by other gases | $[a_0(m)][1 - a_w(wm)]$ |
| Fraction scattered by air molecules | $[a_0(m)][a_w(wm)][1 - a_a(mp/p_0)]$ |
| Fraction depleted by the aerosol | $[a_0(m)][a_w(wm)][a_a(mp/p_0)][1 - a_d(m)]$ |

breakdown of the aerosol depletion into absorption and scattering is needed. Pending more complete observational evidence, half of the aerosol extinction may be attributed to absorption and the other half to scattering. The scattered fraction may be divided into down and up parts with the aid of table 3.12. The total skylight at the surface is then the aerosol down-scatter plus one-half of the molecular scattering.

The most uncertain steps in this procedure are the selection of an aerosol transmissivity and the division of the aerosol depletion between absorption and scattering. The other approximations are thought to be relatively less important. If clouds are included, the uncertainty of the results becomes greater. For monthly means, the best that can be done, with the data usually available, is to use (3.7) or (3.6). If the type and amount of the cloud cover are known for an individual day, or portion of a day, the cloud transmissivities given in table 3.15 may be used. When clouds are present the radiation passing through them is diffuse. For a somewhat similar model for the computation of solar radiation reaching the surface see Atwater and Brown (1974).

## 3.30 The Disposition of Solar Radiation

A more complete consideration of this subject is given in sections 5.22–5.24, where it is combined with information on the transfer of thermal radiation to give an insight into the total radiative budgets.

The most obvious feature of solar radiation is the annual variation in mid- and high latitudes. This is due primarily to the spherical shape of the earth and to the tilt of the earth's axis. In the annual mean the solar radiation reaching the surface has a small maximum in the subtropics and then decreases toward the pole. In the winter hemisphere, the decrease with latitude is more rapid; in summer there is relatively little change with latitude.

In the global average, roughly one-half of the solar radiation incident on the atmosphere is absorbed at the surface, about 20% is absorbed in the atmosphere, and 30% is reflected or scattered back to space. This planetary albedo of about 0.3 is derived from satellite observations and is appreciably smaller than pre-Sputnik estimates of about 0.35. The difference is concentrated in the tropical and subtropical regions, and it appears that earlier computations overestimated the albedos of both cirrus and cumuliform clouds. Much of the increased absorption evidently occurs at the surface of the earth.

## Problems

1.
Discuss the pros and cons of the Smithsonian long and short methods of measuring the solar constant.

2.
What size particles will approximately match the Smithsonian water vapor scattering transmissivities?

3.
Tabulated below are pyrheliometric data taken at Blue Hill, Massachusetts, on 12 April 1969.

| Time | Total on horizontal surface (cal $cm^{-2}$ $min^{-1}$) | Normal to direct beam (cal $cm^{-2}$ $min^{-1}$) |
|---|---|---|
| 0530 | 0 | 0 |
| 0600 | 0.071 | 0.561 |
| 0700 | 0.365 | 0.987 |
| 0800 | 0.645 | 1.162 |
| 0900 | 0.891 | 1.246 |
| 1000 | 1.072 | 1.239 |
| 1100 | 1.189 | 1.278 |
| 1200 | 1.255 | 1.314 |

There was 0.5 cm of precipitable water. The latitude of Blue Hill is 42.3°N.

a. Compute the transmissivity due to the aerosol.

b. Is there any evidence of absorption by the aerosol?

4.
What would be the effect of a doubling of the carbon dioxide on the disposition of the solar radiation? A qualitative or semiquantitative answer is all that is expected.

5.
What is the optimal orientation of a solar-heated building?

# 4

# Principles of Atmospheric Thermal Radiation

## 4.1 Introduction

In the mean, the solar radiation absorbed by planet earth is returned to space as long-wave or thermal radiation emitted by the atmosphere and the surface of the earth. Largely for geometrical reasons the absorbed solar radiation is a maximum near the equator and, in the annual mean, decreases rapidly with increasing latitude. The long-wave radiation to space decreases more slowly with the latitude. Thus there is an excess of absorbed solar radiation in low latitudes and a deficit in high latitudes. These excesses and deficits are redressed by the latitudinal transport of heat by the circulations of the atmosphere and the oceans; in turn, these circulations are driven by the radiational imbalances. The radiational heat source is largely at and near the surface of the earth, while the radiational heat sink, the outgoing long-wave radiation, has its origin primarily in the upper troposphere, thus satisfying the basic requirement of a heat engine. The seasons owe their origin to the annual variation of the solar declination and the consequent changes in the imbalance between the absorbed solar radiation and the outgoing long-wave radiation. The radiation budget exhibits significant changes over smaller time and space scales, which are coupled to the atmospheric circulations through the distribution of cloudiness and water vapor.

It is evident that radiative exchange is basic to an understanding of the atmosphere and its motions, and this has long been recognized. Nevertheless, it is only in the past decade or so that it has become possible to make effective use of the accumulated knowledge of radiative transfer in the atmosphere. The newer numerical models of the global atmospheric circulation include simplified representations of radiative exchange. Concern about possible inadvertent climatic change due to an increasing amount of atmospheric carbon dioxide and a possible increase of atmospheric particulates has stimulated studies of the resultant changes in the radiative budget. The most novel application of radiation is the remote sensing, from earth satellites, of the vertical distribution of temperature, water vapor, and ozone by means of infrared and microwave spectrometers.

This chapter contains a brief historical account of the development of methods for the estimation of thermal radiative transfer in the atmosphere, but most of the chapter is devoted to a simplified account of the far-infrared absorption bands of the principal absorbing gases. Chapter 5 discusses methods and results of computations of thermal radiative fluxes and flux divergence in the atmosphere and the global radiative heat budget. It also includes a section on the remote sensing of atmospheric structure by means of thermal and microwave radiation.

## 4.2 Kirchhoff's Law

This, one of the basic laws of radiation, has stood the test of time through repeated experimental confirmations so long as the basic condition of the law is met, namely, that local thermodynamic equilibrium exists.

Consider a cavity enclosed by opaque walls that may have different radiative properties in different regions. When thermodynamic equilibrium is attained, the entire cavity will have a uniform temperature in accordance with the second law of thermodynamics. At a given wavelength, the radiative flux within the cavity must be the same everywhere, since none can escape. Each unit area of the wall emits radiation at the wavelength in question and also absorbs some fraction of the uniform flux of radiation within the cavity. To maintain the prescribed equilibrium the radiation absorbed on each unit area of the wall must be equal to the radiation it emits. Thus

$$E_1 = a_1 B, \qquad E_2 = a_2 B, \qquad \ldots, \qquad E_m = a_m B, \tag{4.1}$$

where $B$ is the uniform flux within the cavity, $E_1, E_2, \ldots, E_m$ are the emissions per unit area of different portions of the wall as designated by the subscripts, and $a_1, a_2, \ldots, a_m$ are the absorptivities of these wall sections, all for the selected wavelength and the equilibrium temperature. (Absorptivity is the fraction of the incident radiation absorbed.) This leads to

$$\frac{E_1}{a_1} = \frac{E_2}{a_2} = \cdots = \frac{E_m}{a_m} = B = \text{constant}. \tag{4.2}$$

This is one form of Kirchhoff's law. In words, at a given temperature and wavelength, under conditions of local thermodynamic equilibrium, the ratio of the emission to the absorptivity of any substance is a constant. The maximum possible value of $a_i$ is unity and, by definition, a blackbody is one that has an absorptivity of unity at all wavelengths. Thus the constant flux $B$ is the blackbody flux at the prescribed temperature and wavelength. It follows that a blackbody emits the maximum possible amount of radiation at any given temperature and wavelength. Thus a blackbody is often called a "perfect" radiator or absorber and is used as the standard against which other radiators or absorbers are compared. We define the emissivity of any body as the ratio of its emission at a selected temperature and wavelength to that of a blackbody under the same conditions. Inserting this definition in (4.2) leads to an alternative form of Kirchhoff's law,

$$\frac{\varepsilon_1}{a_1} = \frac{\varepsilon_2}{a_2} = \cdots = \frac{\varepsilon_m}{a_m} = 1, \tag{4.3}$$

or

$$\varepsilon_1 = a_1, \qquad \varepsilon_2 = a_2, \qquad \ldots, \qquad \varepsilon_m = a_m. \tag{4.4}$$

As already noted, the flux within the cavity designated by $B$ in (4.2) is the blackbody flux at the equilibrium temperature. Advantage is taken of this in the construction of laboratory sources of blackbody radiation. A cavity of any convenient shape is heated to a uniform temperature, and a small fraction of the blackbody radiation within the cavity is allowed to escape through a small hole.

Experiment has shown that Kirchhoff's law is valid so long as the conditions for thermodynamic equilibrium are fulfilled. In the case of gaseous absorbers, as in the atmosphere, what is required is that the molecules have an opportunity to exchange energy with their neighbors by a sufficient number of collisions to approach thermal equilibrium during the lifetime of the excited state responsible for the emission. As the altitude increases, the rate of molecular collisions decreases with the air density (and temperature), whereas the characteristic time of the emission process remains nearly the same. Evidently a level is finally reached at which the density is so low that the rate of collisions will no longer suffice to maintain kinetic equilibrium. Goody (1954) and others have stated that Kirchhoff's law is probably valid up to 50–70 km, depending on the particular excited state. The lifetime of a rotational state is typically shorter than that of a vibrational state.

## 4.3 Blackbody Radiators

By using observations of Tyndall on the radiation emitted by a laboratory blackbody of the kind described, Stefan found empirically that the total radiation emitted was proportional to the fourth power of the absolute

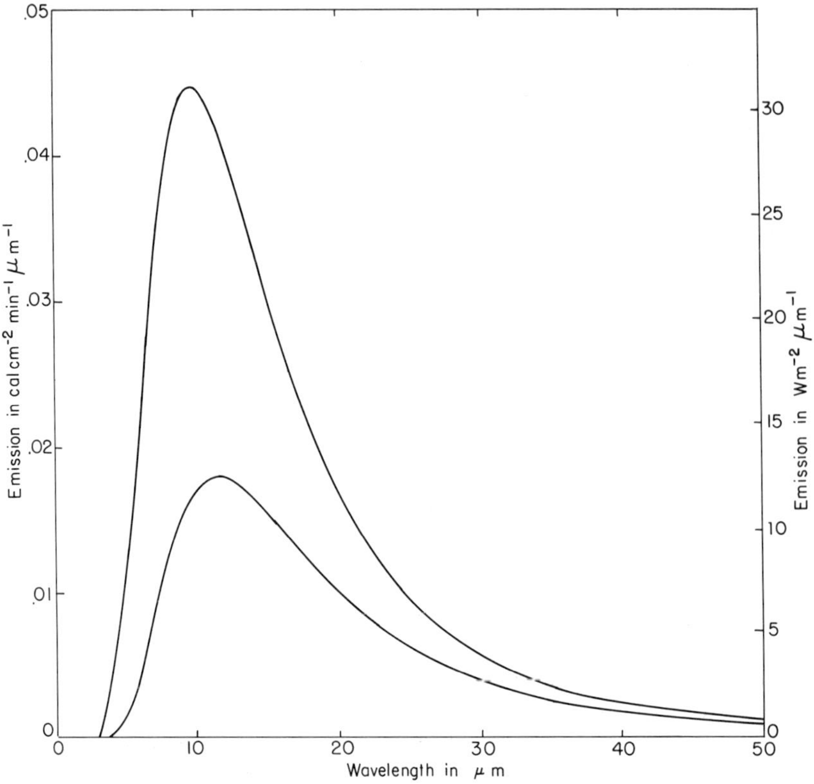


Figure 4.1 The emissions of black bodies at 250 and 300°K.

temperature. Later, Boltzmann derived this relation from thermodynamical considerations. The result, $E = \sigma T^4$, where $T$ is the absolute temperature, is called the Stefan-Boltzmann law and $\sigma$ the Stefan-Boltzmann constant.

Further experimentation established the spectral distribution of the radiation emitted by a black or ideal radiator. Figure 4.1 shows the spectral emission of blackbodies of temperature 300 and 250°K. Each curve has a single maximum and is asymptotic to zero as the wavelength approaches zero or infinity. The wavelength of maximum emission increases with decreasing temperature, and the emission at any given wavelength increases with temperature.

## 4.4 The Ultraviolet Catastrophe

Attempts to explain the experimental results on the basis of classical physics led Planck to develop his famous distribution law. Efforts to invest this law with physical significance by Planck and other physicists, including Boltzmann, Lorentz, and Einstein, finally led to the quantum theory and the beginning of modern physics. A brief outline of the predictions of classical theory that ultimately led to the development of quantum theory is in order here because of its historical interest.

The classical theory of radiation was based on the theory of elastic vibrations. Standing waves are set up in any finite length of medium as a result of constructive interference between direct and reflected waves. The simplest example is a stretched wire or string. The fundamental vibration is at a wavelength of twice the length of the string. The frequencies of the overtones are 2, 3, 4, . . . times the fundamental, and in principle, an infinite series of overtones of increasing frequency is formed. In a solid the series is terminated when the wavelength approaches twice the separation of the atoms, but no such limit is imposed when this principle is applied to radiation.

If the simple theory of a vibrating string is extended to the three-dimensional case, as is appropriate to the standing waves within the blackbody cavity, it is found that

$$dN = \frac{8\pi V\nu^2}{c^3}\,d\nu, \tag{4.5}$$

where $dN$ is the number of overtones in the frequency interval from $\nu$ to $(\nu + d\nu)$, $V$ is the volume of the cavity, and $c$ is the speed of light. The classical theory of linear oscillators requires that the energy of each overtone be $kT$, where $k$ is Boltzmann's constant and $T$ is temperature. The total energy density is obtained by multiplying $dN$ by $kT$ and then dividing by $V$ and $d\nu$. This gives

$$\text{energy density} = 8\pi kT\nu^2/c^3. \tag{4.6}$$

This is the Rayleigh-Jeans radiation law. It correctly explains the blackbody emission curve well to the right of the maximum, but predicts an increase in the energy density without limit as the frequency increases. This is the so-called ultraviolet catastrophy. It leads to the impossible result that the total radiation emitted by a blackbody at any finite temperature is infinite.

Planck postulated that the radiation is emitted only in finite packets or quanta and that the energy of a quantum is given by $h\nu$, where $h$ is Planck's constant and $\nu$ is the frequency. Thus no radiation whatever can be emitted at a given frequency $\nu$ unless an atom or molecule has available for emission at least the energy $h\nu$. At any temperature the energies of the molecules follow a Boltzmann distribution. At higher and higher frequencies the number of molecules with energy states equal to $h\nu$ becomes smaller and smaller. Although the number of modes or overtones increases without limit with frequency, the population of the requisite energy states decreases toward zero. Thus the quantum theory correctly predicts the observed decrease of emission toward higher frequencies (shorter wave-lengths) and ultimately to zero as the wavelength approaches zero.

### 4.5 Planck's Distribution Law

The quantum theory predicts that the average energy of a linear oscillator is

$$\frac{h\nu}{\exp(h\nu/kT) - 1} \quad \text{rather than } kT. \tag{4.7}$$

Note that this reduces to $kT$ when $h\nu/kT$ is small enough to permit the representation of the exponential by the first two terms of the series expansion. This applies when the quantum energy $h\nu$ is small compared with the thermal energy $kT$ and corresponds to the low-frequency (or long-wave-length) region of the blackbody curve, where the Rayleigh-Jeans law is satisfactory.

If the energy of a linear oscillator as predicted by quantum theory, (4.7), is used instead of $kT$, the energy density in the cavity becomes

$$\frac{8\pi h\nu^3}{c^3}\,\frac{1}{\exp(h\nu/kT) - 1} = \text{energy density.} \tag{4.8}$$

If cgs units are used, the dimensions of the energy density are ergs per unit frequency interval per cubic centimeter. The radiant power escaping per unit area per unit frequency interval per unit solid angle may be obtained by multiplying the energy density by the speed of light and dividing by $4\pi$ steradians:

$$I_\nu = \frac{2h\nu^3}{c^2[\exp(h\nu/kT) - 1]}, \tag{4.9}$$

The radiative flux emitted by a blackbody of unit area into a hemisphere is easily shown to be

$$E_\nu = \frac{2\pi h\nu^3}{c^2[\exp(h\nu/kT) - 1]}, \tag{4.9a}$$

where $E$ is the radiant power per unit frequency interval per unit area. This may be readily written in terms of the wavelength by noting that $\lambda = c/\nu$, where $c$ is the velocity of light:

Table 4.1
Numerical values of the first and second radiation constants

| Unit of $E_\lambda$ | $C_1$ | $C_2$ |
|---|---|---|
| cal cm$^{-2}$ min$^{-1}$ ($\mu$m)$^{-1}$ | $5.3616 \times 10^5$ | $1.4388 \times 10^4$ |
| mW cm$^{-2}$ ($\mu$m)$^{-1}$ | $3.7418 \times 10^7$ | $1.4388 \times 10^4$ |
| W m$^{-2}$ ($\mu$m)$^{-1}$ | $3.7418 \times 10^8$ | $1.4388 \times 10^4$ |

$$E_\lambda = \frac{2\pi hc^2}{\lambda^5[\exp(ch/k\lambda T) - 1]}, \tag{4.10}$$

where $E$ is now the radiant power in a hemisphere per unit wavelength per unit area. The several physical constants are usually combined into two:

$$E_\lambda = \frac{C_1}{\lambda^5[\exp(C_2/\lambda\theta) - 1]}, \tag{4.11}$$

where $C_1 = 2\pi hc^2$ and $C_2 = ch/k$. It is customary to measure $\lambda$ in $\mu$m. In most of the older literature $E_\lambda$ is given in cal cm$^{-2}$ min$^{-1}$ ($\mu$m)$^{-1}$, but the units now used are mW cm$^{-2}$ or W m$^{-2}$. Values of $C_1$ and $C_2$ are given in table 4.1 for these three units of radiant power.

The wavelength of maximum emission may be obtained by setting the derivative of (4.11) with respect to the wavelength equal to zero. This leads to the transcendental expression

$$\exp[-(C_2/\lambda T)] = 1 - (C_2/5\lambda T), \tag{4.12}$$

the numerical solution of which is

$$C_2/\lambda_m T = 4.9651, \tag{4.13}$$

where the subscript m denotes maximum. On inserting the numerical value of $C_2$,

$$\lambda_m = 2898/T, \tag{4.14}$$

where $\lambda_m$, the wavelength of maximum emission, is in micrometers and $T$, the blackbody temperature, is in degrees Kelvin. This is Wien's displacement law, which he obtained by other means prior to Planck's development of the distribution law.

From (4.14) and figure 4.1 it is clear that thermal radiation at terrestrial temperatures peaks around 10–15 $\mu$m and extends roughly from 4 to 50 $\mu$m.

The integral of (4.11) over wavelength yields

$$E(T) = \frac{\pi^4 C_1}{15 C_2^4} T^4 = \sigma T^4. \tag{4.15}$$

This will be recognized as the Stefan-Boltzmann equation, the initial development of which has already been noted. The numerical value of $\sigma$, the Stefan-Boltzmann constant, is $8.125 \times 10^{-11}$ cal cm$^{-2}$ min$^{-1}$ °K$^{-4}$, $5.6701 \times 10^{-8}$ W m$^{-2}$ °K$^{-4}$, and $5.6701 \times 10^{-9}$ mW cm$^{-2}$ °K$^{-4}$.

## 4.6 Real Radiators or Absorbers

All real substances emit and absorb less radiation than a blackbody at any given temperature and wavelength. That is, their absorptivities (or emissivities) are less than unity and generally vary more or less rapidly with wavelength. Gases in atomic form absorb only in widely spaced spectral lines that result from the quantized changes in the electronic energy states of the particular atoms. Molecular gases exhibit band spectra, each band consisting of a large number of closely spaced lines; several well-separated bands are commonly found. Thus the spectra of gases appear to be discontinuous, although the absorption does not quite vanish between the lines or bands. In solids and liquids the atoms and molecules are so closely bound by their neighbors that they do not show the discontinuous spectra typical of gases. Emission of radiation results from the thermal vibrations that produce a continuous spectrum like that of a blackbody except that the emissivity is less than unity and varies with wavelength.

## 4.7 The Gray Absorber

To avoid the complexities of the spectra of gases, the concept of a gray absorber was introduced. A gray absorber is one that has a constant absorptivity less than unity at all wavelengths. No real gray absorbers are found, although some substances approximate one over a relatively narrow wavelength interval. This concept has been abandoned in studies of radiation in the earth's atmosphere, but it is still used in estimating radiative exchanges in planetary atmospheres, where, in general, knowledge of the distribution of radiationally active gases is limited.

## 4.8 The Temperature of the Stratosphere after Humphreys

A simple example of the use of the gray atmosphere approximation is given here because of its historical importance. This particular study was made by Humphreys (1929) in 1909 to explain the temperature of the nearly isothermal stratosphere that had been found from balloon soundings

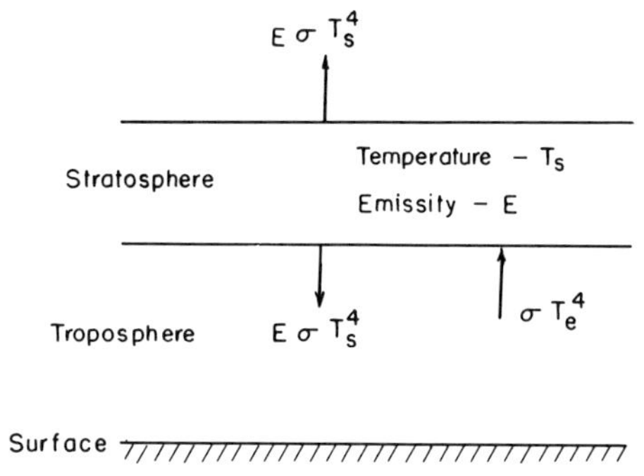


Figure 4.2 The model used to compute the radiative equilibrium temperature of the stratosphere by Humphreys (1929).

around the turn of the century. As shown in figure 4.2, the stratosphere was considered to be a gray absorber of absorptivity $a_s$. The stratosphere is heated through absorption of the upcoming thermal radiation from the surface and the lower atmosphere and cools by radiating both up and down with an emissivity equal to the absorptivity.

The radiation balance is

$$2\varepsilon\sigma T_s^4 = \varepsilon\sigma T_e^4, \tag{4.16}$$

where $T_e$ is the temperature of a blackbody that yields the same radiation as that upwelling from the earth and lower atmosphere at the tropopause. For the earth as a whole, over a year, this upwelling radiation must equal the similarly averaged absorbed solar radiation. Thus

$$4\pi R^2\sigma T_e^4 = \pi R^2(1 - P)E_0, \tag{4.17}$$

where $R$ is the radius of the earth, $P$ is the planetary albedo of the earth, and $E_0$ is the solar constant. Combining this with (4.16) gives

$$T_s = [(1 - P)E_0/8\sigma]^{1/4}. \tag{4.18}$$

By using the values of $P$, $E_0$, and $\sigma$ accepted at the time (0.32, 1.94, and $8.22 \times 10^{-11}$ respectively, $T_s$ was found to be 212°K, in good agreement with the then available measurements of the stratospheric temperature, and it was assumed that the problem was essentially solved. Nearly the same result will be obtained if present-day values of the parameters are used. Note that this is not an explanation of an isothermal stratosphere, that no account is taken of nonradiational heat exchanges, and that the stratosphere as a whole is assumed to be in long-wave radiational equilibrium. A little numerical exploration of (4.18) will show how insensitive it is to variations of the parameters within reasonable limits.

## 4.9 Simpson's Contributions

G. C. Simpson (1927–1928) made an attempt to compute the outgoing long-wave radiation at the top of the atmosphere, assuming that the atmosphere could be considered a gray absorber. He found that from latitude 50°N to 50°S the outgoing radiation was both uniform and nearly independent of the surface temperature. These are his final two sentences: "As the outgoing radiation is practically independent of the temperature of the surface, the problem arises as to how the temperature of the atmosphere readjusts itself to changes in solar radiation. This problem has been considered in detail, but no solution is found." In a later paper Simpson (1928–1930) abandoned the gray absorber assumption and developed a simplified representation of the spectral variation of atmospheric absorption due to water vapor and carbon dioxide. After inspection of the available data on the absorption of these two gases, he assumed that the atmosphere was black from 5.5 to 7 $\mu$m and for $\lambda > 14$ $\mu$m, that it was completely transparent from 8.5 to 11 $\mu$m, and that it was intermediate between black and transparent from 7 to 8.5 $\mu$m and from 11 to 14 $\mu$m. With these assumptions it is a simple matter to compute the outgoing radiation, as illustrated in figure 4.3. This figure contains two blackbody curves, the upper one at the surface temperature and the lower one at the temperature of the highest atmospheric layer containing enough water vapor or carbon dioxide to be black in the regions 5–7.5 and $>14$ $\mu$m. Simpson placed this layer in the lower stratosphere. The vertical dashed lines represent Simpson's divisions of the spectrum. The outgoing radiation is then the cross-hatched area on the figure. In the case of a cloud cover, a blackbody curve at the temperature of the cloud top is substituted for that for the surface temperature. Simpson showed that the outgoing radiation computed in this way varied in a reasonable fashion with latitude and further that the mean emission for the entire earth was within 3% of his independent estimate of the total absorbed solar radiation. The latter was based on a planetary albedo of 0.43, in contrast with the currently accepted value of about 0.30. It is also known that the upper black layer taken by Simpson to be the lower stratosphere is instead in the upper troposphere and is therefore warmer. These modifications are compensatory and could possibly lead to a balance as good as that obtained by Simpson.

Although other workers were well aware of the general nature of the absorption spectrum of the atmosphere and had commented on its probable effects a decade or more earlier, Simpson's two papers finally delivered the "coup de grace" to the gray atmosphere assumption in studies of terrestrial radiation. Spurred on by detailed experimental and theoretical studies of the absorption spectra of the atmospheric gases, the quantitative knowledge of radiative transfer has grown rapidly during the past few decades. It is no longer possible to be satisfied with approximations of the

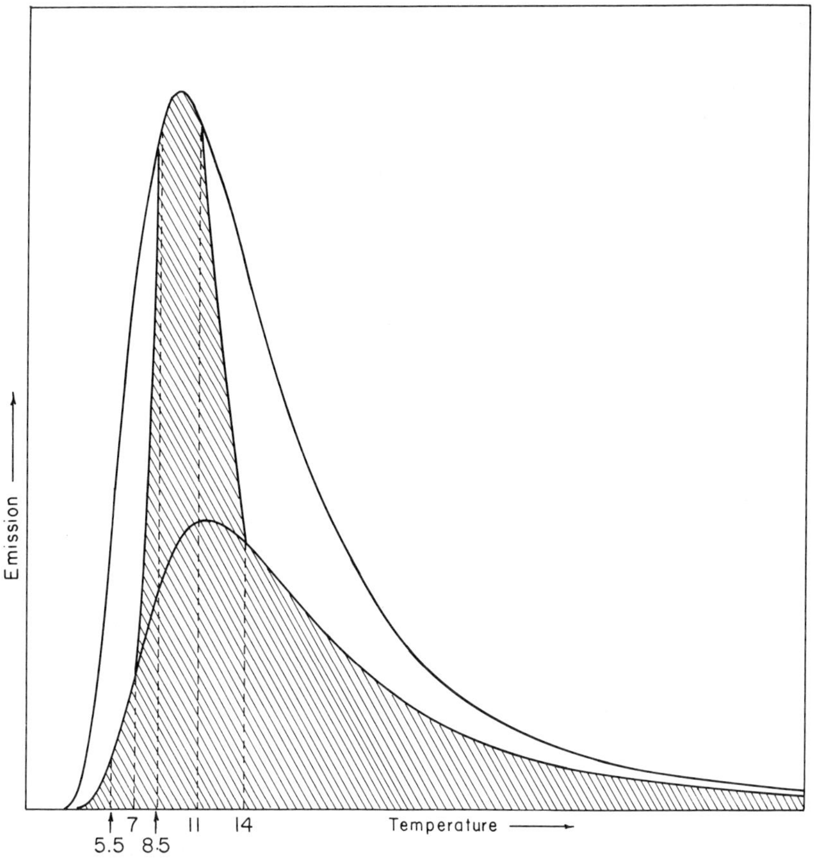


Figure 4.3 The method for evaluating the thermal radiation leaving the atmosphere to space devised by Simpson (1928–1930). The blackbody curves are at the surface temperature and the temperature of the highest black layer, respectively. The cross-hatched area is the outgoing thermal radiation to space.

kind used by Simpson in spite of their appealing simplicity. As a prelude to the modern approach it is necessary to review the quantum theory of spectra and to abandon hope of finding simple solutions to a problem that is, in truth, extremely complex.

## 4.10 The Absorption Spectra of Gases

As already noted, the absorption spectra of gases are essentially discontinuous. The simplest example is the absorption spectrum of a gas in atomic form, for example, hydrogen. Its spectrum consists of a relatively small number of widely spaced lines in the ultraviolet and visible portions of the spectrum. These lines correspond to electronic energy transitions characteristic of the particular atomic species. The energy of a quantum in the visible spectrum is about $4 \times 10^{-12}$ erg, which is to be compared with the thermal kinetic energy, $3kT/2$. At atmospheric temperatures this is about $6 \times 10^{-14}$ erg, and consequently very few of the atoms will be at energy levels above the ground state. The absorption lines correspond to transitions from the ground state to higher excited states. In order to obtain an atomic emission spectrum it is necessary to raise the gas to a very high temperature so as to populate the higher energy states. The resulting emission spectrum consists not only of the lines resulting from transitions to the ground state but also lines resulting from transitions between excited energy states. The emission spectrum therefore contains more lines than the absorption spectrum.

When the spectrum of a molecular gas is examined it is found that the absorption occurs in bands; with sufficient spectral resolution, the bands are seen to consist of a rather large number of closely spaced spectral lines. Molecular spectral bands in the ultraviolet and visible regions are usually due to electronic energy transitions. The reason for the bands is that every electronic transition is typically accompanied by vibrational and rotational changes, and thus each of the electronic lines is subdivided into many lines. These electronic bands are not important for radiative transfer at atmospheric temperature because the population of these energy states is exceedingly small.

The important molecular bands for the present purposes are due to the quantized vibrational and rotational energy transitions and their interactions. These energy transitions are very much smaller than the electronic transitions and are consequently well populated at atmospheric temperatures. For orientation table 4.2 gives, for each of several spectral regions, the principal type of energy transitions and the approximate magnitude of a quantum. The limits of the regions are arbitrary, and for some molecules, the type of energy transition is that of the line above or below that in the table. Since the translational kinetic energy at a temperature of 200–300°K is about $5 \times 10^{-14}$ erg, it is clear that the vibrational and rotational energy

Table 4.2

Type of transitions responsible for lines in various spectral regions

| Wavelength region ($\mu$m) | Quantum magnitude (erg) | Type of transition |
|---|---|---|
| Ultraviolet: <0.4 | $10^{-11}$ | Electronic |
| Visible: 0.4–0.7 | $4 \times 10^{-12}$ | Electronic |
| Near infrared: 0.7–2. | $10^{-2}$ | Electronic/vibrational |
| Infrared: 2.–$10^2$ | $10^{-14}$ | Vibrational/rotational |
| Far infrared: $10^3$–$10^5$ | $10^{-15}$ | Rotational |
| Microwave: $10^3$–$10^5$ | $10^{-17}$ | Rotational |

states will be well populated at atmospheric temperatures, whereas the electronic energy states will not be. For this reason we will be concerned only with the rotational and vibrational spectra of the molecular species found in the atmosphere.

## 4.11 Rotational Spectra

Diatomic molecules have a rotational spectrum only if the rotation results in an oscillating electric dipole moment. For this reason, homonuclear molecules, such as $N_2$ and $O_2$, have no rotational bands. Other species of diatomic gases are rare in the atmosphere, and in most cases, their rotational bands lie in the microwave region. The principal radiationally active atmospheric gases are $H_2O$, $CO_2$, and $O_3$, all of which are triatomic. Of these only $H_2O$ has a pure rotational band in the spectral region responsible for atmospheric radiative exchanges. The band center lies around 65 $\mu$m, but it extends to 14–15 $\mu$m at the high-frequency end of the band. $CO_2$ is a linear molecule with no permanent dipole moment and hence has no rotational band. The $O_3$ rotational band lies in the microwave region. The $H_2O$ rotational band is centered at a relatively high frequency because the light hydrogen atoms lead to relatively small moments of inertia and the rotational energies are inversely proportional to the moment of inertia.

The spectrum of a single rotator consists of a considerable number of lines evenly spaced on a frequency scale. The energies of the rotational states are small compared with $kT$, so that a large number of rotational states are well populated. More spectral lines are commonly observed than the number of rotational states. Contributors are the rotations of the atomic nuclei, the variation of the rotational moments of inertia with the quantum number $J$, overtones, and Coriolis effects.

The atoms of the water molecule are arranged in the form of an isoceles triangle with the oxygen atom at the apex, as shown in figure 4.4. There are

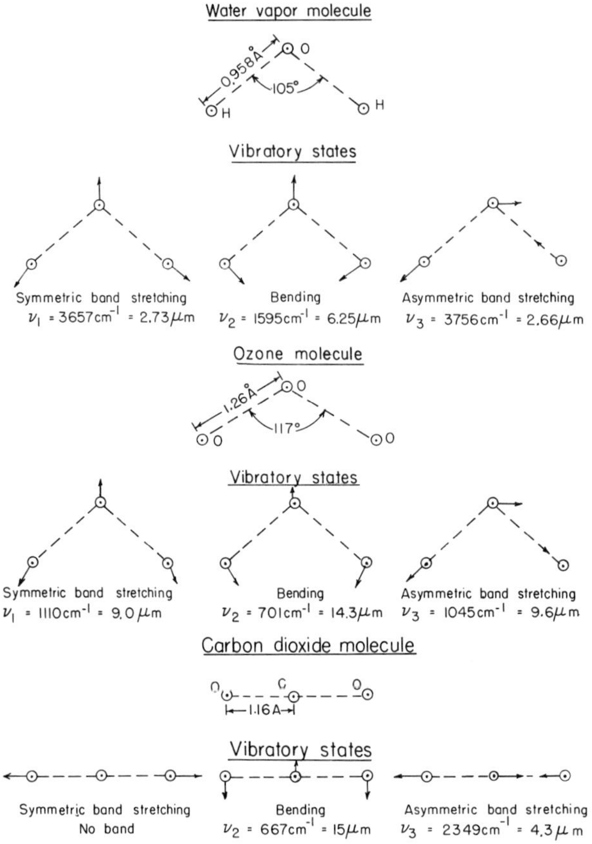


Figure 4.4 The structures of the water vapor, ozone, and carbon dioxide molecules and their vibratory states.

three mutually perpendicular axes of rotation passing through the center of mass of the molecule. The moments of inertia about these axes are all different, and the $H_2O$ molecule is therefore an asymmetric top. The interactions of the rotational states about the three axes and with the vibratory states lead to a spectrum comprised of many lines with no discernable regularity. A portion of the rotational spectrum of water vapor is shown in figure 4.9.

## 4.12 Vibrational Bands

Vibrational bands occur at higher frequencies (higher energies) than rotational bands. In general, a molecule has $3n - 6$, or $3n - 5$ if it is linear, vibrational degrees of freedom, where $n$ is the number of atoms in the molecule. The structure and principal vibrational modes of the three triatomic molecules of greatest interest here are shown in figure 4.4. By the standard spectroscopic terminology, frequency is given in inverse centimeters ($cm^{-1}$), which is referred to as the wave number. It is simply the reciprocal of wavelength and may be converted to a true frequency by multiplication by the velocity of light.

The vibrational bands of principal interest in the 5–50-$\mu$m range of atmospheric thermal radiation are the $\nu_2$ band of $H_2O$, the fundamental of the 6.3-$\mu$m band; the $\nu_3$ band of $O_3$, the fundamental of the 9.6-$\mu$m band; and the $\nu_2$ band of $CO_2$, the fundamental of the strong 15-$\mu$m band. The $\nu_1$ and $\nu_3$ $H_2O$ bands are of some importance in the solar spectrum, although $\nu_1$ is quite weak. The $\nu_1$ and $\nu_2$ bands of $O_3$ are also relatively weak. There is no band corresponding to $\nu_1$ for $CO_2$ because symmetric band stretching of the linear molecule does not induce an oscillating dipole moment. The $\nu_3$ band of $CO_2$, centered at 4.3 $\mu$m, is not of much importance in radiative transfer, since it falls between the solar and thermal spectra, but it is being used to sense the vertical temperature profile from a satellite. The fourth vibrational state of the linear $CO_2$ molecule turns out to be identical with $\nu_2$ and is therefore not shown separately.

Since the energies of vibrational quanta are some two orders of magnitude larger than those of the rotational quanta, changes in the rotational state accompany changes in the vibrational state. This leads to many lines instead of a single line for each change in vibrational state, as indicated schematically in figure 4.5 for a diatomic molecule. Two vibratory states

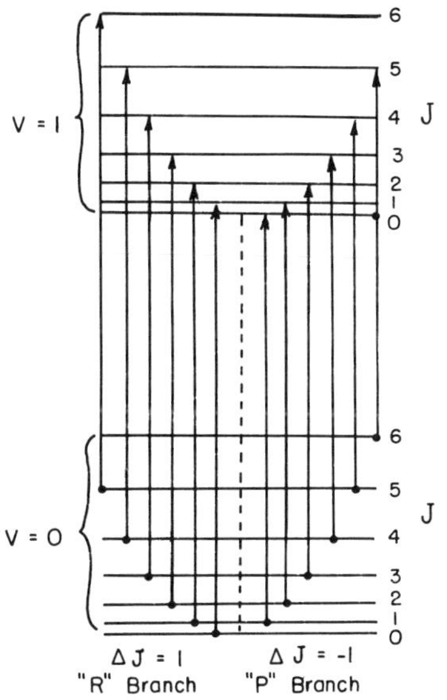


Figure 4.5 Energy levels for a vibrational/rotational band of a diatomic gas. After Barrow (1964).

are shown, $v = 0$ and $v = 1$, and six rotational states. The selection rule is $\Delta J = \pm 1$ and $\Delta J = 0$ is forbidden, where $J$ is the rotational quantum number. This leads to 12 lines instead of the single line that would result from a simple vibrational transition from $v = 0$ to $v = 1$. In a typical case more than the six rotational states will be excited, with a consequent increase in the number of lines. The interactions between the several vibratory and rotational modes lead to a complex spectrum consisting of many lines. Because of the important influence of rotational modes on vibrational spectra, the resulting band is often called a vibrational/rotational band. The 6.3-$\mu$m water vapor band is a good example of such a band.

## 4.13 Other Features of Atmospheric Absorption

Isotopes of the three principal absorbing gases have somewhat different rotational and vibrational bands that add to the complexity of the absorption spectrum of the atmosphere. The natural abundance of the isotopes is generally small enough to neglect their effects on radiative transfer. Other polyatomic trace gases, such as nitrous oxide, methane, ammonia, and carbon monoxide, have absorption bands in the thermal radiation spectrum. Donner and Ramanathan (1980) found that the long-wave bands of $CH_4$ and $N_2O$ contribute about 2°K to the hemispherical mean surface temperature and should therefore be included in climatic models.

Many of the atmospheric gases have lines or bands in the microwave region. Because of the very small thermal radiative flux in the microwave region, these lines and bands are of no importance in the overall transfer of thermal radiation in the atmosphere. However, they are now being used in the remote sensing of the vertical profiles of temperature and humidity from satellites. For example, $O_2$ has a band centered around a wavelength of 5 mm and at about 2.5 mm; $H_2O$ has a band centered around 13.5 mm.

For general orientation, a schematic representation of the principal absorbing bands in the thermal infrared is given in figure 4.6 with a 250°K blackbody curve for reference. Note the "window" in the spectrum between 10.5 and 13 $\mu$m where the absorption is not zero as shown, but is relatively small. Water vapor has a maximum concentration near the surface and generally decreases exponentially with height in the troposphere. Much of the stratosphere is relatively dry, but it still contains a radiationally important amount of water vapor. Carbon dioxide is a uniformly mixed gas up to at least several tens of kilometers and has a present concentration of about 333 ppm (parts per million). Ozone has a maximum concentration at a height of about 20–25 km, and both the total amount of ozone and its vertical distribution are subject to latitudinal and seasonal variations as well as variations on shorter time and space scales. It is a relatively minor constituent of the troposphere and is usually neglected

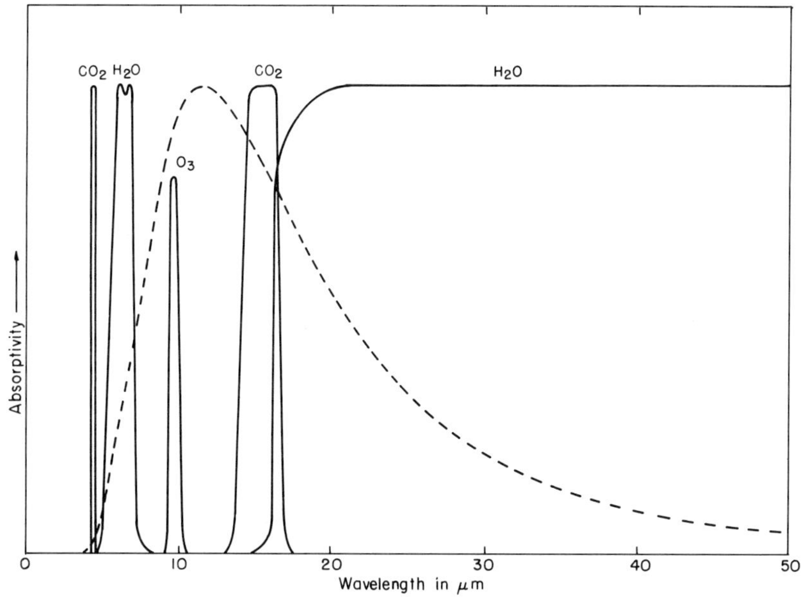


Figure 4.6 Qualitative sketch of the principal atmospheric absorption bands in the thermal infrared. The dashed curve is for a blackbody at 350°K.

in computations of tropospheric radiative exchanges. Thus the principal absorbers in the troposphere are water vapor and carbon dioxide; in the stratosphere all three of the gases are important.

## 4.14 Spectral Lines

A spectral line is not a geometric line but has a finite width on a frequency scale and a characteristic shape. The lines are too narrow to be fully resolved in the infrared, and the best observational data on line shapes have been obtained in the microwave region. Information on line shapes is based largely on theory and indirect experimental evidence.

The emission of a spectral line consists of photons of energy $h\nu$ that are in the form of wave packets. According to Heisenberg's uncertainty principle, the line will occupy a finite, though very small, frequency interval. One may also argue that the finite wave packet can be Fourier transformed into a narrow spectrum around the basic line frequency. This inherent line width is called the natural line width. A typical natural line width is $3 \times 10^{-11}$ cm$^{-1}$, which is so much smaller than the collisional and Doppler line widths, to be discussed, that it may be neglected.

In a real gas at a finite temperature, the molecules are in rapid motion, and numerous collisions occur between molecules. An observed spectral line comes from a large number of molecules, each undergoing the particular transition responsible for the given line. Those molecules moving away from the observer will appear to emit a lower frequency, while those moving toward the observer will yield a higher frequency due to the Doppler effect. The composite line will occupy a widened frequency interval, and this is called Doppler broadening. Since this depends on the mean molecular speed, Doppler broadening increases with the square root of the absolute temperature.

When a molecular collision occurs during the emission of a photon, the interactions between the two molecules result in a phase change of the wave packet or photon. The theory of such interactions is both complex and incomplete. However, there is ample experimental evidence that molecular collisions broaden a spectral line. The observed broadening can be accounted for in terms of the phase shift in the photons emitted from a molecule undergoing a collision. This collisional broadening depends on the rate of collisions, or on the product of the mean molecular speed and the concentration of molecules. In macroscopic terms this product is $\rho\sqrt{T}$ ($\rho$ = density), which, for a perfect gas, may be written $p/(R\sqrt{T})$. Since the range of pressure in the atmosphere is much greater than that of $\sqrt{T}$, the process is often called pressure broadening.

## 4.15 The Shape of a Collision-Broadened Line

The line shape depends on the type of broadening. The simplest and most commonly used expression for the shape of a collisionally broadened line is the Lorentz line shape:

$$k(\nu) = \frac{S}{\pi}\frac{\alpha}{(\nu - \nu_0)^2 + \alpha^2}, \tag{4.19}$$

where $k(\nu)$ is the exponential absorption coefficient, $S$ is the line intensity, sometimes called the integrated absorption, $\nu_0$ is the central frequency of the line, and $\alpha$ is the half-width, which is defined as one-half of the width of the line at the level where $k(\nu)$ is half its value at the line center. See figure 4.7 for a graphical representation of $\alpha$. As already noted, the half-width of a collisionally broadened line, usually called the Lorentz half-width, $\alpha_L$, is directly proportional to pressure and inversely proprotional to the square root of the Kelvin temperature:

$$\alpha_L = \alpha_{L0}\frac{p}{p_0}(T_0/T)^{1/2}, \tag{4.20}$$

where $\alpha_{L0}$ is the Lorentz half-width at a standard pressure $P_0$ and temperature $T_0$.

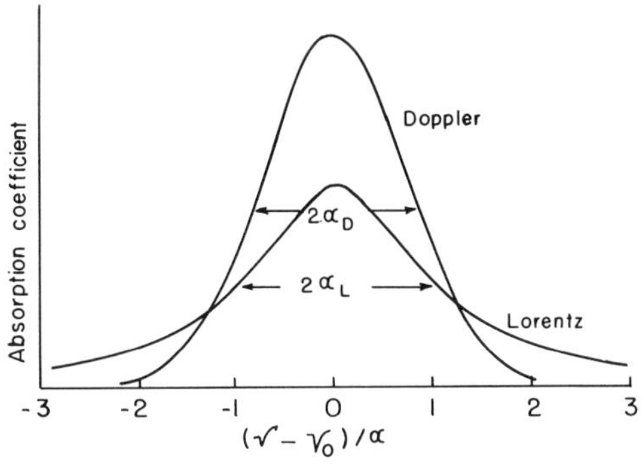


Figure 4.7 Comparison of Lorentz and Doppler line shapes. Pressure and temperature were chosen to make the half-widths nearly the same. After Goody (1964).

It may be readily shown that the integral of (4.19) over frequency from 0 to $\infty$ is $S$. Similarly the integral of (4.19) between $(\nu - \nu_0) = \pm\alpha$ is $S/2$. The Lorentz half-width at standard pressure and temperature depends on the molecular species involved in the collisions and the particular energy transition. The radiationally active substances are trace gases in the atmosphere, and hence the collisions are predominantly between air molecules ($N_2$ and $O_2$) and the absorbing molecule (for example, $H_2O$). In an atmosphere of pure water vapor, the collisonal broadening is greater at a given total pressure than in the usual mix of a small amount of water vapor in air. This is referred to as self-broadening. For this reason experimental absorption data must be taken near the usual concentration of the absorber in air rather than in the more convenient pure absorber atmosphere. The half-width under standard conditions is generally different for each line. Detailed information on the half-widths of individual lines is not readily available, and it is customary to use an average half-width in a given absorption band. This is not very important because it is usually sufficient to know that all of the half-widths vary proportionally with pressure and temperature.

The intensity of a line depends in part on the magnitude of the energy transition, but principally on the population of the excited energy state. This population depends on the temperature through the Boltzmann distribution. Lines are called strong or weak depending on the relative values of $S$ in (4.19). A strong line is strong because the particular excited state is well populated. The absorption by a gas is, in consequence, dependent on the temperature. In general, the effect of temperature on $S$ will be different for different lines. Strong lines are usually less temperature dependent than weak lines. It is possible to compute the effect of temperature on the intensity of a given line by quantum mechanics. Approximate computations of the effect of temperature on entire bands, comprising a mix of strong and weak lines, have been made, and one of these will be discussed later.

## 4.16 The Doppler Line Shape

As shown by (4.19), the Lorentz half-width depends on the total pressure. At a sufficient height the Lorentz half-width will become much smaller than the Doppler half-width, and the Doppler line shape is then the proper representation of the lines. The Doppler line shape is

$$k(\nu) = \frac{S_D}{\sqrt{\pi}\alpha'_D} \exp\left[-\left(\frac{\nu - \nu_0}{\alpha'_D}\right)^2\right], \tag{4.21}$$

$$\alpha'_D = \frac{\nu_0}{c}\left(\frac{2kT}{m}\right)^{1/2}, \tag{4.22}$$

where $\nu_0$ is the frequency at the line center, $c$ is the velocity of light, $k$ is Boltzmann's constant, and $m$ is the mass of the molecule. Equation (4.22) is normalized so that the total area under the line equals the Doppler line intensity $S_D$. The quantity $\alpha'_D$, defined by (4.22), is often called the Doppler half-width. However, the half-width, already defined as half the width of the line at half the maximum ordinate, is given by

$$\alpha_D = \sqrt{\ln 2}\ \alpha'_D \simeq 0.8326\alpha'_D. \tag{4.23}$$

Note that (4.22) gives the half-width as a frequency interval. Division by $c$ will give the half-width in the more customary wave number units.

## 4.17 Comparison of Line Shapes

A schematic representation of the Doppler and Lorentz line shapes for the same line intensity and at a level where $\alpha_L$ is about equal to $\alpha_D$ is shown in figure 4.7, after Goody (1964). It should be noted immediately that the Doppler and Lorentz half-widths are approximately equal at a height of about 30 km. Below this level the Lorentz half-width becomes increasingly larger than the Doppler half-width. At sea level the Doppler half-width is some two orders of magnitude smaller than the Lorentz. Figure 4.7 is intended to facilitate a comparison between the two basic shapes. It is evident that the wings of the Lorentz line extend to a greater distance from the center than those of the Doppler line. As will appear later, the wings of strong lines are very important in vertical radiative transfer.

The Lorentz line shape is appropriate in the troposphere and lower stratosphere, where collisional broadening is more important than

Table 4.3
Typical values of Lorentz line half-widths at standard pressure and temperature

| Band | Authority | Half-width ($cm^{-1}$) |
|---|---|---|
| $H_2O$ (rotational) | Benedict and Kaplan (1959) | 0.087 (0.03–0.11) |
| $H_2O$ (6.3 $\mu$m) | Benedict and Kaplan (1959) | 0.089 |
| $O_3$ (9.6 $\mu$m) | Kaplan (1959) | 0.089 |
| $CO_2$ (15 $\mu$m) | Kaplan and Eggers (1956) | 0.064 |
| $CO_2$ (15 $\mu$m) | Yamamoto et al. (1969) | 0.074 |

Doppler broadening. In the upper stratosphere and lower mesosphere both types of broadening are important. The Voigt line shape is intermediate between the Lorentz and the Doppler shapes and is therefore useful in this region. Only in the upper mesosphere and above is the pure Doppler shape appropriate. Only the Lorentz line shape will be used in all that follows on radiative transfer. The Lorentz line shape is a good approximation to the shape of a collision-broadened line within a few half-widths from the line center. It is not a very satisfactory representation of the far wings of the line. There are other more elaborate formulations that are better in the far wings, but they will not be discussed here. There is also reason to believe that collisional broadening is accompanied by a small shift in the central frequency of the line. This is of negligible importance in atmospheric radiative transfer. A few typical values of the Lorentz half-width at standard temperature and pressure are given in table 4.3 for the principal spectral bands of interest. Doppler half-widths at atmospheric temperatures are on the order of $10^{-3}$ $cm^{-1}$.

## 4.18 Transmissivity Functions

A transmissivity function is a theoretical or empirical relation between the transmissivity, averaged over a spectral interval much larger than a line width, and the absorber path length. The principal spectral bands of interest may each contain some hundreds of spectral lines of various intensities and half-widths. The purpose of a transmissivity function is to reduce this complexity by averaging over relatively wide spectral intervals. Mathematically, a transmissivity function is

$$\tau(\Delta\nu, \mathrm{u}) = \frac{1}{\Delta\nu}\int_{\Delta\nu} \exp\left\{-\left[\int_0^u k(\nu, u)\, du\right]\right\} d\nu, \qquad (4.24)$$

where $\Delta\nu$ is the frequency interval over which the averaging is done and $k(\nu, u)$ is the absorption coefficient, which varies in accordance with the detailed line spectrum in the interval $\Delta\nu$. On the assumption that the lines are of Lorentz shape, (4.20) shows that the absorption coefficient also depends on the pressure and temperature along the path. Since the pressure and temperature can be uniquely related to the path length $u$ through the quantitative description of the path, the absorption coefficient is a function of $u$ as well as of $\nu$, as indicated in (4.24). The relation between $u$ and the pressure and temperature is different along each atmospheric path and is ordinarily known empirically from an atmospheric sounding. In principle, (4.24) may be integrated numerically over a prescribed path and for the specific spectral interval. This requires complete information on line position, intensity, and half-width. Such computations are now readily performed on electronic computers (see Hitschfeld and Houghton, 1961). The result of such a computation is valid only for the particular atmospheric path assumed. The integration over $\nu$ may be computed once and for all if it can be separated from the integration over $u$. There is no exact means for separating the integration over $\nu$ from that over $u$. The procedure usually followed is to assume a homogeneous path, which means that the pressure and temperature are constant along the path. This eliminates the dependence of the absorption coefficient on the path length and hence separates the integrations. Means must then be sought for applying such solutions to nonhomogeneous paths.

## 4.19 Experimental Transmissivities

The basic quantitative transmissivity information comes from experimental determinations made in laboratory chambers with wide-slit spectrometers. The paths are invariably homogeneous, since there is no way in which a continuous variation of pressure and temperature along the path can be achieved in a laboratory chamber. By wide slit is meant that the bandwidth over which the transmissivity is measured is typically 10 $cm^{-1}$, which is some two orders of magnitude larger than a line half-width. Thus the experimental wide-slit spectrometric data yield solutions of (4.24) for specific homogeneous paths and various spectral intervals. The absorbent concentration and total pressure in the chamber can be controlled, although they are constant for a given run. Variations of path length are ordinarily achieved by varying the concentration and by a system of mirrors whereby the radiation may traverse the chamber once to several times. In most cases the temperature has not been controlled and is ordinarily at laboratory ambient temperature. Figure 4.8 is an example of a wide-slit spectrogram, taken by Burch et al. (1962) in the 15-$\mu$m $CO_2$ band. The curves are for different path lengths and pressures. The integrated absorption over the entire band is independent of slit width (see Neilsen et al., 1944). Results are often presented in this form. A sample of such data for the 15-$\mu$m $CO_2$ band from Burch et al. (1962) is given in table 4.4. The path length in atmospheric centimeters is the geometric path length in

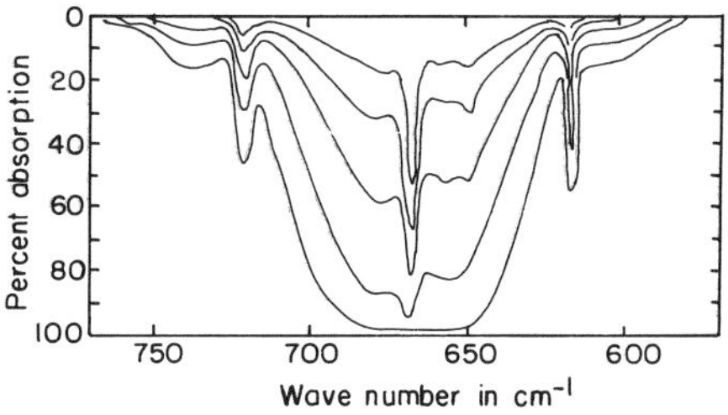


Figure 4.8 Wide-slit spectrogram of the 667-$cm^{-1}$ (15-$\mu$m) band of $CO_2$. After Burch et al. (1962).

| Path length (atmospheric cm) | Equivalent total pressure (mm mercury) |
|---|---|
| 0.187 | 132 |
| 0.371 | 263 |
| 0.756 | 537 |
| 1.55 | 1,100 |
| 3.09 | 2,200 |

Table 4.4
Integrated absorption of $CO_2$ (495–875 $cm^{-1}$)

| Path length (atmospheric cm) | Integrated absorptivity at indicated pressure (mm Hg) | | |
|---|---|---|---|
| | 300 | 760 | 1,000 |
| 0.186 | 0.042 | 0.052 | 0.055 |
| 0.748 | 0.087 | 0.112 | 0.119 |
| 1.55 | 0.126 | 0.156 | 0.166 |
| 3.14 | 0.172 | 0.206 | 0.214 |
| 5.73 | 0.215 | 0.251 | 0.253 |
| 11.5 | 0.262 | 0.295 | 0.299 |
| 46.4 | 0.337 | 0.371 | 0.376 |
| 96.6 | 0.389 | 0.432 | 0.434 |
| 193 | 0.442 | 0.482 | 0.484 |
| 388 | 0.479 | 0.508 | 0.511 |

centimeters multiplied by the partial pressure of the absorbing gas, corrected to standard temperature. The pressure is the equivalent pressure $p_e = p + 1.30p(CO_2)$, where $p$ is the pressure of the diluent (here $N_2$) and $p(CO_2)$ is the partial pressure of the absorber. This recognizes that self-broadening is more effective than foreign gas broadening. Table 4.4 may be considered to represent an empirical transmissivity function, although it is usual to collapse the three right-hand columns into one by using a "reduced" path length, as will be described later.

It has already been noted that the integrated transmissivity over an entire spectral band is independent of slit width, whereas this will not be true for narrower intervals within the band. By itself this would suggest integration over entire bands, which has the further advantage of minimizing the number of different spectral intervals needed. In using these functions to evaluate radiative transfer, the emissivity $1 - \tau$ is multiplied by the Planck distribution law, which is a function of temperature. To minimize this effect of temperature, the frequency interval $\Delta\nu$ should be small enough so that the Planck distribution function is nearly constant within it. In practice, narrow intervals of about 10 $cm^{-1}$ are used when maximum accuracy is desired, and wide intervals, including entire bands, when lesser accuracy is adequate.

## 4.20 Band Models

Band models are idealized representations of spectral intervals containing one or more spectral lines of prescribed shape, usually Lorentzian. They are suggested, in part, by actual spectra, but primarily so that the form of the absorption coefficient permits the analytic integration of (4.24). The purpose of band models has changed with time. The purpose of the first band model was probably to account for the failure of experimental band transmissivities to follow the simple exponential law. Somewhat later, the purpose was to provide a means for extending the very few experimental transmissivity "fixes" to the entire thermal radiation spectrum (see Elsasser, 1942). Now, with the abundance of data, the purpose is somewhat different. Fitting a band model to experimental data provides a useful means of interpolating and extending the experimental data. More important, the band models are invaluable in assessing the ways in which pressure and temperature affect the band transmissivity and, through this, devising approximate means for applying homogeneous path transmissivities to nonhomogeneous paths.

Before proceeding to the discussion of three typical band models it is well to examine the spectra of the absorbers. Portions of the rotational band of $H_2O$ and the 15-$\mu$m $CO_2$ band are shown in figure 4.9 and 4.10. These spectra are of moderate resolution and do not show much of the fine structure. The relative regularity of the $CO_2$ spectrum and the random

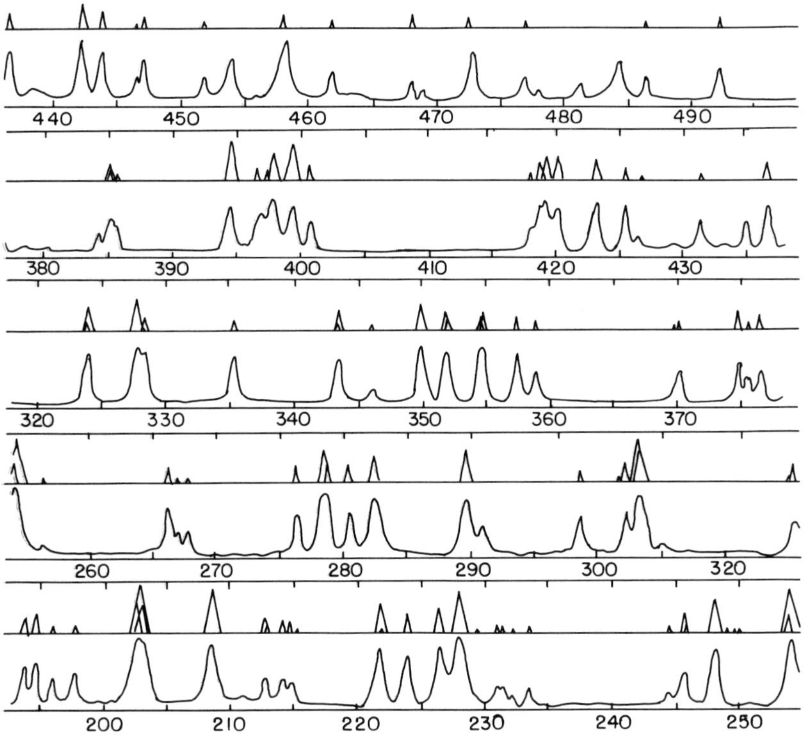


Figure 4.9 Portion of the absorption spectrum of water vapor in the rotational band. Triangles show computed positions and relative intensities of the principal lines. After Randall et al. (1937).

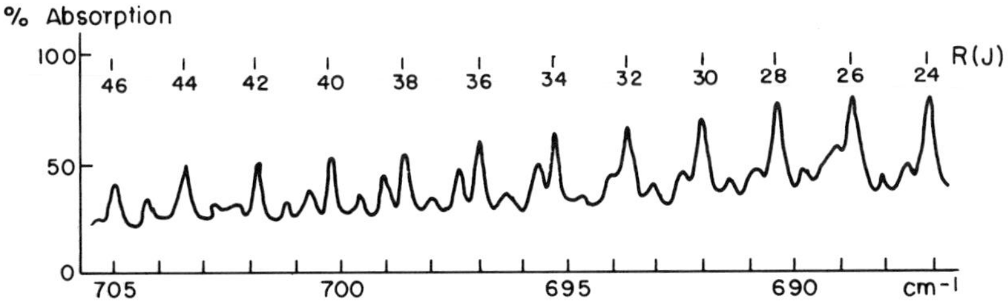


Figure 4.10 Part of the 15-$\mu$m $CO_2$ band. After Rossman et al. (1956).

nature of the $H_2O$ spectrum are readily apparent. In some regions the $H_2O$ lines are well separated, and in other regions they overlap strongly. In the portion of the 15-$\mu$m $CO_2$ band that is shown the lines are of fairly uniform spacing and intensity and there is moderate overlap in the wings of the lines. These features are pertinent to the several band models to be discussed.

The exponential law of absorption holds only for a frequency interval within which the absorption coefficient is essentially constant. In the case of line absorbers this interval must be small compared to the half-width and certainly much smaller than the half-width near line centers. It is not permissible to use a linear mean absorption coefficient over a band that is broad compared with the line width because of the highly non-linear character of the exponential absorption law. Consider a single line within a frequency interval that is very large compared with the half-width. Suppose that radiation of uniform intensity within this spectral interval passes through this absorber. Because of the very large absorption coefficient near the line center, the radiation in this narrow interval will be almost completely absorbed in a very short path length. This will be a very small part of the total incident radiation simply because the line width is so small compared with the spectral interval being considered. As the path length is increased there is little more radiation left to be absorbed at the line center, and the further absorption occurs in the tails of the line. Here the absorption coefficients are much smaller than at the line center, but there is much more radiation available, and it will be absorbed if the path is sufficiently long. For path lengths long enough to yield substantial absorption in the tails of the line the total radiation absorbed is so much greater than that absorbed by the narrow line center that the latter may be neglected. It should be clear that the relation between average transmissivity in such a spectral interval and the path length is not a simple exponential, but that it will depend on the line shape.

Of a number of spectral band models that have been proposed it will suffice to review three. The first assumes a frequency interval very large compared with the half-width that contains a single spectral line. This is the case discussed qualitatively earlier and is also applicable to the case of several nonoverlapping lines within a frequency interval. The second model to be discussed is the Elsasser (1942) band which consists of overlapping lines of uniform intensity and spacing. This model is suggestive of the $CO_2$ spectrum. The third model is the Goody random model, which assumes a normal distribution of line intensity and a random position of the lines. This was suggested by the $H_2O$ spectrum. In each case the Lorentz line shape is assumed and the integration is performed over a homogeneous path.

## 4.21 Band Containing a Single Line

It is assumed that a single Lorentz line lies within the frequency interval $\Delta\nu$, where $\alpha \ll \Delta\nu$. The expression for the transmissivity of this band in a homogeneous atmosphere is

$$\tau = \frac{1}{\Delta\nu}\int_{\Delta\nu} \exp\left[-\frac{S\alpha u}{\pi[(\nu-\nu_0)^2+\alpha^2]}\right] d\nu. \tag{4.25}$$

The integration is straightforward but lengthy, and only the result, first derived by Ladenburg and Reiche (1913), will be given. The only assumption made in the integration is that the limits may be extended to infinity, which is justified on the basis that the interval $\Delta\nu$ was chosen to be very large compared with $\alpha$ and hence includes essentially all of the line in any event. The result is

$$\tau(\Delta\nu) = 1 - \frac{1}{\Delta\nu} e^{-x}[J_0(ix) - iJ_1(ix)], \tag{4.26}$$

where $x = Su/2\pi\alpha$ and $J_0(ix)$ and $J_1(ix)$ are the Bessel functions of the zeroth and first orders. Both terms in the brackets are real and are represented by the following series:

$$J_0(ix) = 1 + x^2/4 + x^4/64 + \cdots \to \frac{e^x}{\sqrt{2\pi x}} \text{ for large } x,$$

$$iJ_1(ix) = -x/2 - x^3/16 - x^5/384 - \cdots \to \frac{-e^x}{\sqrt{2\pi x}} \text{ for large } x.$$

These series rather rapidly approach the limits shown on the right as $x$ becomes larger; already for $x \geq 3$ the asymptotic sums are an adequate representation. It is difficult to visualize the form of this transmissivity expression by inspection. It is instructive to look at two special cases, the first for very small $x$ (small $u$) and the second for $x$ large enough so that the asymptotic expressions for the Bessel functions are valid.

If $x$ is very small,

$$\tau = 1 - Su/\Delta\nu. \tag{4.27}$$

This means that for very short paths the absorptivity $1 - \tau$ increases linearly with $u$. For very short paths all of the absorption occurs in the center of the line and is given adequately by the first (linear) term of the series expansion of the exponential. This approximation is equally valid for very small $S$ and is usually called the weak line approximation.

If $x$ is moderately large,

$$\begin{aligned}\tau &= 1 - \frac{2Su}{\Delta\nu\sqrt{2\pi x}} \\ &= 1 - \frac{2}{\Delta\nu}\sqrt{\alpha Su}.\end{aligned} \tag{4.28}$$

This is the well-known square root law for band absorption. It has been found to fit observational data quite well in spectral regions where there is not extreme overlapping of the lines. Note that for a frequency interval that contains several widely spaced lines of equal $\alpha$ we may write

$$\tau = 1 - \frac{2}{\Delta\nu}\sqrt{\alpha u}\sum\sqrt{S}, \tag{4.29}$$

where the summation is taken over all of the lines in the interval $\Delta\nu$. It is clear that the square root law must fail for sufficiently large $u$, since it does not meet the obvious requirement that $\tau$ should approach zero as $u$ approaches infinity. Equation (4.26) also fails to meet this test. This is due to the error introduced by extending the limit of integration to infinity. Nevertheless, these expressions are found to fit observed transmissivities surprisingly well even when there is some line overlap.

For moderate path lengths and lines with relatively large $S$ most of the absorption will occur in the part of $\Delta\nu$ outside of the line center. As already stated, this is because the center of the line occupies such a small fraction of $\Delta\nu$ that even complete absorption there represents a very small fractional absorption over the entire band. As soon as the path length is great enough that significant absorption occurs in the tails the absorption there rapidly becomes much greater than that absorbed by the line center. One can recognize this at the outset by neglecting $\alpha^2$ in comparison with $(\nu - \nu_0)^2$ in (4.25). This assumption restricts the absorption to the line wings. If this is done at the start the integration is much simpler, and the result is again that

$$\tau = 1 - \frac{2}{\Delta\nu}\sqrt{\alpha Su}.$$

Physically this assumption may be seen to be of the same kind as the use of the asymptotic expressions for the Bessel functions in (4.28). This is called the strong line approximation since it requires a fairly large value of $Su$. Evidently the terms strong and weak are relative and are dependent on the path length as well as on the line intensity.

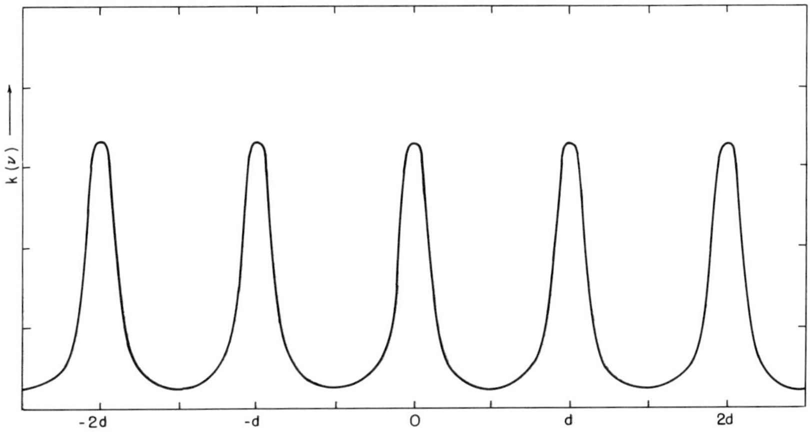


Figure 4.11 The Elsasser band model. After Elsasser and Culbertson (1960).

## 4.22 The Elsasser Band Model

The Elsasser band consists of a series of lines of uniform intensity and spacing. It provides means for studying the effects of line overlap and can be expected to be applicable to regular spectra, perhaps that of $CO_2$. The general appearance of the Elsasser band is shown in figure 4.11. Elsasser showed that the addition of an indefinite number of Lorentz lines of uniform intensity and half-width with their centers separated by a constant frequency interval $d$ led to the following expression for the absorption coefficient:

$$k(\nu) = \frac{S}{d} \frac{\sinh \beta}{\cosh \beta - \cos(2\pi\nu/d)}, \tag{4.30}$$

where $\beta = 2\pi\alpha/d$ and is therefore a measure of the extent to which the lines overlap. Using this absorption coefficient and integrating over a frequency interval large compared to $d$ for a homogeneous atmosphere, Elsasser and Culbertson (1960) found, after somewhat lengthy mathematics, that

$$\tau = \int_y^\infty \exp[-(y \coth \beta)] J_0(iy/\sinh \beta)\, dy, \tag{4.31}$$

where $y = Su/(d \sinh \beta)$.

Kaplan (1953) has shown that (4.31) may be evaluated by expanding it into an infinite series. His numerical results have been extended by Work and Walk (1960). The behavior of this function can be judged only through such computations, and a few examples are shown in figure 4.12. These curves show that the transmissivity correctly approaches unity for very

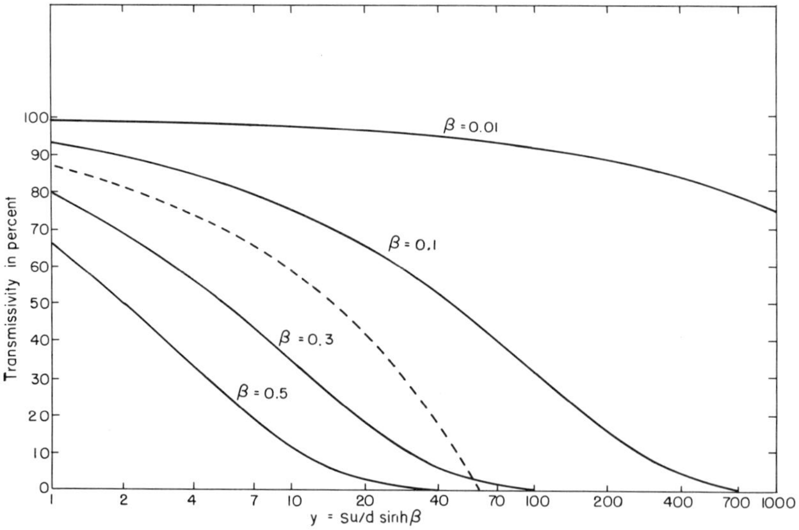


Figure 4.12 The transmissivity of an Elsasser band as a function of the path length parameter $y$ for the indicated values of $\beta$. The dashed curve is the square root law for comparison.

small paths (small $y$) and zero for large paths. For comparison, a square root curve has been drawn to represent the strong line approximation of the Ladenburg and Reiche model. Note that it would be possible to fit a square root expression to the Elsasser results for small $\beta$ and moderate path lengths. This is to be expected, since small $\beta$ corresponds to minimum line overlap and the square root expression is known to fail at large path lengths.

An approximation to (4.31) for small $\beta$ can be obtained by using the first two terms of the series expansions for the hyperbolic functions and using the asymptotic expansion for the Bessel function. As shown by Elsasser, this leads to

$$\tau = 1 - \operatorname{erf}\sqrt{\frac{y\beta}{2}} = 1 - \operatorname{erf}\sqrt{\frac{\beta Su}{2d}} = 1 - \operatorname{erf}\frac{\sqrt{\pi\alpha Su}}{d}, \tag{4.32}$$

where erf stands for "error function of", which is defined by

$$\operatorname{erf}(x) = \frac{2}{\sqrt{\pi}}\int_0^x \exp(-x^2)\,dx$$

$$= \frac{2}{\sqrt{\pi}}\,(x - x^3/3 + x^5/10 - x^7/42 + x^9/216\cdots).$$

The error function or probability integral is tabulated in Peirce (1929) and elsewhere. The assumption of small $\beta$ leads to equation (4.32) and is equivalent to the neglect of core absorption in comparison with absorption in the line wings. This is the strong line assumption discussed earlier.

If the argument is small, the error function may be approximated by the first term of the series:

$$\tau = 1 - \frac{2}{\sqrt{\pi}} \sqrt{\frac{\beta S u}{2d}}. \tag{4.33}$$

Substituting for $\beta$ reduces this to

$$\tau = 1 - \frac{2}{d}\sqrt{\alpha S u}, \tag{4.34}$$

which will be recognized as the square root expression previously derived for the case of nonoverlapping lines. The assumption made here is that $\beta$ is small and this corresponds to small overlap.

Elsasser called $\beta S/d^2$ the generalized absorption coefficient $L$. If it is assumed or established empirically that the error function relation has the same form as experimental data, $L$ may be determined numerically as a relatively slowly varying function of frequency or wavelength. This was the procedure he used in developing the data for his radiation chart discussed in section 5.9.

When $\beta \geq 1$, $\tau$ becomes a simple exponential within a few percent, as would be expected when there is nearly complete overlap. For $\beta \leq 0.1$ the error function expression for $\tau$ approximates the exact integral expression very closely except for very long paths. In the strong absorption bands of $H_2O$ and $CO_2$, $\beta$ is more usually between 0.1 and 1.0, where the error function approximation is not very satisfactory. As would be anticipated the Elsasser transmissivity function fits the more regular spectrum of $CO_2$ better than it does the $H_2O$ spectrum.

The weak line approximation for the Elsasser band may be obtained by evaluating (4.31) for very small $y$, and it is then found that $\tau = 1 - Su/d$. This will be recognized as the same as the weak line approximation for the single-line model.

## 4.23 The Statistical Band Model

The statistical band model proposed by Goody (1952) has proved very useful. This model was suggested by the apparent randomness of the line positions and intensities in the water vapor spectrum. Let there be $n$ lines, with average spacing $d$ in a spectral interval $\Delta\nu$ that is large compared to the line width and $\Delta\nu = nd$. It is assumed that each line has an equal probability of being anywhere within the interval $\Delta\nu$ and that the intensities of the lines follow a Poisson distribution. If $P(S_i)$ is the probability that a line has the intensity $S_i$

$$P(S_i) = \frac{1}{S_0} \exp[-(S_i/S_0)], \tag{4.35}$$

which is normalized so that $\int P(S)\, dS = 1$.

The mean transmissivity over the interval $\Delta\nu$ does not depend on $\nu$ so long as we are far enough from the ends of the interval. The mean transmissivity over the interval due to a single line is

$$\tau = \frac{1}{\Delta\nu} \int d\nu \int P(S_i) e^{-k_i u} ds_i \tag{4.36}$$

The mean transmissivity due to the $n$ lines in the interval is the product of the transmissivities of the separate lines. Since all of the integrals are of the same form, this may be written

$$\begin{aligned} \tau &= \left[\frac{1}{\Delta\nu} \int d\nu \int P(S) e^{-ku}\, dS\right]^n \\ &= \left[1 - \frac{1}{\Delta\nu} \int d\nu \int P(S)(1 - e^{-ku})\, dS\right]^n. \end{aligned} \tag{4.37}$$

Since $\Delta\nu = nd$, this expression is seen to approach an exponential as $n$ becomes large:

$$\tau = \exp\left[-\frac{1}{d} \int P(S)(A_i \Delta\nu)\right] ds, \tag{4.38}$$

where

$$A_i = \frac{1}{\Delta\nu} \int (1 - e^{-ku})\, d\nu \tag{4.39}$$

is the absorptivity of a single line averaged over $\Delta\nu$. The strong line case of the Goody model is obtained by introducing the Lorentz line shape, which leads to

$$\tau = \exp\left\{-\left[\frac{\alpha S_0 u}{d(\alpha^2 + \alpha S_0 u/)^{1/2}}\right]\right\}. \tag{4.40}$$

It is permissible to neglect $\alpha^2$ in the denominator in the strong line case:

$$\tau = \exp\left\{-\left[\frac{(\pi\alpha S_0 u)^{1/2}}{d}\right]\right\}. \tag{4.41}$$

The weak line approximation may be obtained by letting $S_0 u$ become small, giving

$$\tau = \exp\left[-\frac{S_0 u}{d}\right]. \tag{4.42}$$

An important feature of the Goody model is that it gives the transmissivity of a band containing many lines of different strengths in terms of the transmissivity of a single line. Also, the statistical design of the band model seems more realistic than either of the other two models considered.

## 4.24 Summary of Band Models

The strong and weak line approximations for the three models discussed are summarized in table 4.5 to facilitate comparisons and to draw some important conclusions. The similarities are evident, and if the error function and the exponential are approximated by the linear terms in the series expansions, they are identical (assuming that $d$ and $\Delta\nu$ are comparable). In part, the similarity of the results is due to the common use of the Lorentz line shape, but it also confirms the experimental finding that all transmissivity functions have a similar form. Lest we go too far in stressing the similarities, it should be pointed out that there are differences in the intermediate regions, where neither the weak nor the strong approximation is valid.

It is to be noted that all of the strong line expressions contain $\sqrt{\alpha S}/d$, while all of the weak line expressions contain $S/d$. This reflects the physical differences between tail absorption and core absorption. The effects of total pressure on the absorption are introduced through the half-width,

Table 4.5
Summary of the strong and weak approximations of the three band models

| Band model | Strong approximation | Weak approximation |
|---|---|---|
| Ladenburg-Reiche (single Lorentz line) | $\tau = 1 - \frac{2}{\Delta\nu}\sqrt{\alpha S u}$ | $\tau = 1 - \frac{Su}{\Delta\nu}$ |
| Elsasser | $\tau = 1 - \operatorname{erf}\frac{\sqrt{\pi\alpha S u}}{d}$ | $\tau = 1 - \frac{Su}{d}$ |
| Goody random | $\tau = \exp\left[-\frac{\sqrt{\pi\alpha S_0 u}}{d}\right]$ | $\tau = \exp\left[-\frac{S_0 u}{d}\right]$ |

and hence strong, but not weak, absorption is pressure dependent. Spectral bands normally contain both strong and weak lines, and it is to be anticipated that the effects of pressure broadening on the band as a whole will be intermediate between the effects on strong and weak lines. Temperature affects not only $\alpha$ but $S$, and changes in $S$ will be relatively more important in the weak than in the strong line approximation.

## 4.25 Fit of Experimental Data to Band Models

It is important to note that the band models cannot be successfully fitted to experimental absorptivities, integrated over entire spectral bands such as those in table 4.4. The basic reason is that mean line strengths are orders of magnitude larger near the band center than in the band wings. In many bands the mean line spacing is smaller in the band center than in the wings. If a fit is essayed at short path lengths, where the band center is dominant, the spectral parameters found will fail to reproduce the transmissivity at long path lengths, where the band wings are of greater importance. Consequently, good fits with band models can be expected only in spectral intervals that are small compared with the width of the entire spectral band or by introducing a variation of the mean line intensity with the frequency, measured from the band center.

Howard et al. (1956) found a good fit of the Goody random model to observational transmissivities for the 6.3-$\mu$m $H_2O$ band and several other higher-frequency water vapor bands by assuming that the mean line intensity $S_0$ varied inversely with the path length required to yield a transmissivity of one-half. This means that $S_0$ is large in the band center, where the path length for a transmissivity of one-half is small, and conversely in the wings. The range of $S_0$ from band center to band wing was nearly five orders of magnitude, which emphasizes the impossibility of obtaining a good fit of an entire band to the band models with a single value of $S_0$.

With the possible exception of the single-line band model, one does not use the transmissivity functions of the band models by inserting experimental or computed values of $\alpha$, $S$, and $d$. The multiline models are only an approximation to real bands, and therefore these spectral parameters are considered to be adjustable constants to achieve a fit to experimental transmissivities. The usual purpose of obtaining such a fit is to provide a convenient means for extending and interpolating the experimental data.

## 4.26 Empirical Band Models

Now that there exist substantial quantities of experimental transmissivity data of good quality, empirical transmissivity functions are usually derived directly from the data, using the band models only as a guide. A typical set of experimental data will consist of transmissivities taken at various values of the total pressure, and partial pressure of the absorber, for a wide range of absorber path lengths and a number of spectral intervals. It is desirable to summarize these data in the form of a single transmissivity function for each spectral interval, that is, to remove the effect of pressure. The band models tell us that the reduced path length varies linearly with pressure in the strong line case and not at all in the weak line approximation. In a typical band, the overall effect of pressure will be intermediate between these limits, suggesting an effective or reduced length $u_* = u(p/p_0)^\kappa$, where $p_0$ is standard pressure, usually 1,000 mbar, and $\kappa$ is expected to have a value between zero and one. By standard methods, the value of $\kappa$ is found that brings all of the data into a single curve that is independent of pressure. Experience has shown that a better fit of the data to a single curve can usually be obtained by using the equivalent pressure $p_E$, definded as

$$p_E = p_N + ap_A \tag{4.43}$$

where $p_N$ is the partial pressure of the neutral gas (air or nitrogen), $p_A$ is the partial pressure of the absorber, and $a$ is an empirical constant greater than one. The justification is that self-broadening is more effective than foreign gas broadening. More realistically, this provides another adjustable constant because the empirical values of $a$ are sometimes unrealistically large. This does not really matter if the basic purpose is to get a fit as good as possible to the experimental data.

When the procedure outlined is complete, the result will be a series of transmissivity (or absorptivity) curves, one for each spectral interval covered by the data. The difference between the several curves will reflect the spectral properties of the several bands. If the curves are of similar shape, one can determine a generalized absorption coefficient for each band. If successful, this procedure will result in a single transmissivity function for all of the bands plus a tabulation of the generalized absorption coefficient for each spectral interval. It is found that it is convenient to express the transmissivity as a function of $(\log u_* + \log L)$, where $L$ is the generalized absorption coefficient, which is determined from the data.

## 4.27 Effect of Temperature

Temperature affects the line intensity through the temperature dependence of the rotational partition function and the Boltzmann distribution. For water vapor (and ozone) the temperature dependence of the line intensity for a given spectral line is

$$\frac{S(T)}{S(T_0)} = \left(\frac{T_0}{T}\right)^{1.5} \exp\left\{-\left[\frac{E''}{k}\left(\frac{1}{T} - \frac{1}{T_0}\right)\right]\right\}, \tag{4.44}$$

where $E''$ is the energy of the lower state of the transition responsible for the line, $k$ is the Boltzmann constant, and $T_0$ is a standard temperature. For $CO_2$ the exponent on $T_0/T$ is one. The factor $(T_0/T)^{1.5}$ and the exponential vary oppositely with $T$, and it is easily shown that the derivative of (4.44) with respect to $T$ is negative when $E''/kT < (T_0/T)^{0.5}$ and positive when $E''/kT > (T_0/T)^{0.5}$. For $CO_2$ the derivative changes sign at $E''/kT$ equal to one. The strong lines in the center of a spectral band correspond to well-populated transitions and $E''$ is small; the converse is true of the lines in the band wings. Consequently, there is a tendency for $S(T)$ to increase with decreasing temperature in the band center and to decrease with decreasing temperature in the band wings. The net effect of temperature on the integrated absorptivity of the band may be of either sign depending on the detailed structure of the band.

The temperature dependence of an entire spectral band may be obtained by applying (4.44) to each line and then integrating over the band. Elsasser and Culbertson (1960) have developed an approximate method of evaluating the effect of temperature by fitting a parabola to the variation of $E''$ with temperature. With this assumption (4.44) becomes

$$\frac{S(T)}{S(T_0)} = \left(\frac{T_0}{T}\right)^{1.5} \exp\left\{-\left[2.3026a(\nu - \nu_0)^2\left(\frac{T_0 - T}{T}\right)\right]\right\}, \tag{4.45}$$

where $a$ is a constant obtained by fitting a parabola for each band and $\nu_0$ is the frequency of the band center. Elsasser and Culbertson give the values of $a$ in $cm^2$ for $\nu$ and $\nu_0$ in $cm^{-1}$ in table 4.6. The asymmetry of the bands is recognized by different values of $a$ on either side of the band center.

Burch et al. (1962) obtained the integrated absorptivities of the 15-$\mu$m $CO_2$ band (also the 9.4-$\mu$m and the 10.4-$\mu$m bands of $CO_2$) as a function of temperature from room temperature to about 70°C. They were unable to cool their chamber to obtain more typical atmospheric temperatures. In all cases they found that the integrated band absorption increased with temperature. This is not necessarily true for other bands or other gases.

Table 4.6
Values of the constant $a$[a] in equation (4.45)[b]

| Band | $\nu_0$ ($cm^{-1}$) | High-frequency side | Low-frequency side |
|---|---|---|---|
| 6.3-$\mu$m $H_2O$ | 1,595 | $1.75 \times 10^{-5}$ | $2.4 \times 10^{-5}$ |
| 9.6-$\mu$m $O_3$ | 1,043 | $1.4 \times 10^{-3}$ | $4 \times 10^{-4}$ |
| 15-$\mu$m $CO_2$ | 667 | $3.4 \times 10^{-4}$ | $4.6 \times 10^{-4}$ |
| Rotational $H_2O$ | 155 | $9.8 \times 10^{-6}$ | — |

a. Values of $a$ are in $cm^2$.
b. After Elsasser and Culbertson (1960).

## 4.28 The Water Vapor Continuum

Emphasis has been placed on the principal absorption bands of $H_2O$, $CO_2$, and $O_3$ in the spectral region of interest in atmospheric radiation (about 5–50 $\mu$m). There is a well-known region of high transmissivity, called the "window," lying between the 6.3-$\mu$m water vapor band and the 15-$\mu$m carbon dioxide band (about 10.5–13 $\mu$m), interrupted by the narrow 9.6-$\mu$m ozone band. This region is used for satellite sensing of surface and cloud top temperatures and is an important region for the escape of surface long-wave radiation to space, lying as it does near the maximum of black-body curves for terrestrial temperatures.

It has been known for some time that there is finite absorption in the window appearing as a continuum rather than as a line spectrum. Since there are no important water vapor lines in this region, it has been hypothesized that the absorption is mainly due to the far wings of the strong lines in the adjoining bands. Measurements are difficult because long paths are required and there is possible interference from the aerosol. If it were a true continuum, the transmissivity would be expected to follow the simple exponential law. One of the more complete discussions of the continuum is that of Bignell (1970), who gave new measurements and summarized much of the earlier data. He found evidence for two types of absorption. One is that due to the foreign broadening of the wings of the strong water vapor lines in the adjacent water vapor band. The other is an absorption that is dependent on not only the water vapor path length but also the partial pressure of the water vapor. This was earlier attributed to self-broadening of the same strong water vapor lines mentioned earlier. However, this second type of absorption becomes as important as the first at a vapor pressure of 10 mbar, and this is too much to be accounted for by self-broadening. Further, there is a negative temperature dependence that is larger than that of any water vapor line. Bignell made the tentative suggestion that this second type of absorption might be due to the dimer molecule $(H_2O)_2$, which would be expected to exhibit very broad lines, resulting in a near continuum.

Platt (1972) found that his observations in the 10–12-$\mu$m interval agreed well with Bignell's findings. Lee (1973) observed the downcoming atmospheric radiation as a function of height. He compared his results with a line wing model and a water dimer model and found that the latter model gave the better fit. Tomasi et al. (1974) made observations in the atmospheric window that were in good agreement with Bignell's findings. This and other evidence has confirmed Bignell's suggestion that there are two types of absorption in the window region, one due to the tails of adjacent strong bands and the other to the water dimer. Bignell writes the absorption coefficient in the window as

$$k(p, \theta, e) = k_1(p, T) + k_2(e, T), \qquad (4.46)$$

where $k_1$ is due to the pressure broadening of the adjacent strong bands and $k_2$ is due to the water dimer. In (4.46) $p$ is the total pressure and $e$ is the water vapor pressure. Bignell also found that $dk_1/dT$ was about $5 \times 10^{-3}$°C$^{-1}$ and that $dk_2/dT$ was about $-2 \times 10^{-2}$°C$^{-1}$.

## 4.29 Nonhomogeneous Paths

Both the theoretical and experimental transmissivity functions discussed earlier are for homogeneous paths in which both pressure and temperature are constant along the path. Most atmospheric paths of interest are nonhomogeneous; in particular, a vertical path through the atmosphere covers a large range of total pressure and a considerable range of temperature. It is obvious that nonhomogeneous paths of this kind cannot be reproduced in the laboratory. In principle, the transmissivity in a nonhomogeneous atmosphere can be determined by the integration of (4.24) for a particular band model and for a prescribed nonhomogeneous path. An exact analytic solution is possible only for very special nonhomogeneous atmospheres. An example is the case in which the mixing ratio of the absorbent is independent of total pressure. The mathematical difficulty in the integration over the Lorentz line shape resides in the presence of $\alpha^2$ in the denominator.

The simplest means of adapting homogeneous path data to nonhomogeneous paths is based on the band models discussed earlier and is called pressure scaling. As shown in table 4.5 the strong line transmissivity is dependent on the linear product $\alpha S u$, while the weak line transmissivity depends on $Su$. The collision-broadened half-width $\alpha$ varies directly with the total pressure. In simple pressure scaling, the reduced path length $u^*$ is obtained by multiplying the actual path length by the ratio of the actual pressure to a standard pressure of 1,000 mbar. This assumes strong line absorption, but may be modified for the usual mix of strong and weak lines by using an exponent of less than unity in the pressure ratio. If information is available on the temperature dependence of the absorption, it may be used to estimate the effect of the varying temperature along the nonhomogeneous path. All of the information needed for pressure scaling is available from the reduction of experimental data to one or more universal curves, as described earlier. The basic assumption of pressure scaling is that the influence of pressure and temperature on the transmissivity over homogeneous paths can be applied without change to nonhomogeneous paths. This cannot be strictly correct, since it implies that (4.24) may be integrated with an absorption coefficient independent of the path and the result used for the case in which the absorption coefficient varies along the path. Pressure scaling should be considered an empirical approach, although some justification for it will be given. It has the virtue of simplicity and seems to yield fairly good results.

A more justifiable procedure for applying homogeneous path data to nonhomogeneous paths is the Curtis-Godson approximation. This was first suggested by Curtis (1952) and independently by Godson (1953). The mathematical problem of a variable half-width results from the $\alpha^2$ in the denominator of the Lorentz line expression. This difficulty can be overcome if a constant value of $\alpha$, say $\bar{\alpha}$, can be used in the denominator. Curtis noted that almost any reasonable value of $\bar{\alpha}$ would suffice for strong lines. We have already seen that even setting $(\bar{\alpha})^2 = 0$ is adequate for the strong line case. The Curtis suggestion was to select a value for $\bar{\alpha}$ that would also be quite satisfactory for weak absorption. It may be anticipated that a value of $\bar{\alpha}$ that is good for both weak absorption and strong absorption will be at least moderately satisfactory in between. For weak absorption the exponential may be closely approximated by the first two terms of the series

$$\int_{\Delta\nu} \int_{u} \exp[-(k\,du)]\,d\nu = 1 \int_{\Delta\nu} \int_{u} k\,du\,d\nu, \tag{4.47}$$

where

$$k = \frac{S}{\pi} \frac{\alpha}{(\nu - \nu_0)^2 + \alpha^2},$$

the Lorentz line shape. The condition for the Curtis approximation is

$$\int_{\Delta\nu} \int_{u} \frac{S}{\pi} \frac{\alpha\,du}{(\nu - \nu_0)^2}\,d\nu = \int_{\Delta\nu} \int_{u} \frac{S}{\pi} \frac{\bar{\alpha}\,du}{(\nu - \nu_0]^2}\,d\nu. \tag{4.48}$$

This can be rewritten in the form

$$\frac{1}{\pi} \int_{u} S\,du \int \frac{dx}{x^2 + 1} = \frac{1}{\pi\,\bar{\alpha}} \int S\alpha\,du \int \frac{dx}{x^2 + 1}. \tag{4.49}$$

The value of the mean half-width is then

$$\bar{\alpha} = \frac{\int S\,\alpha\,du}{\int S\,du}. \tag{4.50}$$

It is seen that $\bar{\alpha}$ is the mean value of $\alpha$ over the particular path in question. The integrations must usually be performed numerically and, if possible, should take account of the variation of $S$ with temperature as well as the variations of $\alpha$ with pressure and temperature. Once $\overline{\alpha}$ has been determined it may be used to specify a corresponding mean pressure $\overline{p}$ through the relation $\overline{p} = \overline{\alpha}\overline{p}_0/\alpha_0$. The transmissivity over the nonhomogeneous path is then taken as the measured transmissivity over a homogeneous path at a pressure equal to $\overline{p}$.

If the variations of $S$ and $\alpha$ with temperature are neglected, (4.49) becomes

$$\overline{\alpha} = \frac{\int \alpha\, du}{\Delta u}, \tag{4.51}$$

where $\Delta u$ is the path length defined by the limits of the integration. By noting that $\overline{\alpha} = \overline{\alpha}_0 \overline{p}/p_0$ and $\alpha = \alpha_0 p/p_0$, this may be written

$$\overline{p} = \frac{\int p\, du}{\Delta u}. \tag{4.52}$$

This shows that $\overline{p}$ is the average pressure along the path weighted in proportion to $u$. Note that $\overline{p}$ as defined by (4.51) is in the same ratio to $p_0$ as the reduced path length is to the actual path length with linear pressure scaling. In view of the linear product of $\alpha$ and $u$ in the transmissivity relations, this means that the Curtis-Godson approximation leads to the same result as linear pressure scaling if $\alpha$ is linearly dependent on the pressure and is independent of temperature.

Simple pressure scaling represents the absorption along an inhomogeneous path in terms of a single parameter, a mean pressure. The mix of strong and weak lines in a band changes from the band center to the wings and also with the path length. Pressure scaling is best for relatively narrow spectral intervals and when the homogeneous path data used are taken at a pressure close to the mean pressure of the inhomogeneous path. The Curtis-Godson approximation represents the absorption in terms of two parameters: the mean pressure and the mean temperature. Since it is exact for both weak and strong absorption, it may be used for an entire spectral band.

Walshaw and Rodgers (1963) and Rodgers and Walshaw (1966) have estimated the errors resulting from the Curtis-Godson approximation in radiative heating rate calculations. They concluded that the Curtis-Godson approximation is reasonably satisfactory in the four major bands in the thermal infrared in most cases of meteorological interest.

## Problems

1.
Where and why in the atmosphere does Kirchhoff's law fail?

2.
In chapter 4 it is stated that for a square root law transmissivity function, the transmissivity for diffuse radiation is equal to the parallel beam transmissivity at normal incidence in a layer containing 16/9 as much absorber. Show that this is true.

3.
Simpson (1928–1930) selected a temperature for the ground surface, one for the cloud top temperature, and a third for the temperature of the uppermost radiating layer as a basis for calculating the outgoing thermal radiation to space. Select current values of the these temperatures and recompute the outgoing radiation to space. How do the results compare with Simpson's? A comparison at one latitude will suffice.

4.
Why is it that the transmissivity of a band containing a single line does not increase exponentially with path length?

5.
There are two sealed chambers each of which has the same water vapor path length and the same temperature. One contains only water vapor, while the other holds a mixture of water vapor and nitrogen. Which has the smaller transmissivity?

# 5

# Radiative Transfer and the Radiation Budget of the Atmosphere

## 5.1 Introduction

In chapter 4, the basic principles of the absorption of thermal radiation by gases have been applied to the development of transmissivity functions for homogeneous paths. This permits the effective use of experimental wide-slit spectrometric data over a wide range of pressures and temperatures. The band models have suggested approximate but fairly satisfactory means for applying the transmissivity functions for homogeneous paths to the nonhomogeneous paths that are of principal interest in the atmosphere.

It remains to apply the transmissivity functions to the computation of radiative transfer in the atmosphere. In view of Kirchhoff's law, an absorbing layer of gas also emits radiation at the same frequency. This is represented by Schwarzschild's equation:

$$dE_\nu = -A_\nu E_\nu \, du + A_\nu B(\nu, T) \, du, \tag{5.1}$$

where $E_\nu$ is the radiation at the frequency $\nu$, $A_\nu$ is the absorptivity (also the emissivity in view of Kirchhoff's law), $u$ is the absorber path length, and $B(\nu, T)$ is the blackbody function at the given frequency and temperature $T$. The first term on the right-hand side represents the reduction in $E$, due to the absorption, while the second term is the emission of the element of path that acts to increase $E_\nu$. Evidently the sign of $dE_\nu$ may be either positive or negative depending on the magnitudes of $E_\nu$ and $B(\nu, T)$. Schwarzschild's equation, (5.1), is the basic equation of radiative transfer. However, we shall develop another form of this equation that explicitly incorporates the transmissivity function.

Both the experimental and theoretical (band model) transmissivity functions are for a parallel or collimated beam of radiation. Long-wave thermal radiation in the atmosphere originates in the atmosphere and at the surface and is diffuse rather than in a parallel beam. Fortunately, a relatively simple geometric transformation may be used to convert parallel beam transmissivities to diffuse transmissivities.

Transmissivity functions are for a given absorbing gas and tacitly assume that it is the only absorber in the spectral interval of the transmissivity function. In fact, there is overlap between certain bands of the three principal absorbing gases, notably, between the 15-$\mu$m $CO_2$ band and the $H_2O$ rotational band. If this is ignored, a total absorptivity greater than one will result for sufficiently long paths in the overlapped region. Ways must be found to avoid this physically unrealistic result.

Various graphical and tabular schemes have been devised for simplifying the repetitive numerical integration of the radiative transfer equation. Although the high-speed computer has now minimized this problem, the simpler methods are still useful, and the graphical ones offer the further advantage of illustrating the effects of clouds and different atmospheric structures on radiative transfer. This chapter therefore includes a brief review of some of these computational aids.

It is impractical to summarize adequately the results of the many computations of radiative transfer that have been made. A few examples must suffice, but emphasis has been placed on the global radiative budgets, including both the long-wave thermal radiative transfer and the absorption of solar radiation.

Finally there is included a discussion of the principles of the remote sensing of atmospheric structure from satellites by means of long-wave thermal radiation and microwave radiation. These techniques offer the promise of providing what was heretofore economically impractical: securing global observations of the atmosphere in three dimensions.

## 5.2 Transmissivity of Diffuse Radiation

Both the band models and the experimental transmissivities discussed in chapter 4 are for parallel beams. Atmospheric thermal radiation is diffuse, and means must be found to relate diffuse transmissivity to the transmissivity of a parallel beam. This relation depends on the angular distribution of diffuse radiation. It is conventional to assume that the radiation is perfectly diffuse, or, in other words, that the intensity (flux per unit solid angle) varies with the cosine of the angle of incidence. This is the Lambert cosine law, which, in the visible spectrum, causes the apparent brightness of a perfectly diffuse radiator or reflector to be independent of the angle at which it is viewed. Real sources and reflectors are usually not perfectly diffuse, but many are a close approximation.

A further assumption is that the atmosphere is horizontally stratified. In figure 5.1 the upper horizontal surface is taken to be a perfectly diffuse source. First we shall find the irradiance on the lower plane when there is

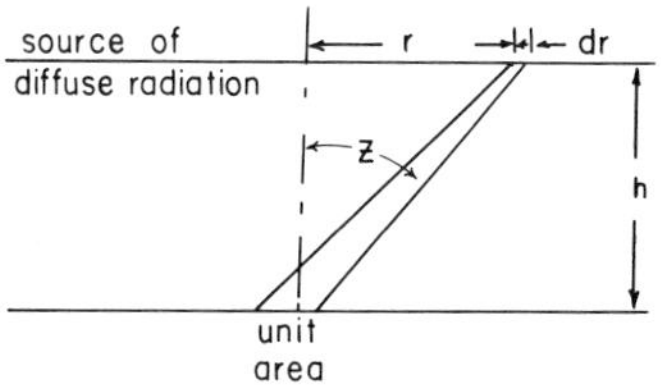


Figure 5.1 Illustrating the procedure for determining the transmissivity of a layer for diffuse radiation.

neither a source nor a sink of radiation in the layer between the two planes. Let the intensity per unit area of the source in the vertical direction be $I_0$. Consequently, the intensity at zenith angle $Z$ is $I_0 \cos Z$. The radiation reaching a unit area of the lower plane at the angle $Z$ comes from an annulus in the source plane of radius $r$ and width $dr$. The infinitesimal flux reaching the unit area at angle $Z$ is then $I_0 \cos Z$ multiplied by the area of the annulus and by the solid angle subtended by the unit horizontal area on the lower plane:

$$dF(Z) = 2\pi I_0 \cos Z\, r\, dr \cos Z/(h \sec Z)^2, \tag{5.2}$$

where $h$ is the vertical distance between the planes. From the geometry of 5.1, $r = h \tan Z$ and $dr = \mathrm{h} \sec^2 Z\, dZ$:

$$dF(Z) = 2\pi I_0 \sin Z \cos Z\, dZ. \tag{5.3}$$

Integration of (5.3) over the hemisphere yields the flux, $F = \pi I_0$.

Now assume that the layer contains an absorber, the parallel beam transmissivity of which is known as a function of the path length. The infinitesimal irradiance given by (5.3) is parallel radiation, since $Z$ is constant. The path length is $u \sec Z$, where $u$ is the path length in the vertical. The infinitesimal irradiance on the lower plane at angle $Z$ is

$$dF'(Z) = 2\pi I_0 \tau(u \sec Z) \sin Z \cos Z\, dZ, \tag{5.4}$$

where $\tau(u \sec Z)$ is the parallel beam transmissivity for a path length $u \sec Z$. The diffuse transmissivity of the layer, sometimes called the slab transmissivity, is given by the ratio of $F'(Z)$ to $F(Z)$:

$$\tau^*(u) = 2 \int_0^{\pi/2} \tau(u \sec Z) \sin Z \cos Z\, dZ. \tag{5.5}$$

Alternative forms of (5.5) are

$$\tau^*(u) = \int_0^1 \tau(u \sec Z)\, d(\sin^2 Z) \tag{5.6}$$

$$= 2\int_1^\infty \frac{\tau(uy)}{y^3}\,dy, \tag{5.7}$$

where $y = \sec Z$. Equation (5.7) is one of Gold's functions and has been tabulated.

## 5.3 Diffusivity Factors

Empirical transmissivity functions from laboratory data may be converted to diffuse transmissivities by the numerical integration of (5.5), (5.6), or (5.7). Analytic transmissivity functions may, in some cases, permit direct integration of (5.5) or (5.6). For example, if the simple square root model is taken as the transmissivity function, it is readily found that $\tau^* = 1 - (16u/9)^{1/2}$, where $u$ is the vertical path length of the absorber. This shows that the diffuse transmissivity is equal to the normal incidence parallel beam transmissivity of a layer 16/9 as thick. One can then multiply all of the actual path lengths by this factor, which is rather unfortunately called the diffusivity factor, and then treat the radiation as a normal incidence parallel beam. Elsasser and Culbertson (1960) found that the diffusivity factor for his error function model was 1.66.

These diffusivity factors apply only to strong absorption and even then only for transmissivites of intermediate magnitude. Plass (1952) has cautioned against the uncritical use of a constant diffusivity factor. He points out that for weak absorption the factor approaches two and that it approaches one as the transmissivity approaches zero. Nevertheless, a factor of 1.6–1.7 has been used frequently. Fortunately, in most cases, the use of an incorrect diffusivity factor has relatively small effect on computations of radiative transfer.

The basic assumption in the derivation of (5.5)–(5.7) is that the radiation is perfectly diffuse. Unsworth and Monteith (1975) have examined this assumption on the basis of experimental determinations of the angular distribution of the down coming atmospheric thermal radiation at the surface. They found that the radiation was not perfectly diffuse but that the diffuse transmissivity computed from the empirical data differed from that given by (5.5) by only about 1%.

## 5.4 Radiative Transfer Equation

It is desired to compute the upgoing radiation at an arbitrary reference level that comes from the atmospheric layer extending from the ground surface to the reference level as illustrated in figure 5.2. By convention, the absorber path length is measured, positive-down, from the reference level. If the downcoming radiation were desired, the absorber path length would be measured positive-up from the reference level.

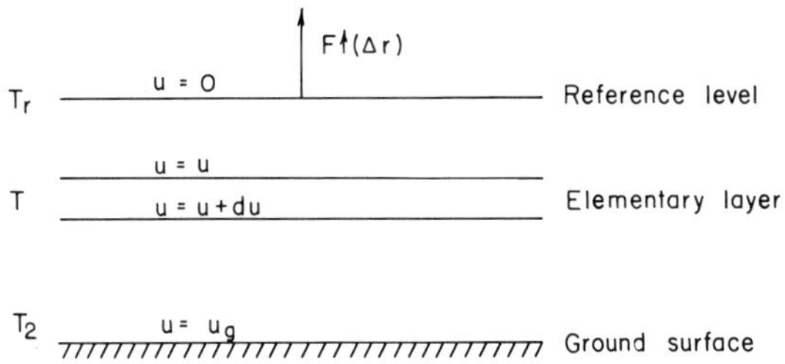


Figure 5.2 Illustrating the alternative derivation of the radiative transfer equation.

The emissivity of the elementary layer extending from $u = u$ to $u = u + du$ is[1]

$$\varepsilon\, du = \frac{\tau^*(u) - \tau^*(u + du)}{\tau^*(u)} = -\frac{1}{\tau^*(u)} \frac{d[\tau^*(u)]}{du} du, \tag{5.8}$$

where $\varepsilon$ is the emissivity for unit $u$ and $\tau^*$ is the diffuse transmissivity. The contribution of the elementary layer to $F\uparrow$ at the reference level is

$$d[F\uparrow(\Delta\nu)] = \varepsilon\, du\, B(T, \nu)\tau^*(u), \tag{5.9}$$

where $B(T, \nu)$ is the blackbody radiation at temperature $T$ and at a frequency in the center of the frequency interval $\Delta\nu$. Combining (5.9) with (5.8) gives

$$d[F\uparrow(\Delta\nu)] = -B(T, \nu)\frac{d\tau^*(u)}{du} du. \tag{5.10}$$

Integrating from the reference level to the atmosphere at ground level gives the basic radiative transfer equation:

$$F\uparrow(\Delta\nu) = -\int_{u=0}^{u=u_g} B(T, \Delta\nu)\frac{d\tau^*(u)}{du} du. \tag{5.11}$$

The transmissivity $\tau^*(u)$ depends also on pressure and temperature, both of which can be written as a function of $u$ in the particular atmosphere under consideration.

---

1. Note that for unit downward radiation incident at the reference level, the numerator of (5.8) is the absorption in the elementary layer and the denominator is the radiation reaching the elementary layer. Their ratio is the absorptivity or emissivity.

An alternative form of (5.11) is obtained by integrating by parts, which yields

$$F\uparrow(\Delta\nu) = B(T_r, \Delta\nu) - B(T_g, \Delta\nu)\tau^*(u_g) + \int_{T_r}^{T_g} \tau^*(u)\frac{dB(T, \Delta\nu)}{dT}dT. \quad (5.12)$$

This form is useful because the derivative of $B$ is an analytic function and is therefore to be preferred to the derivative of $\tau^*$, which is usually an empirical function.

The total upgoing flux at the reference level includes the contribution of the ground surface. If the ground is black and at the same temperature as that of the surface atmosphere, this contribution is $B(T_g, \Delta\nu)\tau^*(u_g)$. Note that the inclusion of this in (5.12) results in the cancellation of the second term on the right-hand side. If the ground is not completely black or is not at the temperature $T_g$, a small remainder will result.

The downcoming radiative flux $F\downarrow$ at the reference level is given by (5.11) or (5.12), with different limits of integration, $u = 0$ to $u = u_t$ or $T_r$ to $T_t$, where the subscript t stands for the top boundary, which may be a cloud layer or the level at which the absorber concentration falls to a negligible value. Only if the upper boundary is a black cloud layer of temperature $T_t$ will the second term on the right of (5.12) cancel.

## 5.5 Flux Divergence

When the thermal radiative flux increases with height (positive divergence), the radiation is acting as a heat sink, and conversely. Knowledge of the divergence of the radiation is essential to a determination of the total heat budget of the atmosphere. What is wanted is the divergence of the net flux, the algebraic sum of the upgoing and downcoming fluxes. In (5.11) and (5.12) and their counterparts for the downcoming flux, $F\uparrow$ and $F\downarrow$ are both positive numbers, a result of the convention that $u$ is measured down and up, respectively, from the reference level. In order to get the net flux it is necessary to adopt a sign convention, and it is logical to call the upgoing flux positive:

$$F_{net} = F\uparrow - F\downarrow. \quad (5.13)$$

In most cases the net flux will be up (positive) because the temperature usually decreases upward. Both $F\uparrow$ and $F\downarrow$ are given by the same integrals, but with different limits. It is tempting to subtract the two to get the net radiation, but it must be remembered that $u$ has a different meaning in $F\uparrow$ and $F\downarrow$. Because of the confusion that can result, this subtraction will not be made here. Rather, from (5.11) we write

$$F_{net} = -\int_{u=0}^{u=u_g} B(T, \nu)\frac{d\tau^*(u)}{du}du + \int_{u=0}^{u=u_t} B(T, \nu)\frac{d\tau^*}{du}du. \tag{5.14}$$

The vertical divergence of the net flux may be obtained by differentiating (5.14), still keeping the two integrals separate. That is to say, the divergence of the net radiation is the algebraic sum of the divergence of $F\uparrow$ and of $F\downarrow$. With the necessary changes of the variables and taking $h$ as the vertical height, positive up,

$$\frac{dF_{net}}{dh} = -\int_{h_r}^{h=0} \frac{dB}{dT}\frac{dT}{dh}\frac{d\tau^*}{dh}dh + \int_{h_r}^{h_t} \frac{dB}{dT}\frac{dT}{dh}\frac{d\tau^*}{dh}dh, \tag{5.15}$$

where the subscripts have the same meanings as before.

Often the divergence of the net flux is determined for a moderately thick layer by computing the up and down fluxes at the top and bottom of the layer. The net flux at the top and bottom is found by subtraction, and then a further subtraction is made in computing the divergence in the layer. This numerical procedure of obtaining the difference of the differences emphasizes the effects of any uncertainties in the transmissivity function on the divergence. It is for this reason that relatively thick layers are commonly used. This problem is somewhat camouflaged in (5.15), but resides in the derivative of the transmissivity. In the regions of interest this derivative may range over several orders of magnitude, and it is difficult to determine the slope of empirical transmissivity functions with the necessary accuracy. This matter was discussed in some detail by Brooks (1950).

It has been assumed so far that $\Delta\nu$ is small enough that the Planck function at the mean frequency may be considered to be a function only of temperature. It is often desired to use larger frequency intervals, such as complete bands. In such cases, as noted by Yamamoto (1952), the band transmissivity should be weighted by the Planck function at some average temperature. This is clearly an approximation. In the transfer equations developed earlier, the necessary integration over frequency has not been indicated in the interests of simplicity.

It has become standard practice to express the radiative divergence in terms of the equivalent cooling (heating) rate in degrees Celsius per day. The cooling rate is

$$\frac{dT}{dt} = \frac{1.44 \times 10^3 \operatorname{div} F_{net}}{c_p \rho} \,°\mathrm{C\,day^{-1}}, \tag{5.16}$$

where the divergence is measured in $\mathrm{cal\,cm^{-3}\,min^{-1}}$, $c_p$ is the specific heat of air in $\mathrm{cal\,g^{-1}\,°C^{-1}}$, and $\rho$ is the air density in $\mathrm{g\,cm^{-3}}$. If the divergence is in $\mathrm{W\,m^{-3}}$, the specific heat in $\mathrm{J\,kg^{-1}\,°C^{-1}}$, and the density in $\mathrm{kg\,m^{-3}}$, the numerical factor in (5.16) is $8.64 \times 10^4$ instead of $1.44 \times 10^3$.

## 5.6 Overlapping Absorption Bands

The principal bands of water vapor, carbon dioxide, and ozone in the spectral region of interest are reasonably well separated, but there is significant overlap of the 15-$\mu$m $CO_2$ band and the $H_2O$ rotational band (see figure 4.6). Neglect of this overlap over paths of moderate length will result in overestimates of the absorption in the region of overlap and may even yield absorptivities greater than unity, a physical impossibility.

The exact approach to the absorption in the region of overlap is to take the product of the transmissivities of the two absorbers at intervals small compared with the line width and then to integrate over the region of overlap. This line-by-line approach is very time consuming and must be repeated in each case because there is no unique relation between the amounts of the two absorbers in an atmospheric path. A more reasonable approach is to take the product of the band transmissivities of the two absorbers in the region of overlap. This procedure was adopted by Yamamoto (1952) and incorporated in his radiation chart and has also been used by Manabe and Strickler (1964) and Manabe and Wetherald (1967). The error introduced by this approximation has been examined by Braslau (1972) by comparing it with line-by-line computations. He reports that the error due to the approximation does not exceed 10% and concludes that the detailed spectral calculation is an unnecessary refinement.

The simplest of all overlap approximations was proposed by Elsasser (1942). Arguing that most atmospheric paths of interest contain enough $CO_2$ so that the region of overlap is black due to the $CO_2$ absorption, he eliminated the absorption by $H_2O$ in the overlapped region. This means that the overlapped region is taken as black for all paths, independent of the water vapor concentration. Obviously this approximation is not valid for very short paths. The selection of the width of the overlapped region is somewhat arbitrary, since the width of the black interval tends to increase with the path length. Elsasser chose to set the radiation in the region of overlap as 18.5% of the total blackbody radiation. Other workers have chosen somewhat different fractions.

## 5.7 Radiation Charts

A number of radiation charts have been devised to facilitate the evaluation of thermal radiation in the atmosphere. These charts are designed to solve the radiative transfer equation (5.11), given the properties of a particular atmospheric path. The integration over the thermal radiation spectrum is performed once and for all in the construction of the chart. It is evident that significant approximations are involved, including integration over the entire spectrum and separation of the integration over frequency from that over the absorber path in (4.24). Significant improvements could be

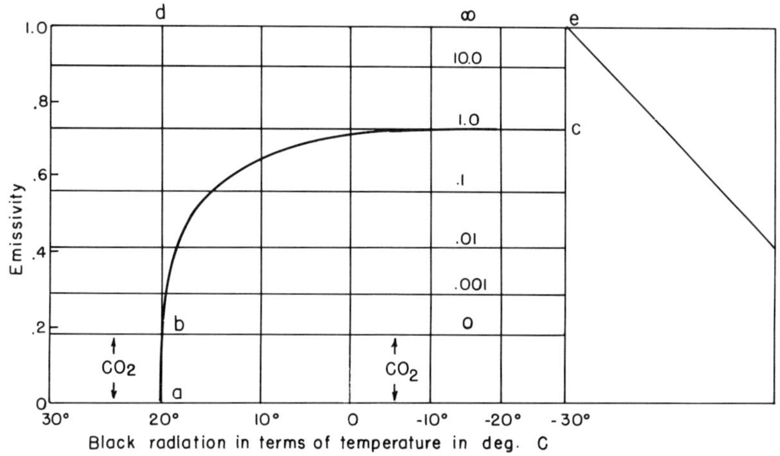


Figure 5.3 The Kew radiation chart with a typical path plotted. The column of numbers between −10 and −20°C is water vapor path length in centimeters of water. After Robinson (1950).

achieved by constructing separate charts for each major band and for the region of band overlap, but this would come at the expense of their principal advantage of simplicity. These refinements and others are now provided by numerical models run on the ubiquitous computer. The radiation charts are of historical interest and are useful for visualizing the effects of atmospheric structure and clouds on radiative transfer. A brief account of three radiation charts is give in sections 5.8–5.10.

## 5.8 The Kew Chart

This chart, devised by Robinson (1950), is not the first chart, but it is simple both to construct and to understand. It is illustrated in figure 5.3. The ordinates are emissivity on a linear scale, and the abscissas are blackbody flux, also on a linear scale, but labeled in temperature. Area on the chart is therefore radiative flux. The horizontal lines of constant path length of water vapor in centimeters of precipitable water are located from the empirical emissivity data from Robinson's measurements. Overlap between the $CO_2$ and $H_2O$ bands is handled by the Elsasser approximation described earlier. The area between the baseline of the chart and the horizontal line for $u = 0$ represents the assumed black emission of $CO_2$ in the interval from 13.3 to 17.1 $\mu$m which is taken to be 18.5% of blackbody emission. In use, the observed data for an atmospheric path in the form of path length $u$ versus temperature are plotted on the chart. The hypothetical example shown in the figure represents a cloudless atmosphere with a surface temperature of 20°C and a total precipitable water content of 1 cm.

The reference level is taken at the surface. The downcoming radiation is given by the area to the right of the curve *abc*, extending to zero blackbody emission. The chart is truncated at $-30°C$, but an accessory curve, shown as the sloping line in the upper right corner, yields the area to the right of $-30°C$. The top of the chart, marked $u = \infty$, is unit emissivity, and therefore area to the right of any isotherm is the blackbody flux at that temperature. It follows that the area *adecba*, extended to the zero abscissa beyond *ce*, represents the net outgoing radiation at the surface if the ground radiates as a blackbody at 20°C.

The Kew chart and all similar radiation charts permit the graphical evaluation of total upgoing or downcoming radiation and the net radiation at any level in the atmosphere. If cloud layers are considered to radiate as blackbodies, their effects on radiative transfer may also be found. Note that a black cloud can be represented as an isothermal layer of infinite thickness. The flux divergence in finite layers may also be evaluated by finding the differences in the net fluxes between the top and bottom of the layer. Three schematic examples are given in figure 5.4 to show how the chart permits one to visualize the process of radiative transfer.

## 5.9 The Elsasser Chart

The best-known radiation chart in the United States is the Elsasser (1942). Like the Kew chart, in this chart area is proportional to the radiative flux. Elsasser chose to use an equal area transformation. He defines

$$Q(u, T) = \int_0^\infty \tau^*(\nu, u) \frac{dB(\nu, T)}{dT} d\nu. \tag{5.17}$$

Reference to (5.12) shows that

$$\int Q \, dT = F. \tag{5.18}$$

The coordinates chosen by Elsasser for his chart are $aT^2$ as abscissas and $Q/2aT$ as ordinates, where $a$ is a constant. Area is evidently equal to $\int Q \, dT$. For the evaluation of $Q(u, T)$ Elsasser used the Elsasser band model and estimated the generalized absorption coefficient from the spectral data available in the late 1930s. The dependence of $Q$ on $T$ took account of the variation of the blackbody radiation with temperature and a theoretical estimate of the effect of temperature on the transmissivity. The emission of $CO_2$ and its overlap with the $H_2O$ rotational band was handled as in the Kew chart. Historically the Elsasser chart preceded the Kew chart, and Robinson borrowed the $CO_2$ procedure from Elsasser. Elsasser recommended the use of pressure scaling in which the reduced path length was

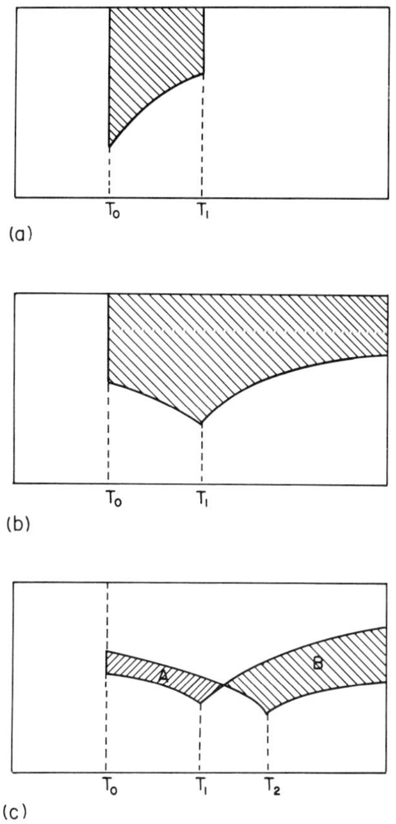


Figure 5.4 Schematic representations of the cross-hatched areas on the Kew chart that yield the indicated fluxes: (a) net upward flux at surface with a cloud of base temperature $T_1$, surface temperature $T_0$; (b) net upward flux at a level of temperature $T_1$, surface temperature $T_0$ (cloudless); (c) flux divergence of layer extending from $T_0$ to $T_2$ ($T_1 > T_2$) given by area $B - A$, which is positive, divided by vertical depth of the layer.

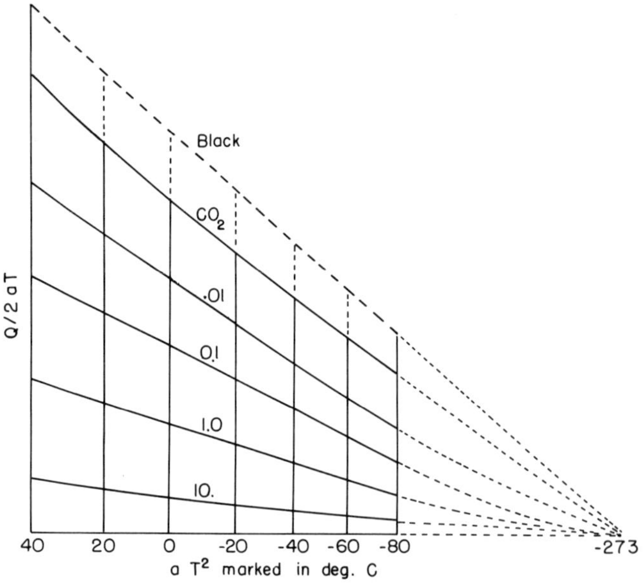


Figure 5.5 The Elsasser radiation chart. Sloping lines are isopleths of precipitable water, marked in centimeters. The area between black and $CO_2$ is the flux due to $CO_2$. The dashed portion to the right of $-80°C$ is not included on the chart, but areas in that region may be read from a table printed on the working chart. After Elsasser (1942).

taken as $u(p/p_0)^{0.5}$. The general form of the Elsasser chart is illustrated in figure 5.5. The chart is triangular in shape as a result of the equal area transformation. It is inverted with respect to the Kew chart, with the $CO_2$ contribution at the top and $u = \infty$ at the bottom. It is used in the same way as the Kew chart. The apex, marked 0, is at 0°K; the working chart is truncated at $-80°C$, and the area in the dotted line region is given in a table printed on the chart.

## 5.10 The Yamamoto Chart

The general form of this chart is similar to that of the Kew chart, but Yamamoto's (1952) procedures for obtaining the transmissivities followed Elsasser, while incorporating more recent empirical spectral data. The principal feature of the Yamamoto chart is a more adequate treatment of the $CO_2$ emission and its overlap with the $H_2O$ rotational band. The chart is based on the transmissivity of water vapor alone over the entire spectrum without omission of the region of overlap with the 15-$\mu$m $CO_2$ band. An auxiliary chart is provided that one enters with the path lengths of $H_2O$ and $CO_2$ to obtain an additional flux that is to be added algebraically to the water vapor flux to yield the total. This is based on a product of the band transmissivities of water vapor and $CO_2$ in the region of overlap and is clearly superior to the Elsasser procedure.

Yamamoto made some comparisons of results obtained from his chart with those obtained by London (1957) using the Elsasser chart. He found that his chart yielded downcoming fluxes at the surface 2–11% less that those derived from the Elsasser chart, the larger differences corresponding to the drier atmospheres. These results are similar to those reported by Robinson in his comparison of the Elsasser and Kew charts. The net upgoing flux as given by the Yamamoto chart was found to exceed that given by the Elsasser chart at all levels up to 10 km for all of the test atmospheres investigated.

## 5.11 Nocturnal Radiation

The net upward radiative flux at the surface has long been called the nocturnal radiation because of its importance in cooling the surface during the night; it occurs during daylight as well, but is overbalanced by the absorbed solar radiation. Several simple empirical expressions have been proposed for the computation of the downcoming flux at the surface in terms of quantities observed at the surface. Those of Ångstrom (1929a, b) and Brunt (1932) are given in (5.19) and (5.20), respectively:

$$F\downarrow = \sigma T_0^4 [a - b(10)^{-\gamma e}], \tag{5.19}$$

$$F\downarrow = \sigma T_0^4 (c + d\sqrt{e}), \tag{5.20}$$

where $T_0$ and $e$ are the Kelvin temperature and the water vapor pressure in mbar at the surface and $a$, $b$, $c$, $d$, and $\gamma$ are numeric constants that vary from place to place. The constants $a$, $b$, *and* $\gamma$ have been reported to have the ranges 0.71–0.80, 0.24–0.325, and 0.04–0.07, respectively. Similarly, $c$ and $d$ have been found to have the ranges 0.43–0.62 and 0.029–0.082, respectively.

The form of (5.19) and (5.20) reflects the fact that a substantial portion of $F\downarrow$ comes from the strong water vapor bands in the lower atmosphere, where the surface vapor pressure is a fair measure of the water vapor path length. The correlation coefficient between measured values of $F\downarrow$ and (5.20) is about 0.7. These expressions are valid only for cloudless skies, and attempts to include the effect of clouds have not been very satisfactory. Better results in the computation of the nocturnal radiation can be obtained from radiation charts, most of which are at their best near the surface. To use the charts it is necessary to have vertical moisture and temperature profiles. The effect of low or middle clouds can be incorporated if the height of the cloud base is known.

## 5.12 Flux Divergence from the Charts

Rodgers and Walshaw (1966) have compared the flux divergences derived from their very detailed computational method with those obtained from the Elsasser chart and the Yamamoto and Onishi (1953) chart for a series of model atmospheres. The latter chart was derived from the same data as the Yamamoto chart, but is designed to yield the flux divergence directly. It was found that the flux divergences from the Elsasser chart were consistently low at all tropospheric levels and high in the stratosphere. The flux divergences from the Yamamoto and Onishi chart were found to be too large in the lower troposphere, but reasonably good in the upper troposphere and in the stratosphere.

## 5.13 Tabulated Emissivities

It is now customary to determine thermal radiation in the atmosphere by numerical integration of the radiative transfer equation, using tabulated emissivities. This has proved to be faster and more accurate than the radiation charts. The simplest kind of tabulated emissivities are those for the entire spectrum, omitting the region overlapped by $CO_2$. Robinson (1950) found the emissivities in the second column of table 5.1 from measurements of the downcoming radiation at the surface during cloudless nights. These are the emissivities he used in the construction of the Kew chart. The emissivities in the third column were derived by Brooks (1950) from Robinson's data and show the latitude permitted by the data. The Elsasser emissivities in the fourth column of table 5.1 were computed from the tabular data on the Elsasser chart at a temperature of 0°C. Elsasser included a theoretical temperature correction with the result that the emissivity increases with decreasing temperature; the choice of 0°C made here is arbitrary.

In using the data in table 5.1, the flux due to water vapor, omitting the $CO_2$ overlapped region, is computed from the radiative transfer equation after the reduced path lengths have been found from pressure scaling or otherwise. Black radiation in the $CO_2$ region at the temperature of the reference level is then added to get the total flux. Robinson found that the Kew chart gave results for the downcoming radiation at the surface that agreed better with observations than the Elsasser chart. For a surface or near-surface reference level, the Robinson emissivities in table 5.1 are probably better than the other two sets. The sole virtue of these emissivities is the ease with which they can be used to compute radiative transfer. The results should be much the same as those obtained with the charts and must be considered only as a first approximation.

Elsasser and Culbertson (1960) have given a much more complete tabulation of emissivities based on the detailed spectral data available in the late

Table 5.1
Diffuse emissivity of water vapor omitting $CO_2$ band emissivity

| Path length (cm) | Robinson | Brooks | Elsasser | Path length (cm) | Robinson | Brooks | Elsasser |
|---|---|---|---|---|---|---|---|
| 0.0001 | 0.023 | 0.024 | 0.030 | 0.2 | 0.420 | 0.448 | 0.487 |
| 0.001 | 0.105 | 0.116 | 0.117 | 0.5 | 0.487 | 0.519 | 0.556 |
| 0.002 | 0.138 | 0.153 | 0.159 | 0.7 | 0.512 | 0.547 | 0.580 |
| 0.005 | 0.189 | 0.205 | 0.218 | 1 | 0.540 | 0.575 | 0.604 |
| 0.007 | 0.208 | 0.225 | 0.242 | 2 | 0.589 | 0.631 | 0.637 |
| 0.01 | 0.225 | 0.246 | 0.267 | 5 | 0.657 | 0.702 | 0.704 |
| 0.02 | 0.262 | 0.288 | 0.316 | 7 | 0.680 | 0.727 | 0.725 |
| 0.05 | 0.329 | 0.347 | 0.382 | 10 | 0.703 | 0.754 | 0.747 |
| 0.07 | 0.352 | 0.370 | 0.407 | $\infty$ | 0.815 | 0.815 | 0.815 |

1950s. They derived and tabulated the emissivities of each of the major bands separately and also a correction for band overlap. One of the "bands" included is the water vapor continuum in the window region. The effects of temperature on the emissivities were computed from the theoretical approximation given by (4.45). It appears that a mathematical error was inadvertently introduced in deriving the band emissivities from the generalized absorption coefficients as a function of wave number. This has been corrected by Staley and Jurica (1970), who have presented revised tables of the emissivities at standard pressure for the 6.3-$\mu$m water vapor band, the window, the water vapor rotational band, the 15-$\mu$m $CO_2$ band, the 9.6-$\mu$m ozone band, and a water vapor-$CO_2$ overlap correction. These tables, in abridged form, are given in table 5.2. The emissivities are all in percent of blackbody flux, so that they may be added algebraically (the overlap correction is negative) to yield the total gaseous emissivity. In a later paper Staley and Jurica (1972) have given new water vapor emissivities on the basis of more recent laboratory data, but they did not include the new tables. They report that the computed downcoming radiation at the surface is closer to, but still somewhat smaller than, that observed. This may be due to neglect of the contributions of the aerosol and trace gases or to errors in the measurements or in the emissivities.

All of the emissivities in table 5.2 are for standard sea level pressure. For the reasons discussed in section 4.26 it is expected that the reduced path length at other pressures will be proportional to $(p/p_0)^\kappa$, where $\kappa$ is less than unity. The exponent $\kappa$ for the water vapor bands ranges from about 0.7 for small optical depths to nearly 1 for large optical depths. Staley and Jurica (1970) used $\kappa = 1$ for the water vapor bands. In their second paper (1972) they were able to obtain more complete information on the effects of pressure and temperature on the reduced path length from the newer laboratory data. In effect this amounted to smaller, though variable, values of $\kappa$ and consequently to an increase in the downcoming flux at the surface. The water vapor emissivities in table 5.2 can be used with $\kappa = 1$ as intended by Staley and Jurica or with $\kappa = 0.8$–$0.9$ as an empirical correction that will increase the computed downcoming radiation at the surface. It is not known whether this empirical adjustment would be desirable in the computations of the upgoing radiation or at higher levels in the atmosphere.

Staley and Jurica say that the exponent for the Elsasser transmissivities in the 15-$\mu$m $CO_2$ band is 0.88 but that Elsasser adjusted his tabulations to correspond to $\kappa = 1$. The emissivity in the 9.6-$\mu$m band of $O_3$ is so temperature dependent that the tabulated values are quite uncertain, and it is pointless to consider a preferable value of $\kappa$. Fortunately the contribution of $O_3$ to the total emissivity in the troposphere is small, and no large error will result from the assumption that the emissivity due to ozone is 5% at all tropospheric levels.

Table 5.2
Flux emissivities at standard pressure (%)[a]

Water vapor emissivities (%)

| | Rotation band (°C) | | | Window (°C) | | | 6.3-$\mu$m band (°C) | | |
|---|---|---|---|---|---|---|---|---|---|
| log $u^b$ | −40 | −10 | 20 | −40 | −10 | 20 | −40 | −10 | 20 |
| −4.0 | 5.57 | 4.65 | 3.98 | 0 | 0 | 0 | 0.59 | 1.07 | 1.67 |
| −3.0 | 14.5 | 12.3 | 10.7 | 0.01 | 0.01 | 0.01 | 1.23 | 2.27 | 3.58 |
| −2.7 | 18.1 | 15.5 | 13.6 | 0.01 | 0.01 | 0.02 | 1.48 | 2.75 | 4.34 |
| −2.3 | 23.2 | 20.2 | 17.8 | 0.03 | 0.04 | 0.07 | 1.83 | 3.41 | 5.42 |
| −2.0 | 27.1 | 23.7 | 20.9 | 0.06 | 0.09 | 0.18 | 2.08 | 3.90 | 6.21 |
| −1.7 | 30.8 | 27.2 | 24.1 | 0.13 | 0.21 | 0.41 | 2.37 | 4.46 | 7.10 |
| −1.3 | 35.6 | 31.7 | 28.1 | 0.34 | 0.57 | 0.96 | 2.76 | 5.18 | 8.24 |
| −1.0 | 39.1 | 34.9 | 31.0 | 0.70 | 1.10 | 1.69 | 3.04 | 5.69 | 9.03 |
| −0.7 | 42.4 | 37.9 | 33.8 | 1.40 | 2.04 | 2.89 | 3.32 | 6.19 | 9.79 |
| −0.3 | 46.4 | 41.6 | 37.1 | 3.29 | 4.40 | 5.72 | 3.68 | 6.83 | 10.7 |
| 0 | 49.2 | 44.2 | 39.2 | 5.94 | 7.57 | 9.34 | 3.96 | 7.30 | 11.4 |
| 0.3 | 51.8 | 46.4 | 41.0 | 10.2 | 12.5 | 14.7 | 4.22 | 7.72 | 12.0 |
| 0.7 | 54.7 | 48.8 | 42.8 | 18.5 | 21.8 | 24.5 | 4.52 | 8.18 | 12.6 |
| 1.0 | 56.6 | 50.1 | 43.6 | 25.5 | 29.5 | 32.4 | 4.72 | 8.45 | 12.9 |
| 1.3 | 58.1 | 51.1 | 44.1 | 30.8 | 35.3 | 38.2 | 4.89 | 8.64 | 13.1 |

Carbon dioxide emissivities (%)

| log $h^c$ | −40°C | −10°C | 20°C | log $h^c$ | −40°C | −10°C | 20°C |
|---|---|---|---|---|---|---|---|
| −4.0 | 0.119 | 0.115 | 0.108 | −0.3 | 6.36 | 6.54 | 6.49 |
| −3.7 | 0.171 | 0.167 | 0.158 | 0 | 7.99 | 8.26 | 8.23 |
| −3.3 | 0.262 | 0.260 | 0.248 | 0.3 | 9.61 | 10.0 | 10.0 |
| −3.0 | 0.354 | 0.352 | 0.338 | 0.7 | 11.7 | 12.3 | 12.4 |
| −2.7 | 0.471 | 0.471 | 0.454 | 1.0 | 13.3 | 13.9 | 14.1 |
| −2.3 | 0.691 | 0.693 | 0.672 | 1.3 | 14.8 | 15.6 | 15.8 |
| −2.0 | 0.928 | 0.934 | 0.907 | 1.7 | 16.7 | 17.7 | 18.0 |
| −1.7 | 1.29 | 1.30 | 1.27 | 2.0 | 18.1 | 19.3 | 19.6 |
| −1.3 | 2.08 | 2.11 | 2.07 | 2.3 | 19.5 | 20.7 | 21.1 |
| −1.0 | 3.03 | 3.08 | 3.03 | 2.7 | 21.3 | 22.7 | 23.1 |
| −0.7 | 4.30 | 4.39 | 4.32 | 3.0 | 22.6 | 24.1 | 24.4 |

Table 5.2 (*cont.*)

Overlap correction (%, $T = -10°C$) for log $u$[d]

| log $u$ | −2.0 | −1.7 | −1.3 | −1.0 | −0.7 | −0.3 | 0 | 0.3 | 0.7 | 1.0 |
|---|---|---|---|---|---|---|---|---|---|---|
| −2.0 | 0.01 | 0.03 | 0.06 | 0.09 | 0.14 | 0.23 | 0.34 | 0.48 | 0.69 | 0.81 |
| −1.7 | 0.02 | 0.04 | 0.08 | 0.12 | 0.19 | 0.33 | 0.48 | 0.67 | 0.96 | 1.13 |
| −1.3 | 0.03 | 0.06 | 0.13 | 0.20 | 0.31 | 0.53 | 0.78 | 1.09 | 1.56 | 1.83 |
| −1.0 | 0.04 | 0.09 | 0.19 | 0.30 | 0.46 | 0.78 | 1.14 | 1.60 | 2.27 | 2.67 |
| −0.7 | 0.06 | 0.13 | 0.27 | 0.42 | 0.65 | 1.10 | 1.61 | 2.26 | 3.22 | 3.80 |
| −0.3 | 0.10 | 0.19 | 0.40 | 0.63 | 0.96 | 1.64 | 2.39 | 3.35 | 4.78 | 5.64 |
| 0 | 0.13 | 0.25 | 0.50 | 0.79 | 1.21 | 2.06 | 3.00 | 4.21 | 6.00 | 7.09 |
| 0.3 | 0.16 | 0.30 | 0.61 | 0.96 | 1.46 | 2.48 | 3.60 | 5.05 | 7.22 | 8.54 |
| 0.7 | 0.20 | 0.38 | 0.75 | 1.17 | 1.79 | 3.03 | 4.40 | 6.17 | 8.82 | 10.5 |
| 1.0 | 0.24 | 0.43 | 0.86 | 1.33 | 2.03 | 3.43 | 4.98 | 6.97 | 9.98 | 11.8 |
| 1.3 | 0.28 | 0.50 | 0.97 | 1.50 | 2.28 | 3.85 | 5.58 | 7.81 | 11.2 | 13.3 |
| 1.7 | 0.33 | 0.59 | 1.13 | 1.74 | 2.62 | 4.41 | 6.37 | 8.90 | 12.7 | 15.2 |
| 2.0 | 0.38 | 0.66 | 1.25 | 1.91 | 2.88 | 4.83 | 6.96 | 9.71 | 13.9 | 16.5 |
| 2.3 | 0.43 | 0.73 | 1.37 | 2.09 | 3.14 | 5.25 | 7.54 | 10.5 | 15.0 | 17.9 |
| 2.7 | 0.49 | 0.82 | 1.52 | 2.31 | 3.46 | 5.77 | 8.27 | 11.5 | 16.4 | 19.6 |
| 3.0 | 0.53 | 0.89 | 1.63 | 2.46 | 3.68 | 6.12 | 8.76 | 12.2 | 17.4 | 20.8 |

Ozone emissivities (%)

| | Flux emissivities (%) | | |
|---|---|---|---|
| log $h$ | −40°C | −10°C | 20°C |
| −4.0 | 0.155 | 0.191 | 0.214 |
| −3.0 | 0.610 | 0.777 | 0.896 |
| −2.7 | 0.872 | 1.12 | 1.29 |
| −2.3 | 1.34 | 1.72 | 1.99 |
| −2.0 | 1.79 | 2.30 | 2.68 |
| −1.7 | 2.31 | 2.99 | 3.51 |
| −1.3 | 3.03 | 3.97 | 4.71 |
| −1.0 | 3.54 | 4.66 | 5.55 |
| −0.7 | 4.01 | 5.30 | 6.33 |
| −0.3 | 4.60 | 6.10 | 7.29 |
| 0 | 5.02 | 6.64 | 7.93 |
| 0.3 | 5.36 | 7.08 | 8.43 |

a. From Staley and Jurica (1970).
b. $u$ is the precipitable water in cm (or g/cm$^2$); the logarithm is to the base 10.
c. $h$ is the path length of the absorbent ($CO_2$) at standard pressure and temperature, 1,013.3 mbar and 0°C.
d. This is a correction to be *subtracted* from the emissivity of $H_2O + CO_2$ to account for the overlap of the spectra. (Similar tables are available in the reference for other temperatures, but this one may be used for other temperatures with usually negligible errors.)

## 5.14 The Use of Table 5.2

A typical nonhomogeneous vertical path in the atmosphere is described by the pressure, temperature, and humidity at a discrete number of elevations. The increment of $u^*$ within each of the layers may then be computed and $u^*$ summed from any desired reference level or levels. The same may be done for $CO_2$ by making use of the fact that $CO_2$ represents about 0.033% of the atmosphere by volume. This is then reduced to the path length in centimeters at standard pressure and temperature and $h^*$ is evaluated in the same fashion as $u^*$. Ozone is a variable constituent, and its path length can be obtained from an observed or assumed vertical distribution. In practice ozone has only a small effect on radiative transfer in the troposphere, but becomes as important as $H_2O$ and $CO_2$ in the stratosphere. Once $u^*$ and $h^*$ are determined, the radiative flux is computed from the finite difference form of the radiative transfer equation, using the flux emissivities in table 5.2.

The Staley and Jurica emissivities can be expected to yield more accurate fluxes than can be obtained from any of the radiation charts or from the total emissivities in table 5.1. These newer emissivities are based on more adequate spectral data. Separation of the emissivities by bands reduces the effect of the variations of the Planck distribution function with temperature, permits a more adequate treatment of band overlap, and allows a better approximation for nonhomogeneous paths including the effects of variable temperature. These advantages are gained at the expense of much greater computational demands, which are probably at the limit of hand computation. They cannot be considered as the last word because they do not incorporate the additional spectral data accumulated since about 1960. The next step in complexity is the use of wideslit spectral data for bands of a width of about 10 $cm^{-1}$, and this certainly requires the use of a high-speed computer.

## 5.15 Observations of Atmospheric Radiation

It should be remarked at the outset that the measurement of atmospheric radiation poses formidable instrumental problems. Among these are the unavailability of window materials that are fully transparent and of surfaces that are black over the entire spectrum. All surfaces, including particularly the housing and shutters of the instruments, radiate in the same region of the spectrum as the atmosphere. All instruments must be calibrated against standard blackbodies, and this poses its own problems. Over the years some fairly satisfactory instruments have been developed for the measurement of the downcoming atmospheric radiation or the net flux at the ground. Valuable results have been obtained with such instruments, but the flux at the ground comes predominately from the lowest layers of the atmosphere. The effects of pressure and the overlap of the spectra are minimal in this case. What is needed is the direct measurement of the upgoing and downcoming fluxes (or their difference) as a function of height at least as far as the lower stratosphere.

Infrared spectrometers have been flown in England by Houghton and Seeley (1960). For the most part these experiments have involved only selected regions of the spectrum, and observations have been taken primarily at high altitudes to reveal more clearly the spectral lines. Brewer and Houghton (1956) have made some measurements of the total spectrum from aircraft in spite of the difficulties of measurement. Some very promising results have been obtained with balloon-borne radiometers that telemeter the data through radiosonde equipment. The instrument used for most of these observations is the economical net radiometer described by Suomi and Kuhn (1958). This instrument measures the upward and downward long-wave radiation and hence the net flux as a function of height. The slope of the curve of net flux versus height is the flux divergence. Comparisons between the measured flux and that computed from laboratory observations by Kuhn (1963a, b) showed good agreement. Suomi et al. (1958) found good agreement between the flux divergence given by this instrument and that computed by Brooks's (1950) method. The best fit between observations and computations was obtained by using less than linear pressure scaling.

## 5.16 Effects of the Aerosol

The aerosol contributes to the long-wave radiation and may explain, in part, the excess of the measured radiation over that computed for the principal gaseous absorbers. The absorption by the aerosol depends on the absorption efficiency factor, which, in turn, depends on the size parameter and the imaginary part of the complex index of refraction. The size parameters are an order of magnitude smaller in the thermal infrared than in the solar spectrum. The imaginary index of refraction is typically two orders of magnitude larger for thermal than for solar radiation. These two effects act in opposite directions, and detailed computations are required to determine the absorptivity of the aerosol in the thermal infrared. Whereas aerosol particles of radius 0.1–1 $\mu$m are the most important in the solar spectrum, particles of 1–10 $\mu$m radius are of greatest importance in the thermal infrared. Thus the absorptivity in the infrared depends strongly on the concentration of the large aerosol particles.

Ackerman et al. (1976) have made calculations of radiative transfer for atmospheres containing aerosol particles. Three different aerosol concentrations were used, categorized as light, average, and heavy, and based on observations in Los Angeles. The aerosol size spectrum was bimodal, one class with a modal radius of 0.2 $\mu$m and the other with a modal radius of

6–8 $\mu$m. They found that the aerosol had little effect within the three major gaseous absorption bands, but was quite noticeable in the window region. The "average" aerosol increased the cooling rate a few tenths to 1°C per day. The "heavy" aerosol caused an increased cooling of about 4°C per day. These excess cooling rates do not differ very greatly from the heating rates that have been computed for the absorption of solar radiation by the aerosol. It also seems that an "average" aerosol could be responsible for the approximate 0–8% excess downcoming radiation found by Robinson (1950). The results of the Ackerman et al. computations are dependent on the assumed bimodal aerosol size distribution, and smaller cooling rates would be anticipated for an aerosol with a smaller concentration of large particles.

A special type of aerosol is the Harmattan haze, which is wind-raised dust from the Sahara and is often observed as far west as the Caribbean. Kuhn et al. (1975) made a number of observations of the thermal radiation in the eastern tropical North Atlantic during visible Harmattan haze and found that the total volume extinction coefficient of the haze was 0.042 $km^{-1}$ and quite uniform. They computed the vertical flux divergence due to the haze and found that it corresponded to a cooling rate of about 0.03°C $hr^{-1}$.

## 5.17 Effects of Cirrus Clouds

Cirrus clouds modify the radiative flux and flux divergence, the magnitude of the effect being dependent on the thickness of the cloud and its optical depth. In some cases cirrus clouds are very thin or tenuous. It has been suggested that such tenuous cirrus might be responsible for the excess radiation observed by Robinson and others because the cloud might be unobserved at night, the time when most of the measurements of long-wave radiation are made. This is a reasonable supposition, but it is probable that, in most cases, the extra radiation observed on clear nights is due to the aerosol.

Cirrus clouds range from scattered wisps to dense cirrostratus layers. The emissivity of cirrus has a similar range and averages well below unity. Measurements of the emissivity are almost always made in the window region to avoid the complications of the gaseous emissions. It has been found that the emissivity of cirrus is a function of the thickness of the cloud. This is shown in figure 5.6 for two sets of observations and a few isolated points. The scatter of the points to which the two curves were fitted is considerable, but is not shown in the figure. This scatter is due to the dependence of the emissivity on the size and concentration of the ice particles as well as on the cloud depth.

Several computations of the emissivity of cirrus clouds have been made by Liou (1974) and Fleming and Cox (1974). The numerical results are

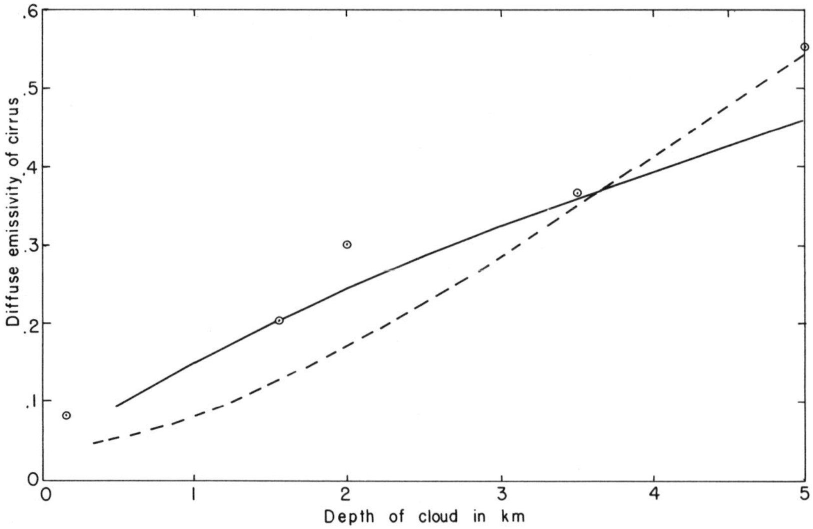


Figure 5.6 Diffuse emissivity of cirrus cloud versus cloud thickness. The solid curve is after Kuhn and Weickmann (1969); the dashed curve is after Davis (1971); circles are from other studies.

limited by the lack of adequate information of the concentration, shape, orientation, and size of the ice crystals in natural cirrus. Initially the particles were assumed to be ice spheres. More realistically, the ice crystals are taken to be cylinders or hexagonal columns; the phase functions of these forms are significantly different from those for spheres. In most cases the hexagonal columns or cylinders are assumed to be oriented with their long axes horizontal and with random orientation along the azimuth. The actual forms of cirrus crystals are thought to be hexagonal columns, bullets, thick hexagonal plates, and assemblages of bullets or of columns, The results of the computations can be adjusted to fit observations by an appropriate selection of crystal concentration, but this does not necessarily show that the numerical models are satisfactory.

## 5.18 Radiative Effects of Middle and Low Clouds

For a long time middle and low clouds were assumed to radiate as blackbodies at the temperature of the cloud base or top. This assumption is still often made because of its simplicity, although it is now known that many clouds are not black.

Allen (1971) on the basis of 248 observations of low clouds at Montreal found that their emissivities ranged from 0.25 to 1.15 with a mean of $1.00 \pm 0.03$ in the window, taken as 8–13 $\mu$m. The unrealistic emissivity of 1.15 is doubtless the result of observational errors and the use of the

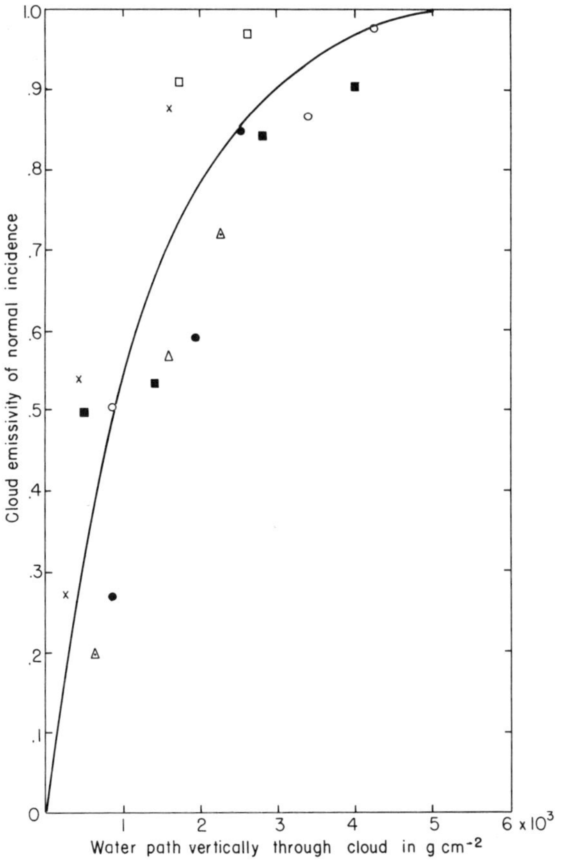


Figure 5.7 Normal incidence emissivity of middle clouds versus water path length. Symbols identify the six separate flights on which data were gathered. After Platt (1976).

temperature at cloud base as the reference temperature. These results, together with others, suggest that low clouds of moderate to large optical depths have emissivities of 0.97–0.99 and may usually be taken as black.

The situation is not so simple for middle clouds, which are mostly alto- and stratocumulus. These cloud layers are often as shallow as a few hundred meters or less, but may have depths of up to several kilometers. They are cellular in structure, reflecting the convective motions that form and maintain them. These clouds may be composed of liquid drops or ice crystals or both.

The observed emissivities of altocumulus and stratocumulus in the window spectral region range from about 0.2 to 1.0 and show no useful relation to cloud depth. From airborne observations Platt (1976) found

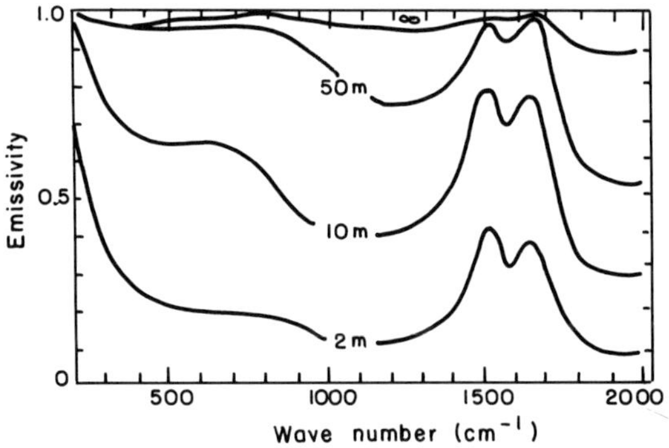


Figure 5.8 Cloud emissivities versus wave number for several cloud depths. After Yamamoto et al. (1970). Minimum drop diameter is 2 $\mu$m, maximum 26 $\mu$m, and mode 9 $\mu$m. Liquid water content is 0.28 gm m$^{-3}$. Cloud drop spectrum is from Diem (1948) for altostratus.

that the cloud emissivity was related to the total liquid water in a vertical column through the cloud, as shown in figure 5.7. The corresponding diffuse emissivities, which were not measured, are larger than the normal incidence emissivities. Even so, this set of data on warm middle clouds just off the Australian east coast shows that many of the clouds are not black, but also that some of the clouds only 100–300 m thick are nearly black.

## 5.19 Theoretical Studies of Cloud Emissivity

The thermal radiative properties of clouds have been examined theoretically by Hunt (1973) and Yamamoto et al. (1970), among others. Hunt considered only the window regions, where he felt that the effects of gaseous absorbers could be neglected. Yamamoto et al. essayed the more difficult task of averaging over the thermal radiation spectrum from 5 to 50 $\mu$m. They therefore included the gaseous absorption as well as the scattering and absorption by the cloud drops. The authors of the two papers chose somewhat different cloud drop size distributions and scattering phase functions as well as different formulations of radiative transfer in a multiple-scattering medium. These differences and the choices they made in presenting their results make it difficult to undertake quantitative comparisons. A rough comparison in the window region at 11 $\mu$m shows that the cloud emissivities of Yamamoto et al. are significantly larger than those of Hunt. The cause of this discrepancy is not known but it may reside in the different assumptions listed earlier. It is not considered useful to present the results of these studies in detail until this discrepancy is resolved.

The variation of the emissivity with wave number for several cloud thicknesses as obtained by Yamamoto et al. is shown in figure 5.8 because

Table 5.3
Measured thermal emissivities of natural surfaces, 8–14 $\mu$m

| Surface | Number of observations | Emissivity | | |
|---|---|---|---|---|
| | | Mean | Maximum | Minimum |
| Ocean | 20 | 0.99 | 1.02 | 0.95 |
| Desert A[a] | 3 | 0.93 | 0.94 | 0.92 |
| Desert B[b] | 3 | 0.90 | 0.91 | 0.90 |
| Melting snow | 4 | 0.99 | 0.99 | 0.98 |

a. 50% sand and 50 % dirt and gravel.
b. 100% sand.

Table 5.4
Normal incidence emissivities in the 8–12-$\mu$m band

| Substance | Emissivity | Substance | Emissivity |
|---|---|---|---|
| Granite | 0.898 | Wet sand | 0.936 |
| Basalt | 0.904 | Concrete | 0.966 |
| Sandstone | 0.909–0.935 | Asphalt | 0.956 |
| Dolomite | 0.958 | Pure Water | 0.993 |
| Dolomite gravel | 0.959 | Melting snow | 0.99 |
| Sand (dry) | 0.914 | Oil film on water | 0.97 |

of its qualitative importance. It will be seen immediately that the emissivity shows a broad minimum in the spectral region from about 900 to 1,400 $cm^{-1}$, which includes most of the atmospheric window. This suggests that the cloud emissivities measured in the window underestimate the emissivity over the thermal radiation spectrum.

## 5.20 Surface Emission

The usual practice of assuming that the surface of the earth radiates as a blackbody at the reported surface air temperature is open to question. Few if any natural surfaces are truly black. Measurements of the emissivity of a few types of natural surfaces in the 8–14 $\mu$m spectral region, as reported by Griggs (1970), are given in table 5.3.

Some measurements of the emissivity of natural substances at normal incidence by Buettner and Kern (1965) are given in table 5.4.

The emissivity of water is of special interest, since it covers some 70% of the earth. The emissivity of 0.993 for water in table 5.4 is one of the highest reported. Other measurements have yielded values as low as 0.96. Most of

Table 5.5
Emissivity of a plane water surface in the window as a function of angle of incidence

| Angle of incidence (°) | Emissivity | Angle of incidence (°) | Emissivity |
|---|---|---|---|
| 30 | 0.98 | 70 | 0.86 |
| 40 | 0.98 | 75 | 0.79 |
| 50 | 0.97 | 80 | 0.64 |
| 60 | 0.93 | | |

the measurements have been made at normal incidence, but McSwain and Bernstein, reported by Buettner and Kern (1965), found that the emissivity of a plane water surface varied with the angle of incidence. Average values for the window region taken from this study are given in table 5.5.

The emissivity of a plane water surface also depends on wavelength, as shown in figure 5.9 from Robinson and Davies (1972). The emissivity is maximum in the window region, suggesting that measurements there overestimate the emissivity averaged over the thermal radiation spectrum. This can be decided only when similar measurements are made in the important region beyond 15 $\mu$m.

From an overview of the still incomplete data, the spectrally averaged emissivity of the oceans may perhaps be taken as 0.97–0.98. Over desert areas the emissivity is about 0.90–0.93. Direct measurements over vegetation are apparently scarce, but there is indirect evidence that the emissivities are close to unity.

## 5.21 Radiative Temperature of the Surface

A simple calculation shows that, for an ambient temperature of around 10°C, an error of about 3.5°C in the radiative temperature has the same effect on the emission of the surface as a 5% error in the emissivity of the surface.

The ocean surface temperature typically differs by a degree or less from the air temperature measured some meters above the surface. Larger differences occur on occasion, as when cold continental air streams over warmer water. Diurnal variations of the sea surface temperature are generally small. The accuracy of computations of the emission of the sea surface are generally limited by errors in the reported air or sea surface temperatures and by inadequate coverage of large ocean areas.

Over land areas substantial differences may exist between the radiative temperature and that taken in the standard thermometer screen. Extreme differences occur in subtropical deserts, where the midday sand surface temperature may be 20–30°C higher than the screen temperature. In

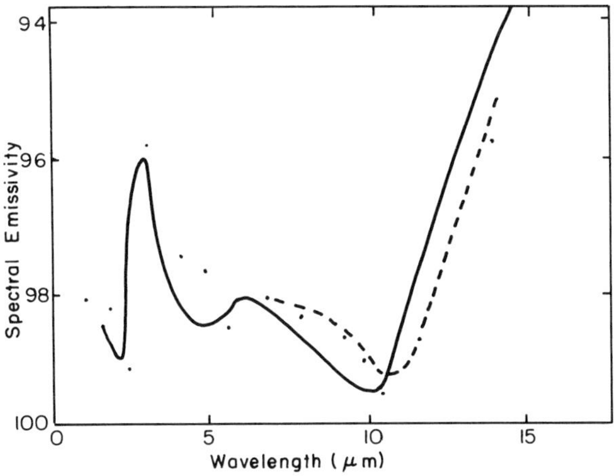


Figure 5.9 Emissivity versus wavelength for pure water. After Robinson and Davies (1972) for dashed curve and including data from Bell (1957) for dotted curve and Kislovskii (1959) for solid curve.

vegetated areas the leaves are the radiators. Robinson (1950) measured the radiation temperature over closely cut grass under conditions of strong insolation and found that it was usually within a degree or less of that of a thermometer in the grass. However, the screen temperature was found to be up to 20°C colder than the surface temperature, the average difference being about 7°C in this particular set of observations. On clear nights with strong radiational cooling, it would be expected that the surface temperature would be much colder than the screen temperature. The data of Rider and Robinson (1951) show the reversal of the difference setting in an hour or so before sunset, but they do not include late night or early morning hours. Similar data for other more complex vegetative covers do not seem to be available.

## 5.22 Survey of Radiative Transfer in the Atmosphere

Before turning to numerical evaluations, it is well to view the subject of radiative transfer in perspective. Much of the solar radiation penetrates the atmosphere and is absorbed at the surface, which is the principal heat source for the atmosphere. The thermal radiation from the surface is largely absorbed by the atmosphere, which then radiates downward as well as upward. The radiative flux from the surface must overbalance the downcoming flux and so provide the net upward flux that, with further absorption and emission, escapes at the top of the atmosphere. In consequence the surface temperature must be much warmer than it would be with a nonabsorbing atmosphere. This is called the "greenhouse effect."

Some have objected to this term because it is not the opacity of the glass to thermal radiation but its suppression of convection that is primarily responsible for the warmer temperature. If this is understood, there would seem to be no reason to change the time-honored term.

In the annual global mean the absorbed solar radiation must closely equal the outgoing thermal radiation to space. In the annual zonal mean both the absorbed solar radiation and the thermal radiation to space decrease with latitude, but the former changes more rapidly than the latter. It is this imbalance that is the basic thermal drive of the general circulations of the atmosphere and the oceans. It is well to keep in mind that the imbalance is a complex function of time and space that have been averaged out in the time and space means.

The outgoing thermal radiation to space originates from the gaseous radiators, the surface of the earth, and clouds. In the strong emission bands of water vapor and carbon dioxide the outgoing radiation comes from the upper layers of the atmosphere because the radiation from lower levels has been absorbed. The upper troposphere is thus the immediate source of the emissions in the 6.3-$\mu$m and rotational bands of water vapor, while much of that from the 15-$\mu$m carbon dioxide band comes from the stratosphere. In the wings of these strong bands, where the lines are weaker and more widely spaced, the contribution to the outgoing radiation comes from lower tropospheric layers. Pressure broadening of the lines permits radiation from the wings at lower (higher pressure) levels to escape almost unhindered to space. The atmospheric window (8–12 $\mu$m) allows a substantial amount of radiation from the surface or the top of a cloud layer to escape to space.

The net thermal radiative flux decreases with height in the troposphere at a rate that is equivalent to a cooling rate of approximately 1–2°C per day. This cooling is partly offset by heating due to the absorption of solar radiation at a rate of, say, 0.5–1°C per day. This radiative imbalance is compensated by latent and sensible heat convected from near the surface. The principal heat source of the stratosphere is the absorption of solar radiation by ozone. The thermal radiative flux due to carbon dioxide, ozone, and water vapor decreases quite rapidly with height and provides a countervailing cooling.

Clouds have a major effect on both solar and thermal radiation. On the average, clouds reflect about a quarter of the total incident solar radiation back to space. The effects of clouds on thermal radiation are of equal importance. Dense low and middle clouds will markedly decrease the net radiation at the surface and, in consequence, nocturnal cooling. Such clouds reduce the outgoing radiation to space primarily by replacing the thermal radiation of the surface in the window by a colder cloud top. The clouds themselves will be heated at their bases and cooled at their tops.

Cirrus cloud usually decreases the radiation to space primarily again through lower emission in the window. When both the solar and thermal fluxes are considered, a cirrus cloud will usually lead to lower surface temperatures, but for certain very high ($>9$ km) and dense ($\varepsilon > 0.5$) clouds, a slight warming at the surface is predicted by Manabe and Strickler (1964).

## 5.23 Numerical Studies of Radiative Effects

Two approaches have been followed in studies of the influence of radiative exchange on the atmospheric structure. One approach is to determine the vertical temperature profile of an atmosphere in radiative equilibrium. An initial temperature profile, usually isothermal, and also the vertical distribution of water vapor are assumed. The first calculation of cooling rates is used to get the second approximation to the equilibrium temperature profile; this procedure is iterated until the cooling rates become equal to, or smaller than, an arbitrary small number. This procedure emphasizes features of the temperature structure that are strongly related to radiative transfer.

The second approach, and the one most often used, is to start with the observed distribution of temperature, pressure, moisture, and cloud and then compute the cooling rate as a function of position. The regions of relatively large cooling rates are those where the nonradiative processes are most important and conversely. It is usual to carry out such computations as a function of latitude and season.

The first approach has been followed by Manabe and Moeller (1961), who adopted a distribution of water vapor in the lower troposphere compiled by London (1957) from radiosonde data; in the upper troposphere they took a constant frost point lapse rate of 6.25°C km$^{-1}$. The effects of clouds were neglected, and many interesting results obtained are not applicable to an atmosphere with a normal cloud cover. The vertical temperature profile for a water vapor distribution appropriate to latitude 35°N in April is shown in figure 5.10. The radiative equilibrium temperature profile is superadiabatic near the surface and only slightly smaller than the dry adiabatic lapse rate in the remainder of the troposphere. A very sharp and cold tropopause is found, suggesting that radiative exchange is important in maintaining the tropopause. Further computations showed that the radiative equilibrium tropopause became warmer and lower as the tropospheric water vapor concentration was decreased and also that the tropopause became warmer with increased ozone and decreased water vapor in the stratosphere. These results are in qualitative accord with some of the well-known features of the real tropopause, such as the decrease of tropopause height with increasing latitude. However, radiation is only one of the

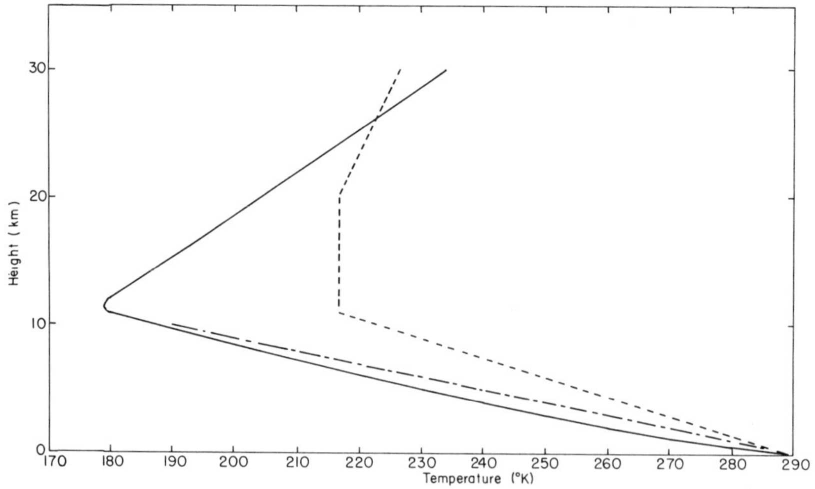


Figure 5.10 Vertical temperature profile (solid curve) for radiative equilibrium. Moisture distribution and solar irradiation appropriate to 35°N in April. After Manabe and Moeller (1961); dash-dot curve is a dry adiabat, and dashed curve is the U.S. Standard Atmosphere of 1962.

controls of tropopause height and temperature, and too much should not be made of similarities between the real and the radiative equilibrium tropopauses.

Manabe and Strickler (1964) introduced a convective adjustment in which the lapse rate is not permitted to exceed a mean tropospheric lapse rate[2] of 6.5°C $km^{-1}$. As noted, pure radiative equilibrium leads to unrealistically steep lapse rates, which would result in convective overturning. The convective adjustment stabilizes the lapse rate by simulating an upward flux of heat by convection. The introduction of the convective adjustment leads to thermal, rather than radiative, equilibrium. Manabe et al. show that the use of the convective adjustment in a one-dimensional model leads to realistic vertical temperature profiles, as shown, for example, in figure 5.11.

Manabe and Wetherald (1967) used this model with convective adjustment to estimate the change in surface temperature due to a doubling of the $CO_2$ concentration in the atmosphere. In this comparison, it was assumed that the relative humidity rather than the absolute humidity was conserved. This was based on a comparison of measured summer and winter water vapor profiles, which shows that the relative humidities are approximately

2. This follows the usual meteorological practice of omitting the negative sign of the lapse rate.

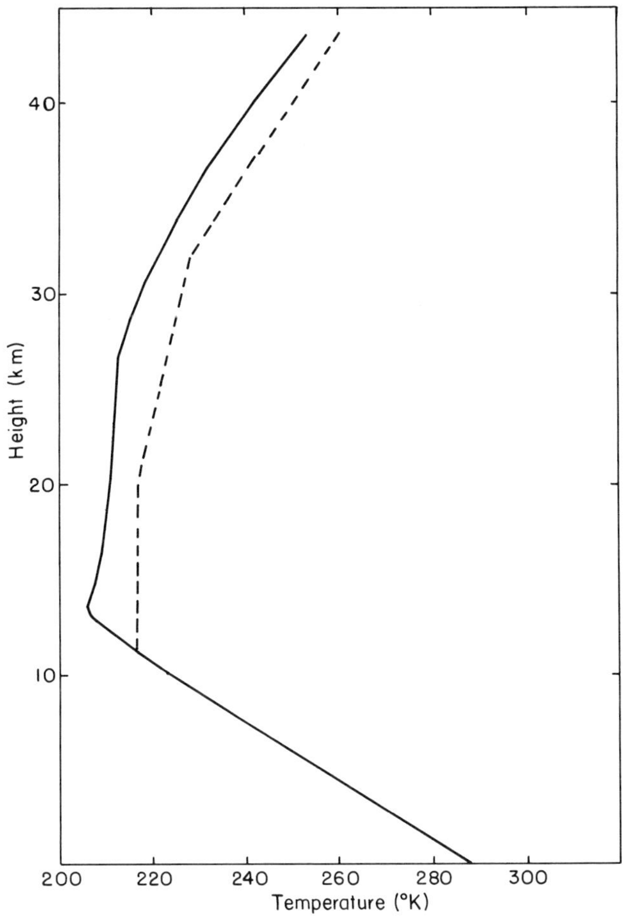


Figure 5.11 Vertical temperature profile (solid curve) with convective adjustment. After Manabe and Strickler (1964); dashed curve is the U.S. Standard Atmosphere of 1962. The curves are practically coincident below 11 km.

the same. This is a matter of some importance, since it increases the precipitable water with temperature and acts as a positive feedback in the temperature rise due to the $CO_2$. In fact, nearly half the heating is due to the increased water vapor.

The convective adjustment procedure has been incorporated in a nine-level primitive equation model by Manabe and Wetherald (1975, 1980). The model domain extends from the equator to the pole and is in the form of two "pieces of pie" of 60° of longitude, one representing continent and the other ocean. There is no coupling to the oceanic dynamics, the ocean serving only as a wet surface. No heat storage is permitted, and there are no seasonal nor diurnal variations of insolation. Radiative transfer and the hydrological cycle are treated in considerable detail. In the 1980 version of this climatic model, clouds are assumed to form wherever condensation of water vapor is predicted, and the clouds, assumed black, modify the radiative exchange. The feedback due to the marked change in surface albedo with snow or ice or their melting is modeled.

The model predictions of cloud cover, zonal mean planetary albedo, and zonal mean surface temperatures are in rough agreement with observations. The model has been used to predict temperature changes due to a doubling and a quadrupling of the $CO_2$ concentration. Some of the results are shown in figure 5.12. The large increase in temperature in high latitudes is due in part to the stable lapse rates there and to the strong positive feedback due to albedo changes as snow and ice melt. Also note the large temperature decreases in the stratosphere, particularly in low latitudes.

There are other climate models, which are based on somewhat different assumptions and yield different numerical results. The models of Manabe and collaborators have been selected for discussion here because of these researchers' extensive work on the problem for two decades. This does not necessarily mean that their numerical results are judged to be more reliable than those of other investigators. In my opinion none of the models is yet capable of predicting the climatic effects of increased $CO_2$ with the confidence required for the serious decisions that must be made in the future.

## 5.24 Warming of the Earth Due to the Increasing Atmospheric Burden of Carbon Dioxide

Although a very few voices were raised earlier, little note was taken of the effects on the global climate of an increase in carbon dioxide, largely because there was no reliable series of observations. The available concentration measurements were of variable quality and served mainly to establish that the $CO_2$ concentration was about 0.03% by volume. In a few cases $CO_2$ was included in tables of the permanent gases.

It may be said that the study of atmospheric carbon dioxide con-

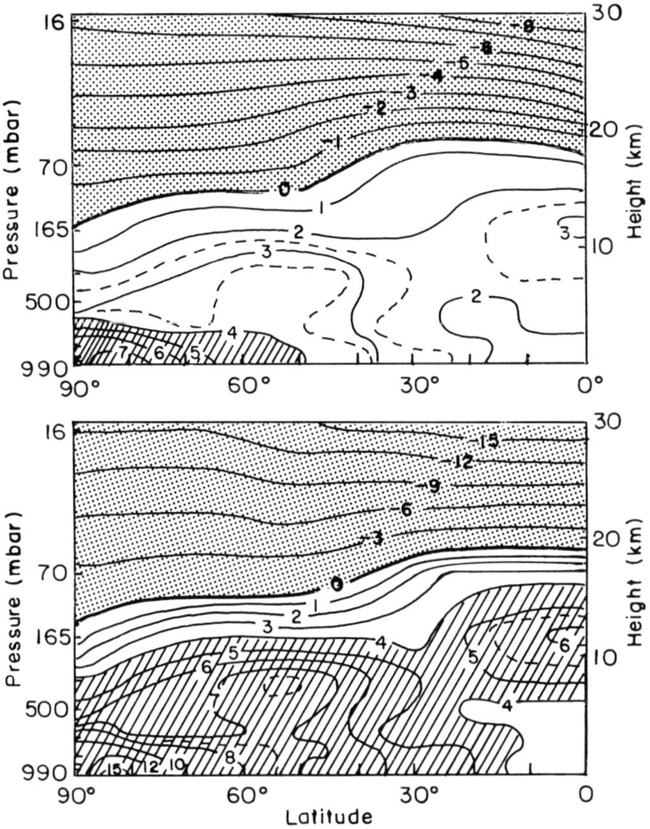


Figure 5.12 Latitude-height distribution of the change of the zonal mean temperature in °K in response to (top) a doubling of the $CO_2$ content and (bottom) a quadrupling of the $CO_2$ content. After Manabe and Wetherald (1980).

centrations began during the International Geophysical Year, 1957–1958, when two series of observations were initiated, one at Mauna Loa Observatory, Hawaii, and the other at the South Pole. As the period of record lengthened, a marked annual increment of nearly 1 ppm $yr^{-1}$ (ppm = parts per million) was revealed, which continued thereafter, as shown in figure 5.15. It was estimated that the rate of release of $CO_2$ from the burning of fossil fuels (coal, oil, gas) was about double that required to produce the observed annual increment of atmospheric $CO_2$. It was generally believed that the excess found its way into the oceans.

The potential effect of a substantial global temperature increase is so far-reaching that every effort must be made to predict the magnitude and distribution of the temperature increases as well as the predicted heating as a function of time.

It is first necessary to estimate the $CO_2$ concentration for a century or more. One is tempted to rely almost entirely on an extrapolation of figure 5.15. However, there is no assurance that this is a valid procedure. The $CO_2$ cycle involves the atmosphere, the oceans, the biosphere, and rocks, and it is probable that there will be changes as other links in the $CO_2$ cycle are found. It has been assumed that the primary source of $CO_2$ is the combustion of fossil fuels, but it now appears that deforestation is almost as large a source. Exhaustion of tropical forests may well occur in the foreseeable future. Other radiatively active gases may be introduced into the atmosphere by humans, as indeed has already occurred. Such trace gases may add significantly to the heating due to $CO_2$.

The complex motions of the atmosphere are driven by nonuniform heating, and the $CO_2$ heating is certain to modify the circulation patterns. In turn, this will feedback to the heating distribution. There are many other potential feedback processes, some positive and some negative. Evidently the $CO_2$ cycle must be super-imposed on a numerical model of the atmospheric circulation.

The preliminary numerical results obtained with the Manabe-Wetherald model yield a global mean warming of 2°C for a doubling of the $CO_2$ concentration and a warming of 4.1°C for a quadrupling of the $CO_2$. They also found that the warming increased strongly at high latitudes, as already noted. These authors warn that the numerical results should be interpreted with caution because of the many simplifications and idealizations that are included. Other numerical estimates of the warming differ by less than a factor of two for the most part. A panel appointed by the National Academy of Science-National Research Council suggested a warming of $3 \pm 1.5$°C. Various estimates of the time required to attain the predicted warming have ranged from a few decades to a century.

All that can be said with confidence is that significant warming is inevitable even if the $CO_2$ concentration does not continue to increase. There is little hope that it will be possible to arrest or reverse the $CO_2$ heating. However, there is reason to believe that the warming will be predicted with increasing confidence as more adequate models are developed and tested.

## 5.25 Cooling Rates of Realistic Atmospheres

Of the many studies in which radiative cooling rates are computed for atmospheres of observed properties, that of Dopplick (1979) will serve as an example. He computed cooling rates due to solar heating and thermal radiation for the four seasons as a function of height and latitude (zonal means). Figures 5.13 and 5.14 have been selected from the many included in the reference and show the total (solar plus thermal) heating rates for winter and summer, respectively. It is seen that there is a net radiative cooling in the troposphere except for small regions of heating at high

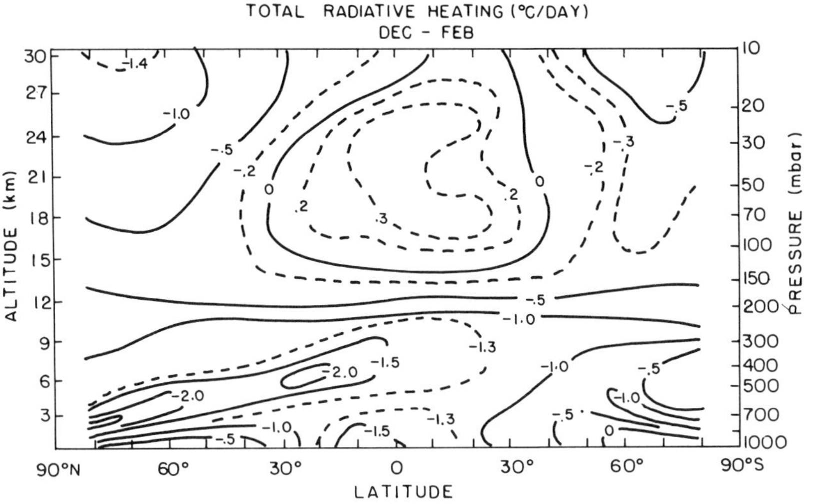


Figure 5.13 Total radiative heating (°C per day) for the December–February quarter. After Dopplick (1979).

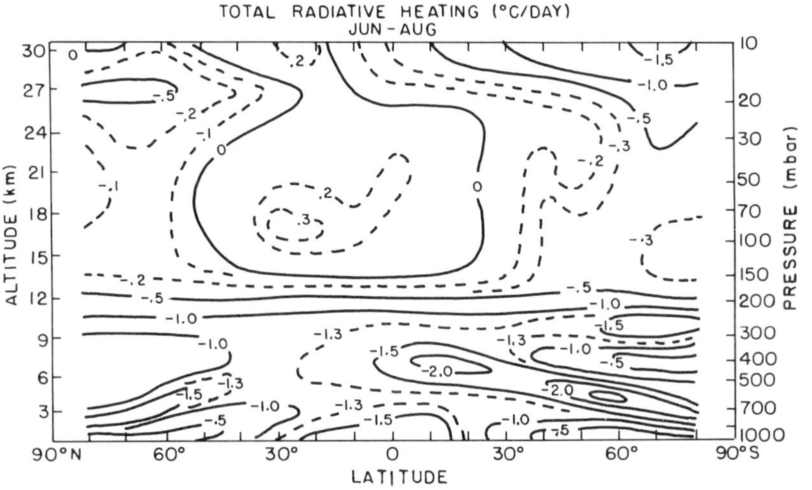


Figure 5.14 Total radiative heating (°C per day) for the June–August quarter. After Dopplick (1979).

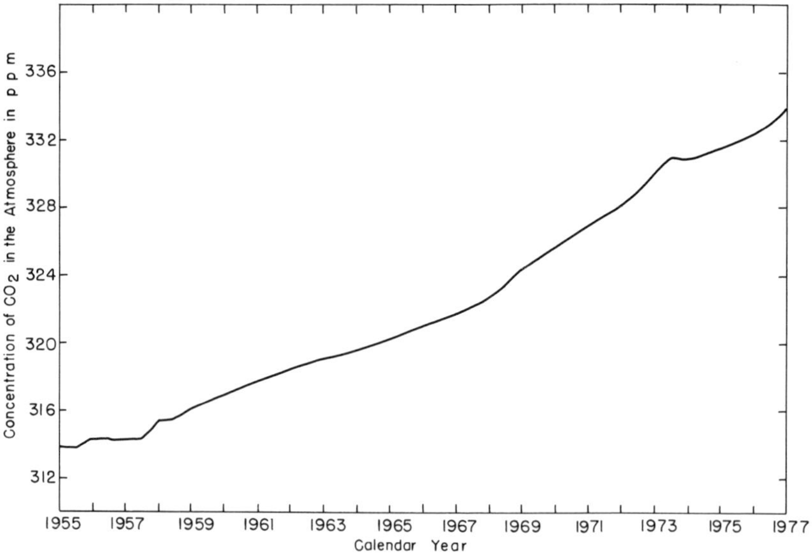

Figure 5.15 Trend of the annual mean concentration of $CO_2$ in the atmosphere with calendar year for Mauna Loa Observatory, Hawaii. Annual means are plotted at mid year. After Keeling (1978).

latitudes in summer. In the stratosphere there is heating in low latitudes and cooling in mid- and high latitudes. The Dopplick results are subject to uncertainties due to incomplete information on upper tropospheric and stratospheric water vapor concentrations and the inadequacies in the treatment of clouds.

## 5.26 Radiative Budgets

In the annual mean the differences between the zonally averaged absorbed solar radiation and the thermal radiation to space must be redressed by the latitudinal transports of latent and sensible heat carried by the meridional circulations of the atmosphere and the oceans. Because of the central importance of this problem, many computations of the radiational imbalances have been made. Until the advent of the artificial earth satellite, both the thermal radiation to space and the back-scattered solar radiation to space had to be estimated from surface-based observations. Now these outgoing solar and thermal radiation fluxes can be measured from satellites.

Direct measurements by satellite are to be preferred to computations from surface-based observations. However, the early measurements from satellites are somewhat uncertain due to degradation of the sensing ele-

Table 5.6
Annual means of absorbed solar radiation and of thermal radiation to space ($W m^{-2}$)

| | Ellis and VonderHaar (1976) | | | London and Sasamori (1971) | | |
|---|---|---|---|---|---|---|
| Latitude (°N) | Absorbed solar | Thermal to space | Albedo | Absorbed solar | Thermal to space | Albedo |
| 0 | 310 | 256 | 0.248 | 308 | 252 | 0.256 |
| 10 | 304 | 255 | 0.251 | 310 | 255 | 0.242 |
| 20 | 288 | 259 | 0.251 | 301 | 257 | 0.231 |
| 30 | 257 | 251 | 0.291 | 266 | 250 | 0.268 |
| 40 | 217 | 230 | 0.333 | 217 | 238 | 0.332 |
| 50 | 175 | 211 | 0.432 | 174 | 224 | 0.380 |
| 60 | 135 | 196 | 0.430 | 135 | 203 | 0.413 |
| 70 | 100 | 185 | 0.498 | 105 | 202 | 0.470 |
| 80 | 78 | 178 | 0.567 | 85 | 189 | 0.525 |
| 90 | 62 | 175 | 0.581 | 80 | 179 | 0.538 |
| Hemispheric mean | 234 | 234 | 0.324 | 238 | 238 | 0.33 |

ments with time and the conversion of the measured intensity to total flux. These problems have been overcome in the newer satellites. A feature of polar-orbiting satellites is their appearance at a given latitude at the same local time, thus failing to average the diurnal variations. This can be mitigated by the use of multiple satellites.

It is important to remember that satellites can measure only the upwelling radiation. For a complete picture, the satellite observations should be supplemented by measurements of the downcoming and upgoing fluxes at the surface. The annual means of the absorbed solar radiation and of the thermal radiation to space are given as a function of north latitude in table 5.6. The Ellis and VonderHaar (1976) data are based on 29 months of satellite data taken from 1964 to 1971. The London and Sasamori (1971) data are based on computations. In both cases, data are also given for the Southern Hemisphere, but the differences between hemispheres are not remarkable.

The Ellis and VonderHaar satellite data were taken from six satellites in the 1964–1971 period. They obtained 29 monthly sets, or 1 to 3 sets for each calendar month. Different satellites crossed the equator at local times from 0830 to 1500. The entries in table 5.6 are mean values for a 10° latitude belt; thus the entry opposite 40° is an average over the 35°–45° belt.

The planetary radiation budget derived from the first 18 months of

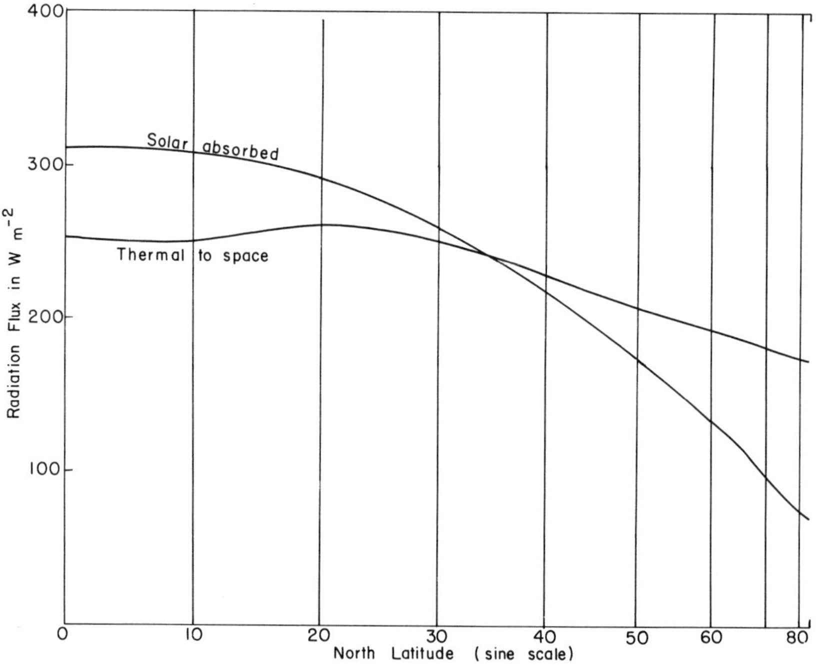


Figure 5.16 Annual means of the solar radiation absorbed and of the thermal radiation to space. After Ellis and VonderHaar (1976).

observations from the ERB experiment on *Nimbus 6* yields annual means of the planetary albedo and the thermal radiation to space of 31% and 234W $m^{-2}$ according to Jacobowitz et al. (1979). These may be compared with the corresponding values in table 5.6.

## 5.27 The Net Radiation

The net radiation is the difference between the absorbed solar radiation and the thermal radiation to space. It is shown graphically in figure 5.16, in which the data of Ellis and VonderHaar in table 5.6 are plotted against the sine of the latitude to reflect the relative areas of the latitudinal belts. The annual means of the net radiation, obtained from table 5.6, are listed in table 5.7. This table also gives the meridional transports of heat required to just balance the radiational excesses and deficits. These quantities are the heat transfers across the entire circles of latitude.

Comparison of the two sets of data in tables 5.6 and 5.7 shows good general agreement, particularly since probably only the first two digits in table 5.6 are significant. The principal difference between the two sets of figures is the generally larger values of the net radiation in the London and

Table 5.7
Net radiation and the meridional heat transports required to balance the radiative surpluses and deficits

| | Ellis and VonderHaar (1976) | | London and Sasamori (1971) | |
|---|---|---|---|---|
| Latitude (°N) | Net radiation ($W m^{+2}$) | Meridional heat transfer ($\times 10^{15}$ W) | Net radiation ($W m^{-2}$) | Meridional heat transfer ($\times 10^{15}$ W) |
| 0 | 54 | 0 | 56 | 0 |
| 10 | 49 | 2.5 | 55 | 2.5 |
| 20 | 29 | 4.3 | 44 | 4.6 |
| 30 | 6 | 4.9 | 16 | 5.8 |
| 40 | − 13 | 4.8 | − 21 | 5.7 |
| 50 | − 36 | 4.1 | − 50 | 4.6 |
| 60 | − 61 | 2.9 | − 68 | 3.1 |
| 70 | − 85 | 1.5 | − 97 | 1.6 |
| 80 | −100 | 0.4 | −104 | 0.4 |
| 90 | −113 | 0 | − 99 | 0 |

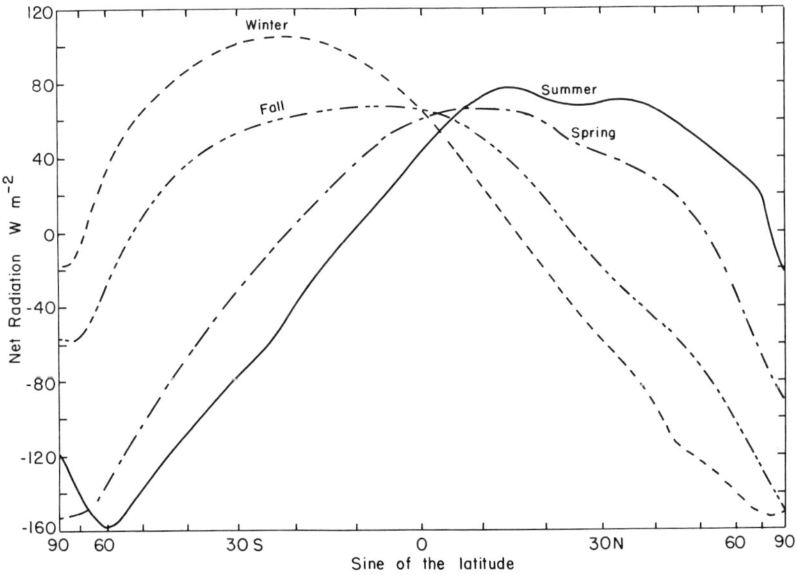


Figure 5.17 The zonally averaged net radiation for the four seasons. Winter includes December, January, and February; spring includes March, April, and May; and so forth. After Ellis and VonderHaar (1976).

Sasamori study; the effects of this are most clearly shown by the meridional heat transports in table 5.7.

To this point only annual means for the Northern Hemisphere have been considered. The very considerable variation of the net radiation with season is shown in figure 5.17 taken from Ellis and VonderHaar (1976). The marked imbalances of the hemispheres, particularly in summer and winter, would seem to demand large transports of heat across the equator. However, most of the excess in the summer hemisphere goes into storage in the oceans, the atmosphere, and the soil; similarly, most of the deficit in the winter hemisphere is compensated by withdrawals from these heat reservoirs. Over a year the net heat storage is very small and not detectable from the available observations.

The most important heat reservoir is the oceans. Efforts to quantify the seasonal storage of heat in the oceans have met with only limited success. Bryan and Schroeder (1960) based a study on some 20,000 bathythermograms (temperature-depth profiles) taken in the North Atlantic between latitudes 20°N and 65°N. The seasonal heat storage from the surface to a depth of 200, calculated from these data, substantially exceeded estimates of the local heating. the authors suggest that this overestimate of the heat storage may result from temperature changes of opposite sign below the relatively shallow depths reached by the bathythermographs.

## 5.28 The Global Heat Balance

In addition to the radiative components, the heat balance includes heat transfer from the surface to the atmosphere by the convection of sensible heat and the evaporation-condensation process. Detailed estimates of the heat transfer terms at the surface have been made by Budyko (1958) in the form of an extensive atlas. No useful summary of this encyclopedic work can be given here. Instead, the annual heat balance of the globe or of the Northern Hemisphere as estimated by several authors is given in table 5.8.

At first sight the agreement between these four estimates is quite good, but it must be remembered that these are averages both over the year and over the earth. There are significant differences in the planetary albedo, the fraction reflected and scattered to space. As noted, the most recent satellite data yield an albedo of about 0.31, which is significantly different from London and Sasamori's 0.33. Further refinements will come from *Nimbus 6* and its successors.

## 5.29 Sounding the Atmosphere by Thermal Radiation

The artificial earth satellite is an observing platform that can provide global coverage, but can receive information only by radiation. The use of televisionlike scanning detectors in the visible or the infrared yield pictorial

Table 5.8
Annual mean heat budget of the hemisphere or the globe (in percent of the incoming solar radiation)

| Source | Houghton (1954) | London and Sasamori (1971) | Budyko (1958) | Kondratyev (1969) |
|---|---|---|---|---|
| Incoming from sun | 100 | 100 | 100 | 100 |
| Reflected by clouds | 25 | 26 | 27 | 24 |
| Back-scattered | 7 | 7 | 8 | 10 |
| Reflected at surface | 2 | 33 | 5 | 3 |
| Planetary albedo | 34 | 33 | 40 | 37 |
| Absorbed by atmosphere | 19 | 22 | 16 | 18 |
| Absorbed at surface | 47 | 45 | 44 | 45 |
| Thermal radiation to space | 66 | 67 | 60 | 63 |
| Net surface thermal radiation | 14 | 15 | 17 | 18 |
| Thermal divergence in atmosphere | 52 | 52 | 43 | 45 |
| Evaporation-precipitation | 23 | 30 | 22 | 23 |
| Sensible heat to atmosphere | 10 | 30 | 5 | 4 |

representations of the clouds and the surface by both day and night. Such pictures have aided meteorologists to construct weather maps in data scarce areas. Of potentially greater value are the vertical sounding of the atmosphere from satellites and the deduction of wind speed and direction. The latter will not be considered here.

Although earlier suggestions were made by others, the first specific proposal for the deduction of the vertical temperature profile from satellite measurements of the upwelling radiation was due to Kaplan (1959a). The radiance in an absorption band of a uniformly mixed gas, such as carbon dioxide, is given by the radiative transfer equation and the Planck function. Since the emissivity is known, the radiance is a measure of the temperature through the Planck function.

At the center of a strong band, such as the 15-$\mu$m $CO_2$ band, the upwelling radiance originates primarily in the stratosphere; the emission from lower layers is nearly completely absorbed and contributes little to the measured radiance. Conversely, in the wing of the band the surface and the lower troposphere are the large contributors to the measured radiance. Radiances at intermediate frequencies can provide the temperatures of intermediate layers.

## 5.30 Formulation of the Problem

The radiative transfer equation for the outgoing radiation at the top of the atmosphere in a narrow spectral interval $\Delta\nu$ is

$$E(\Delta\nu, 0) = \bar{B}(\Delta\nu, T_g)\bar{\tau}(\Delta\nu, p_g) - \int_1^{\tau(\Delta\nu, p_g)} \bar{B}(\Delta\nu, T)\, d\bar{\tau}, \tag{5.21}$$

where the overbars indicate averages over the interval $\Delta\nu$ and the subscript g refers to the surface. It is straightforward to solve (5.21) for $E(\Delta\nu,0)$ given the transmissivity function and the vertical temperature profile. The inverse problem of solving for the vertical distribution of $\bar{B}(\Delta\nu, T)$, and hence of the temperature, from observed radiances $E(\Delta\nu, 0)$ poses some difficulties.

For simplicity, consider a frequency at which the transmissivity from the surface to the top of the atmosphere is very small; then the first term on the right of (5.21), which is the contribution from the ground, may be neglected. If the frequency interval $\Delta\nu$ is small, say, 5–10 $cm^{-1}$, the following linear expression is a good approximation:

$$B(\Delta\nu, T) = \alpha\, \Delta\nu\, B(\nu_r, T) + \beta\, \Delta\nu, \tag{5.22}$$

where $\alpha$ and $\beta$ are constants for any given $\Delta\nu$ and $\nu_r$ is a reference frequency usually at the center of the interval $\Delta\nu$. Use of (5.22) eliminates the frequency dependence of $\bar{B}(\Delta\nu, T)$. Equation (5.21) then becomes

$$\frac{E(\nu, T) - \beta\,\Delta\nu}{\alpha\,\Delta\nu} = -\int_0^{\ln p_g} B(\nu_r, T)\frac{d\tau(\Delta\nu, p)}{d(\ln p)}\, d(\ln p). \tag{5.23}$$

The weighting function or kernel of the integral is $d\tau/d(\ln p)$ and depends on the transmissivity function. Four of these weighting functions, each for a different frequency interval $\Delta\nu$, are shown in figure 5.18. It is the usual practice to measure the radiance at five to seven frequency intervals of 5-$cm^{-1}$ width ranging from the band center to the far wing. A typical set of frequencies is 668.7, 679.8, 692, 701, 709, 734, and 750 $cm^{-1}$. These range from the center of the 15-$\mu$m $CO_2$ band down the high-frequency flank of the band to the wing. Only every other weighting function is shown in figure 5.18 for clarity. These curves show the relative contributions of the atmospheric layers to the measured radiances. A feature of the curves is their broad maxima, which occur at higher levels as the frequency approaches the band center. Channel 1, at the band center, provides information primarily from the stratosphere, while channel 7 in the band wing emphasizes the lower troposphere and the surface. The overlapping of the curves is apparent and would be even more so if all of the channels had been included. Each channel receives some contribution from each of the

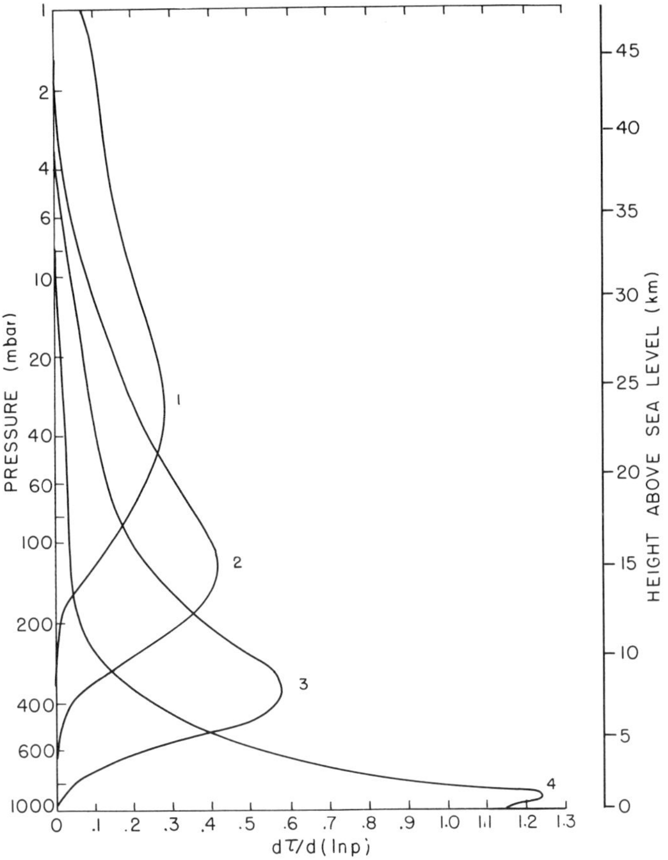


Figure 5.18 Weighting functions for four 10 $cm^{-1}$ intervals on the high-frequency wing of the 15-$\mu$m $CO_2$ band. Intervals are centered at 667 (curve 1), 695 (curve 2), 709 (curve 3), and 750 $cm^{-1}$ (curve 4). After Smith (1972).

atmospheric layers. The redundancy due to the overlap sets a limit on the number of channels that can be effectively used.

## 5.31 The Inversion Process

The retrieval of $B(\nu, T)$, and hence of temperature, requires the inversion of (5.21) or (5.23). Equation (5.21) is (nearly) a Fredholm integral equation of the first kind, and matrix and other forms of the solution are known. At first sight it would seem that the problem reduces to the solution of a number of simultaneous integral equations, one for each channel. It was found that no stable solution could be obtained in this way. The reason is that the redundancy resulting from the overlap of the weighting functions makes the solution indeterminate. Imagine a situation in which there is no overlapping of the weighting functions. The radiance from each channel would then specify a mean temperature of the corresponding atmospheric layer, independently of the other channels.[3]

A stable solution may be obtained by introducing another condition, usually an average observed temperature profile appropriate to the location and season, as shown by Smith et al. (1970). This is a conservative approach, the radiances being used to find the small deviations from the climatological mean profile. There is some bias toward the mean sounding, and the method works best when the departures from the mean sounding are small.

Chahine (1968, 1972) has developed another method that does not depend on a detailed set of mean profiles. He starts with an initial guess that may be a typical observed profile on even an isothermal profile. The radiances are computed for the first-guess profile for each channel. The residuals between these computed radiances and those observed are introduced into a relaxation equation to generate a second-guess profile. This process is repeated until the residuals fall below the noise level of the measured radiances. Convergence is rapid and stable. The method owes its success to Chahine's astute choice of the relaxation equation. In spite of its apparent advantages this method has not been used extensively.

## 5.32 Inversion with Clouds

Clouds in the field of view pose a serious problem in the retrieval of the temperature profile. If there is a solid, dense low cloud layer, the profile may be retrieved down to the cloud top in the same way as in a cloudless case. If the surface temperature is known from another source, the profile may be interpolated to the surface. Broken clouds at one or more levels make it impossible to retrieve the profile from a single set of radiances. Two recent developments in satellite technology provide the additional information needed for profile retrieval in this common situation.

The new features are scanning of the sensors normal to the satellite orbit, so that several contiguous fields are viewed, and a second infrared radiometer, which operates in the 4.3-$\mu$m $CO_2$ band. If the fractional cloud coverage is different in the added fields of view, it is clear that the measured radiances contain information on the effect of cloud cover on the radiances. It has been found that the channel radiances vary linearly with a window channel radiance obtained for a clear path from a high-resolution sensor. This relation can then be used to estimate the radiances of the

---

3. Not quite because of the dependence of the transmissivity on the temperature and pressure.

regular measuring channels in a cloud-free atmosphere. These corrected radiances are then inverted in the usual way to retrieve the temperature profile all the way to the surface.

A more elaborate and more powerful procedure, developed by Chahine (1974, 1977) and Chahine et al. (1977), is only briefly outlined here. By assuming that the radiances for a cloud-free path are the same in contiguous fields of view, one can write the effects of a fractional cloud cover in each field of view as a function of the optical properties of the clouds. Measurements made in adjacent fields of view permit the elimination of the optical properties of the clouds, assumed to be the same in both fields of view. The remaining unknowns are the fractional cloud cover and the cloudless temperature profile. The radiances measured in the two separate $CO_2$ bands are used to get cloud-free radiances that can be used to yield the vertical temperature profile. If the broken clouds are all at the same level, only two fields of view and one window channel are required. If the broken clouds are at three levels, four fields of view and three window channels are needed. In addition to the temperature profile, the method yields the surface temperature and the height and fractional amounts of cloud.

Still largely unsolved are the cases of a solid undercast and tenuous clouds. In view of past progress, it is not unreasonable to anticipate the development of methods capable of dealing with these cases.

## 5.33 Types of Satellite Sounding Instruments

The first instrument, which is still used with modifications, is the satellite infrared spectrometer, or SIRS, described by Hillary et al. (1966). This is a spectrometer with exit slits and sensors for each channel. An on-board blackbody source, the temperature of which is monitored, provides ongoing calibration of the sensors.

Another satellite package developed contemporaneously with SIRS is the infrared interferometer spectrometer, or IRIS, described by Hanel et al. (1972). This is a single detector Michelson interferometer. It provides a continuous spectrum from about 6 to 25 $\mu$m with a resolution of 2.8 $cm^{-1}$ and is continually calibrated against an on-board blackbody. This is a research instrument rather than an operational one, such as SIRS. For example, it may be used to yield the concentrations of variable gases, such as water vapor and ozone, when the temperature profile is known.

Still another instrument is the selective chopping radiometer (SCR), which was designed to sound the upper stratosphere. To achieve this purpose the radiances must be observed with high spectral resolution at the center of the band. This is achieved in part by the use of a cell containing $CO_2$ gas as a filter. Results have been quite satisfactory both in the upper stratosphere and at lower levels as well.

Absorption bands of oxygen, water vapor, and liquid water in the

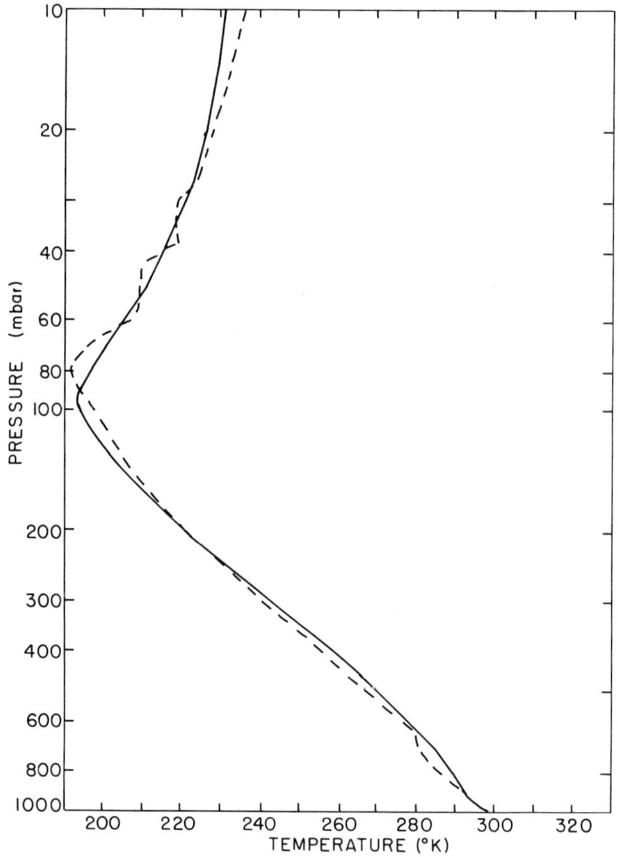


Figure 5.19A Comparison of SIRS (Solid curve) and radiosonde (dashed curve) vertical temperature profiles taken at 16°N, 73°W on 14 August 1969. From Wark and Hillary (1969);copyright 1969 by the American Association for the Advancement of Science.

microwave region have led Staelin et al. (1973) to develop microwave spectrometers to sound the atmosphere. The principal advantages of microwave spectrometers over infrared spectrometers derive from the near transparency of many clouds and the high spectral resolution available and also their ability to distinguish between vapor and liquid water. Their chief disadvantages are that the microwave bands lie in the far wing of the blackbody curves at terrestrial temperatures, leading to unfavorable signal to noise ratios, and that the highly variable emissivity of the ground (about 0.4–1.0) limits most applications to oceanic areas. To a degree, microwave and infrared spectrometers are complementary, and it is advantageous to install both on one satellite.

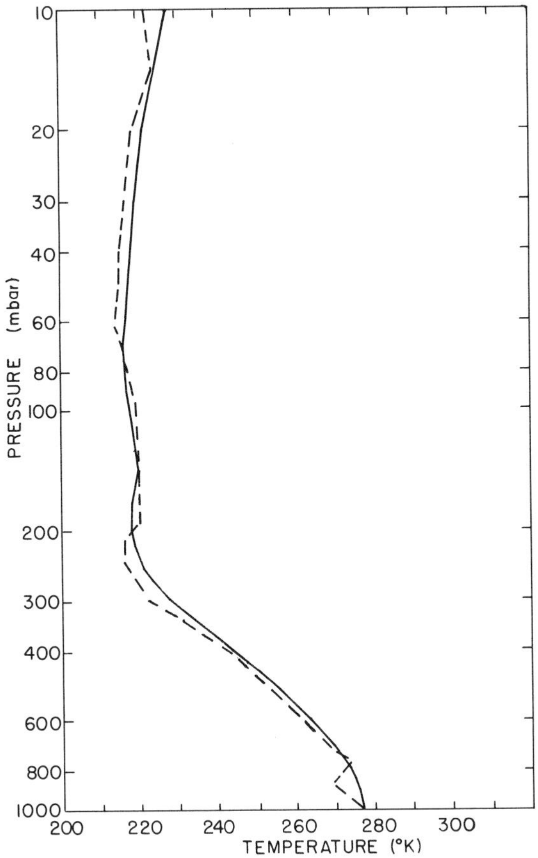


Figure 5.19B Comparison of SIRS (Solid curve) and radiosonde (dashed curve) vertical temperature profiles taken at 44.5°, 86.8° on 19 April 1969. From Wark and Hillary (1969); copyright 1969 by the American Association for the Advancement of Science.

A different geometry is used in the limb-scanning method, in which a detector with a small acceptance angle scans the limb of the earth in the vertical. The ray path is tangent to a sphere concentric with the earth. This geometry provides a long path and is of principal value in the measurement of small absorber concentrations in the upper atmosphere.

## 5.34 Comparisons with Radiosondes

Many comparisons have been made, but only a small number have appeared in the literature. Two examples are given in figures 5.19A and 5.19B, the first of which is the historic first retrieval of a temperature profile from a satellite. The overall agreement is quite good, but SIRS fails to depict sharp inversions and the tropopause structure. This is understandable when it is remembered that the entire SIRS profile is based on only seven measured radiances. Closer agreement has been obtained when two different $CO_2$ bands are used simultaneously.

Statistical comparisons have given root mean square errors of 1–3°C, the range resulting from the choice of the population. The errors are greatest at the tropopause and near the surface and are lowest in the midtroposphere; for comparison, the root mean square errors of radiosonde temperatures are only a fraction of a degree. The presence of cloud in the field of view, which occurs more than half the time, is an important source of errors. It has been estimated that the root mean square error of radiatively sensed temperatures needs to be reduced to 1°C or less if the data are to be of maximum value in numerical weather forecasting.

## 5.35 Improved Radiative Soundings

There are many sources of error in deriving temperature profiles from the measured radiances. One of the limitations is the small number of "data points" that are obtained from the radiances. Due to the overlapping weighting functions, additional channels would provide very little additional information. This is an inherent feature of the transmissivity functions that cannot be changed significantly by using narrower frequency bands. The narrower the band, the smaller the radiance and the poorer the signal to noise ratio.

A more feasible way to get more independent data is to use two or more absorption bands. Carbon dioxide has a strong band centered at 4.3 $\mu$m that can be used in addition to the 15-$\mu$m band. The 4.3-$\mu$m band is more intense than the 15-$\mu$m band, the Planck function changes more rapidly with temperature at 4.3 $\mu$m, but the blackbody radiation is two to three orders of magnitude larger at 15 than at 4.3 $\mu$m. Overall it has been demonstrated that the use of the 4.3-$\mu$m band with the 15-$\mu$m band provides more detail in the troposphere. Still more data can be provided by the addition of a microwave spectrometer. All three bands are used simultaneously in operational satellites.

Errors in the transmissivity function, the effect of overlapping water vapor absorption, and, in particular, the signal-to-noise ratio all contribute to the overall error. It has been estimated that about half of the total error is due to instrumental uncertainties. The other half is presumably due to errors in the transmissivity functions and the inherent lack of detail in the vertical. It will be difficult, but probably not impossible, to reduce the root mean square error to around 1°C. Efforts to this end are certainly justifiable because of the uniform global coverage of the satellite and the economy of the satellite as compared with a radiosonde network of similar global coverage.

## Problems

1.
Construct a Kew radiation chart.

2.
Judge how well (5.19) and (5.20) agree by means of numerical experimentation.

3.
What are the effects of the typical atmospheric aerosol on the upgoing and downcoming thermal radiation in the atmosphere? A qualitative answer is all that is expected.

4.
What are the effects of clouds on the downcoming thermal radiation at the surface and the outgoing thermal radiation to space?

5.
The annual zonal mean of the absorbed solar radiation as a function of latitude usually has a maximum around 20°N rather than near the equator. Why is this?

# 6

# The Nucleation of Water and Ice in the Atmosphere

## 6.1 Introduction

Water vapor is ubiquitous in the atmosphere and is the source of the liquid and solid phases that are the materials of clouds and precipitation. Nucleation is the name given to the process whereby a stable element of a new phase of a substance first appears in an initial phase. The most important nucleation processes in cloud physics are the nucleation of the liquid phase in the vapor, the nucleation of the ice phase in the liquid, and the nucleation of the ice phase in the vapor. In each case, the new phase is one of greater atomic order than the mother phase, and an energy barrier must be surmounted before nucleation occurs.

Some surprise is often evidenced when it is learned that condensation in pure water vapor requires a supersaturation of several hundred percent and that pure water may be supercooled some tens of degrees without the appearance of ice. This may be due to familiarity with phase equilibrium, which is treated in all books on thermodynamics. Obviously, phase equilibrium can have no meaning when there is only a single phase present. Nucleation is a nonequilibrium phenomenon, and it must be approached accordingly.

Two kinds of nucleation are recognized, homogeneous and heterogeneous. In homogeneous nucleation, nothing is present except the pure initial phase (vapor or liquid); heterogeneous nucleation occurs on particles of a foreign substance, called nuclei of condensation, freezing, or deposition.[1] The presence of air seems to have little or no effect, and nucleation in a mixture of air and water vapor can be considered to be

1. Deposition is the formation of ice from the vapor, the reverse of sublimation.

homogeneous. In almost all cases, nucleation in the atmosphere is heterogeneous. However, it is instructive to consider homogeneous nucleation first because it is somewhat simpler in concept and provides a useful basis for consideration of heterogeneous nucleation.

## 6.2 Analytical Discussion of Homogeneous Nucleation

As a result of the thermal motion of individual molecules in a gas, many collisions occur. A small fraction of these collisions is inelastic, leading to the formation of molecular aggregates. Aggregates larger than molecular pairs occur by inelastic collisions between smaller aggregates and by the accretion of individual molecules. The probability of the existence of an aggregate of any given number of molecules decreases rapidly with the number of molecules in the aggregate, but is always finite.

The vast majority of the aggregates have very short lifetimes, since they disintegrate under the continual molecular bombardment. When the largest aggregate attains a size sufficient to survive, nucleation has occurred. When an embryo droplet is formed in this way, the surface free energy must be supplied. The source of this energy, in macroscopic terms, is the latent heat of condensation. As will be shown in section 6.3, a fraction of the latent heat that increases with increasing supersaturation is available to supply the surface energy of the newly born droplet. From an energetic viewpoint the embryo will survive if the available latent heat is equal to or greater than the surface free energy. Since the surface free energy is proportional to the square of the radius and the latent heat is proportional to mass, or the cube of the radius, it is evident that, for any given supersaturation, there must be a critical radius for which the incremental surface energy is just equal to the corresponding increment of latent heat. This critical radius is given by the Kelvin equation, to be developed in section 6.3. The critical radius defines the minimum size of an embryo that will survive; by definition this is homogeneous nucleation. The greater the supersaturation, the greater is the fraction of the latent heat available and, hence, the smaller is the critical radius. As stated, the formation of molecular aggregates is a statistical process, and the probability of the appearance of an aggregate decreases rapidly as its size increases. In principle an aggregate of any arbitrary size will eventually appear. It is therefore necessary to set a somewhat arbitrary limit on the time allowed for nucleation to occur. As will be made clear in section 6.4, the conditions for nucleation are quite insensitive to the choice of this "waiting time." It is clear from what has been set forth that nucleation is favored by a large supersaturation because this decreases the critical radius and thus makes it more likely that an embryo of this size will appear in a reasonable time.

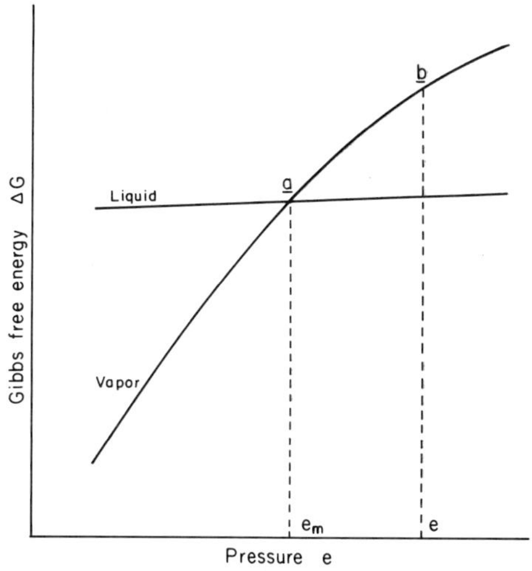


Figure 6.1 Qualitative plot of the liquid and vapor isotherms of a pure substance in the Gibbs free energy-pressure plane.

## 6.3 The Kelvin Equation

The usual derivation[2] of the Kelvin equation is based on the Gibbs free energy. This is also widely used in physical chemistry and is appropriate to the problem at hand because it may be evaluated along irreversible paths; nucleation is an inherently irreversible process. The Gibbs free energy for unit mass is given by

$$\Delta G = U + ev - T\Phi, \tag{6.1}$$

where $U$ and $\Phi$ are the internal energy and the entropy per unit mass, respectively, $e$ is the pressure, and $v$ is the specific volume. This expression applies to either the gaseous or the liquid phase or any mixture of the two phases. For a system in equilibrium $dG$ is zero and $\Delta G$ is at a minimum. In any spontaneous process the Gibbs free energy will decrease.

Differentiating (6.1) and remembering that $T\,d\Phi = dU + e\,dv$ yield

$$dG = v\,de - \Phi\,dT. \tag{6.2}$$

Figure 6.1 shows qualitatively the isotherms of the vapor and the liquid of a pure substance on a plot of $\Delta G$ against $e$. These isotherms are plotted from

2. For Thomson's (Lord Kelvin) own very interesting derivation see, Humphreys (1929).

the integrated form of (6.2) with $T$ constant. The isotherm for the liquid is nearly horizontal because of the very small compressibility of water. It is easily seen from (6.2) that the equation of the vapor isotherm is $\Delta G = R_w T \ln(e) + C$, where $R_w$ is the gas constant for water vapor and $C$ is a constant.

The intersection of the two isotherms defines the equilibrium vapor pressure at the selected constant temperature. The region of the vapor isotherm to the right of this intersection corresponds to supersaturated vapor. The excess free energy of the supersaturated vapor over the equilibrium value at $a$ may be evaluated by integrating (6.2) from $b$ to $a$:

$$\Delta G_v = \int_e^{e_m} v\, de = -R_w T \ln(e/e_m). \tag{6.3}$$

This is the free energy available in the supersaturated vapor to supply the surface free energy of a liquid embryo. As will be shown in this section, it is also the latent heat released by the condensation required to reduce the vapor pressure to the equilibrium value, $e_m$.

Suppose that a spherical embryo of radius $r$ appears spontaneously in the vapor. The change in free energy is the sum of the surface free energy of the embryo and the energy released by the supersaturated vapor. Remembering that the surface energy is given by the area times the specific surface free energy (or the surface tension) and that (6.3) is for a unit mass, we may write

$$\Delta G = 4\pi r^2 \sigma - (4/3)\pi r^3 \rho R_w T \ln(e/e_m), \tag{6.4}$$

where $\sigma$ is the specific surface energy and $\rho$ is the density of the liquid. The equilibrium condition is obtained by setting the first derivative of (6.4) to zero. This leads to Kelvin's equation:

$$r_* = \frac{2\sigma}{\rho R_w T \ln(e/e_m)}. \tag{6.5}$$

In this form, $r_*$ is the minimum radius of the embryo that will persist for a given value of the saturation ratio $e/e_m$ and is therefore the condition for nucleation. Also, (6.5) gives the vapor pressure in equilibrium with a water surface having a radius of curvature $r$. This applies as well to a concave surface, such as a meniscus in a capillary tube, in which case $r$ is negative and $e/e_m$ is less than one. This explains the dehumidifying action of a microporous substance such as silica gel.

Table 6.1 gives values of $r_*$ as a function of $s = e/e_m$ as computed from (6.5) for a temperature of 0°C. Also included is the approximate number of water molecules in spherical drops of the indicated radii.

It will now be shown that the second term on the right-hand side of (6.4)

Table 6.1
The saturation ratio in equilibrium with liquid drops of radius $r_*$ from Kelvin's equation

| $s$ | $r_*$ ($\times 10^{-3}$ $\mu$m) | Number of molecules | $s$ | $r_*$ ($\times 10^{-3}$ $\mu$m) | Number of molecules |
|---|---|---|---|---|---|
| 1.0 | $\infty$ | $\infty$ | 3.5 | 9.59 | 125 |
| 1.5 | 2.96 | 3,650 | 4.0 | 8.66 | 90 |
| 2.0 | 1.73 | 730 | 4.5 | 7.99 | 70 |
| 2.5 | 1.31 | 315 | 5.0 | 7.46 | 60 |
| 3.0 | 1.09 | 180 | 6.0 | 6.70 | 40 |

does represent the latent heat that is available to supply the surface free energy. The Clausius-Clapeyron equation may be written

$$\frac{de_{\mathrm{m}}}{dT} = \frac{L}{vT}, \tag{6.6}$$

where the specific volume of the liquid has been neglected in comparison with that of the vapor. Assuming further that water vapor behaves as an ideal gas and substituting from the equation of state give

$$\frac{de_{\mathrm{m}}}{e_{\mathrm{m}}} = \frac{L}{R_{\mathrm{w}}}\frac{dT}{T^2}. \tag{6.7}$$

It will be remembered that this relation applies only to equilibrium conditions, as is indicated by the use of the subscript m on the vapor pressure. By referring to figure 6.1, integrate (6.7) along an equilibrium path ($\Delta G$ constant) from pressure $e_{\mathrm{m}}$ to $e$:

$$\ln(e/e_{\mathrm{m}}) = \frac{\bar{L}}{R_{\mathrm{w}}}\frac{(T_0 - T)}{TT_0}, \tag{6.8}$$

where $T_0$ is the temperature at which the supersaturated vapor pressure $e$ would be the equilibrium pressure and $\bar{L}$ is the mean value of the latent heat of vaporization. Evidently $T_0$ is the dew point temperature of the supersaturated vapor and $T_0 > T$. If (6.8) is inserted, the second term of (6.4) becomes

$$-\frac{4}{3}\pi r^3 \rho \bar{L}\frac{(T_0 - T)}{T_0}.$$

This shows that this term represents the fraction of the latent heat released

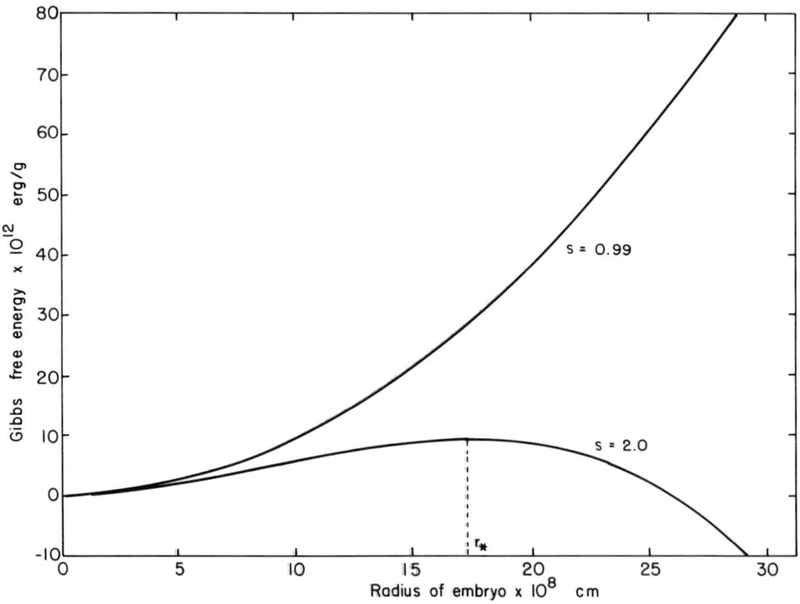


Figure 6.2 Gibbs free energy versus embryo radius for supersaturation ratios less than one and greater than one.

by the condensation required to eliminate the supersaturation and that this is the energy available to supply the surface free energy. This formulation readily leads to an alternative form of the Kelvin equation:

$$r_* = 2\sigma T_0/\rho\bar{L}(T_0 - T). \tag{6.9}$$

This form is most useful in studies of the nucleation of ice in water, to be discussed in section 6.7. For this application $\rho$ is the density of ice and $\bar{L}$ is the mean latent heat of fusion. The variation of the equilibrium temperature of ice in water with pressure is entirely negligible over the small range of pressure involved here, and $T_0$ may be taken as 273°K.

The contrasting behavior of (6.4) for values of the saturation ratio $s$ less than one and greater than one is illustrated in figure 6.2. Only when $s > 1$ is homogeneous nucleation possible, as indicated by the maximum in the lower curve. This maximum occurs at $r = r_*$ and $\Delta G_* = 4\pi\sigma r^2{}_*/3$; $\Delta G_*$ is called the nucleation energy barrier.

## 6.4 Kinetics of Nucleation

The thermal motions of the molecules in a gas lead to a large number of collisions, a very small fraction of which are inelastic. These rare inelastic collisions lead to the continuous formation of molecular aggregates of various sizes, which are, in turn, destroyed by the molecular bombardment. This leads to a stationary statistical distribution of molecular aggregates by size and hence by free energy. This stationary distribution is given by

$$N(r) = N_1 \exp\{-[\Delta G(r)/kT]\}, \tag{6.10}$$

where $N(r)$ is the concentration of aggregate of radius $r$, $N_1$ is the total concentration of unassociated molecules, $\Delta G(r)$ is the free energy of the aggregate, and $k$ is Boltzmann's constant.

Since $\Delta G(r)$ increases with the square of the radius, $N(r)$ decreases rapidly as $r$ becomes larger. Nucleation occurs when aggregates of radius $r_*$ appear. The rate of formation of critical embryos is then

$$J(r_*) = BN_1 \exp\{-[\Delta G(r_*)/kT]\}, \tag{6.11}$$

where $J(r_*)$ is the rate of formation of critical embryos in a unit volume of the gas in unit time and $B$ is the rate at which a critical embryo acquires one molecule. Various estimates of the value of $B$ have been made, but will not be reviewed here. Because of the insensitivity of $J(r_*)$ to the numerical value of $B$, it seems adequate to adopt Fletcher's (1962) suggestion that $BN_1$ be taken as $10^{25}$ cm$^{-3}$ sec$^{-1}$ for the homogeneous nucleation of water in its vapor. Inserting the value of $\Delta G(r_*) = (4\pi\sigma r_*^2)/3$ from (6.5) in (6.11) gives

$$J(r_*) = 10^{25} \exp\left\{\frac{-16\pi\sigma^3}{3k\, R_w^2 \rho^2\, T^3[\ln(s)]^2}\right\}. \tag{6.12}$$

This is plotted in figure 6.3 for a temperature of 273°K. Note that a change of one order of magnitude in $J(r_*)$ corresponds to only a little more than a change of one-tenth in the saturation ratio. This makes the uncertainty in the value of the numerical factor $10^{25}$ in (6.12) of minor importance.

An appropriate value of $J(r_*)$ may usually be selected within an order of magnitude on the basis of the experimental conditions. For example, suppose that a rapid expansion cloud chamber is used to determine the onset of homogeneous nucleation. Assume that the design and operation of the chamber are such that the expansional cooling has a duration of $10^{-2}$ sec, the volume observed is $10^{-1}$ cm$^3$, and some 10 drops must appear in this volume to convince the observer that nucleation has occurred. The appropriate value of $J(r_*)$ in this case would be $10/(10^{-1} \times 10^{-2}) = 10^4$. In a thermal diffusion cloud chamber in which the observation time may be several minutes, the volume larger, and detection of drops easier, $J(r_*)$ may be less than one.

The critical saturation ratio for homogeneous nucleation of the liquid in

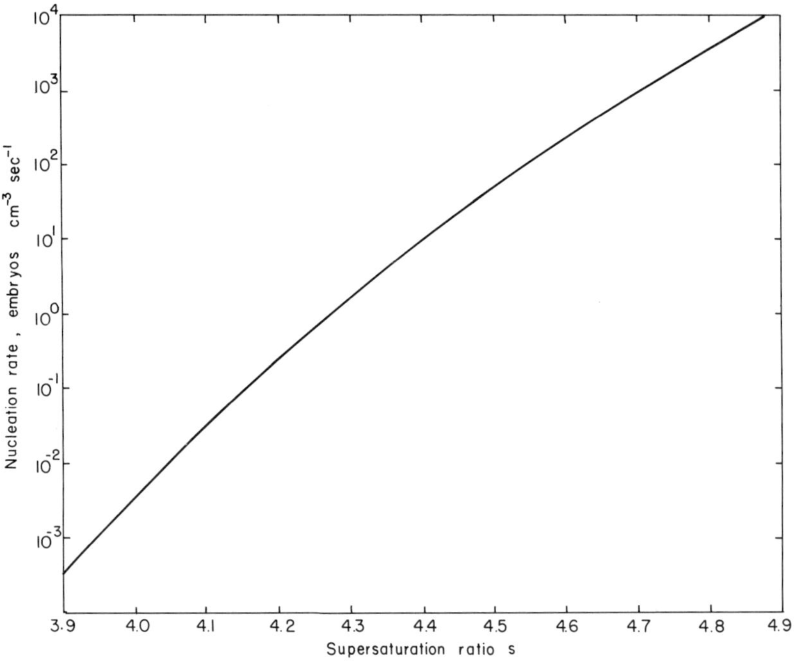


Figure 6.3 The homogeneous nucleation rate of water in its vapor as a function of the supersaturation ratio for an ambient temperature of 273.15°K.

the vapor is a function of temperature because of the temperature dependence of $T$, $\sigma$, and $\rho$ in (6.12). This is shown in figure 6.4 for $J(r_*)$ equal to one together with the experimental observations of several investigators. The selected value of $J(r_*)$ is not appropriate for all of the experiments, but figuure 6.3 shows that this is not an important matter. In general it is seen that there is fair agreement between experiment and theory. The scatter of the observed points at low temperature is thought to be due in part to the increasing experimental difficulties as the temperature is reduced.

## 6.5 Comments on Homogeneous Nucleation Theory

The development just given is called the Volmer or the classical theory of homogeneous nucleation. It captures the essential features of the thermodynamics and kinetics of the problem in readily understandable fashion. Further, its quantitative predictions agree rather well with experiments although there is evidence that the agreement is not so satisfactory for many substances other than water.

The apparent success of the Volmer theory is rather remarkable and perhaps fortuitous, in view of the simplified model on which it is based.

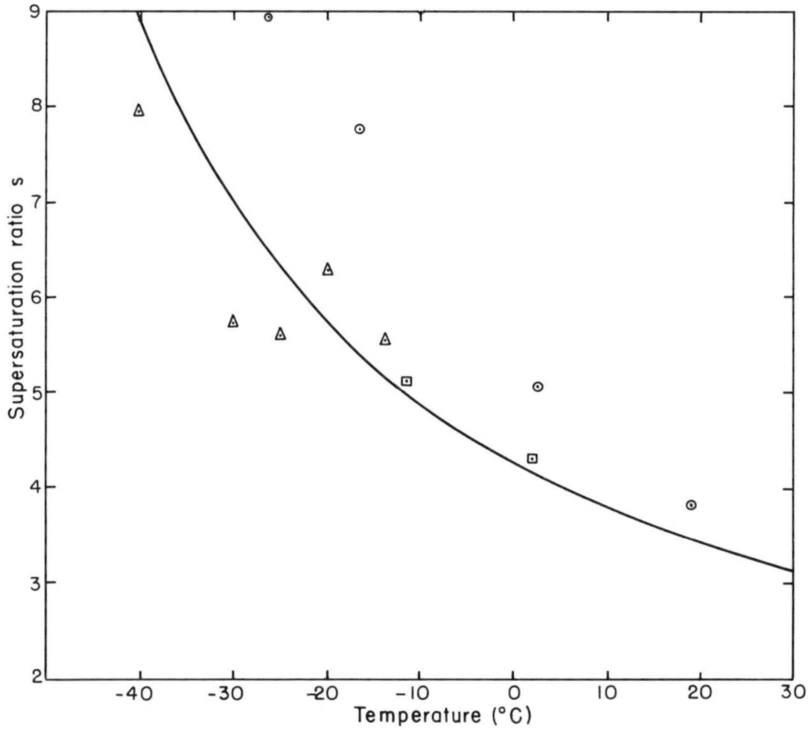


Figure 6.4 The supersaturation ratio required for the homogeneous nucleation of the liquid in the vapor. The solid curve is computed from (6.12) with $J(r_*) = 1$. Key to the experimental points: ○, Powell (1928); △, Pound et al. (1951); □, Volmer and Flood (1926).

The embryos are assumed to be spheres of water even when they are composed of fewer than 100 molecules (see table 6.1). The macroscopic, or bulk, values of the specific surface free energy and density used are of doubtful applicability to the tiny molecular aggregates. It has also been pointed out that not all of the terms involved in the free energy of formation are included. These have been nearly canceled by "replacement terms," at least in the case of water, in such a way that the classical theory gives a nearly correct result.

Hale and Plummer (1974) have used a model that is directly related to molecular properties, such as the binding energy and the internal energy of the inter- and intramolecular vibrations. This would seem to be the proper approach to a more adequate model of nucleation. This is important because it not only provides a more fundamental approach but also offers the hope of a more basic understanding of heterogeneous nucleation. Pending the more complete development and testing of molecular models, the discussion in this chapter is based on the classical Volmer nucleation model, with the understanding that it is an approximation at best.

## 6.6 Homogeneous Nucleation of Ice in the Vapor

In principle, the homogeneous nucleation of ice in water vapor can be described by (6.11) with a different preexponential factor and a suitable value of $\sigma$, which is now the specific surface free energy of ice. The numerical value of this specific free energy has been estimated to be about 100 erg $cm^{-2}$, but is still quite uncertain. Since $\sigma$ appears to the third power in the exponent in (6.12), this uncertainty is so magnified that (6.12) is of no predictive value.

Experimental approaches are hindered by the fact that homogeneous nucleation of the liquid almost certainly involves overcoming a smaller energy barrier than that for ice, at least for temperatures warmer than, say, $-50°C$. As noted in the next section, the homogeneous nucleation of ice in the liquid occurs at a temperature of about $-40°C$. This means that ice will always appear in an expansion chamber in which the temperature falls to $-40°C$ or lower, and no simple measurements can unequivocally distinguish ice nucleated from the vapor from ice formed via the liquid phase. The feeling persists that ice can be nucleated from the vapor at a sufficiently low temperature, but this has yet to be convincingly demonstrated.

## 6.7 Homogeneous Nucleation of Ice in Water

Water has a much more complex structure than water vapor. Models of water based, in part, on observations of x-ray scattering suggest that water consists of transient molecular groupings of three kinds: (a) an icelike structure; (b) a quartzlike structure; and (c) a close-packed, ideal liquid structure. The first of these groupings is the important one for ice nucleation, and it is found only to a small extent and only at temperatures below 4°C. These transient icelike structures play the same role as the molecular aggregates in the homogeneous nucleation of the liquid from the vapor. Nucleation will occur when the supercooling (equivalent of supersaturation) is large enough for the icelike structures to survive. The development follows the same course as that given for homogeneous condensation and will not be repeated here. In the counterpart of (6.4), some investigators prefer to recognize that ice embryos are not spherical. This leads to a dimensionless correction factor in the Kelvin equation. For any reasonable, simple shape this factor is not very different from unity, and it seems unimportant in view of the other uncertainties and the absence of specific information on the shapes of the ice embryos. The Kelvin equation in the form of (6.9) is most useful for ice nucleation. The numerical factor

Table 6.2
Typical homogeneous freezing temperature of water drops

| Drop diameter ($\mu$m) | Freezing temperature (°C) |
|---|---|
| 1 | −40.7 |
| 10 | −37.5 |
| 100 | −35.0 |
| 1,000 | −33.2 |
| 10,000 | −32.0 |

preceding the exponential in the expression for the nucleation rate is uncertain. Fletcher (1962) suggests a value of $10^{30}$ as giving results in reasonable accord with experiments. Thus[3]

$$J(r_*) = 10^{30} \exp\left\{ - \left[ \frac{16\pi\sigma^3 T_0^2}{3\rho_i^2 L_f^2 kT(T_0 - T)^2} \right] \right\}, \tag{6.13}$$

where $\sigma$ is the specific surface free energy of ice in the liquid and $T_0$ is the thermodynamic equilibrium temperature of ice and water (273°K). The numerical value of $\sigma$ is known only approximately and is about 20 erg/cm$^2$. If the numerical factor were more firmly established, experimental data on the supercooling required to nucleate ice in pure water would provide a sensitive evaluation of $\sigma$ through (6.13).

Many experimenters have sought to determine the homogeneous nucleation temperature of water. Only when it was realized that the water must be absolutely free of "motes" were consistent results obtained. The nucleation temperature is found to increase slowly with increasing size of the water volume or drop size. Typical results are given in table 6.2. It is considered that one ice embryo of critical size is sufficient to cause the drop to freeze. The variation of freezing temperature with drop size is then due to the decrease in the nucleation rate, $J(r_*)$, corresponding to the increase in drop volume.

The freezing temperatures in table 6.2 have been read from a smoothed curve to emphasize the variation with drop diameter and are not significant to three digits, as might be implied; in fact, the freezing temperatures are not established to better than about 2°C. On the basis of the experimental evidence, it is generally believed that the homogeneous nucleation temperature of pure water is in the neighborhood of −40°C.

3. The exponent in (6.10) should be $-(\Delta G r_* + \Delta G')/kT$, where $\Delta G'$ is the free energy required to break the intermolecular bonds as molecules diffuse through the water-ice interface. This is neglected here in view of the other uncertainties.

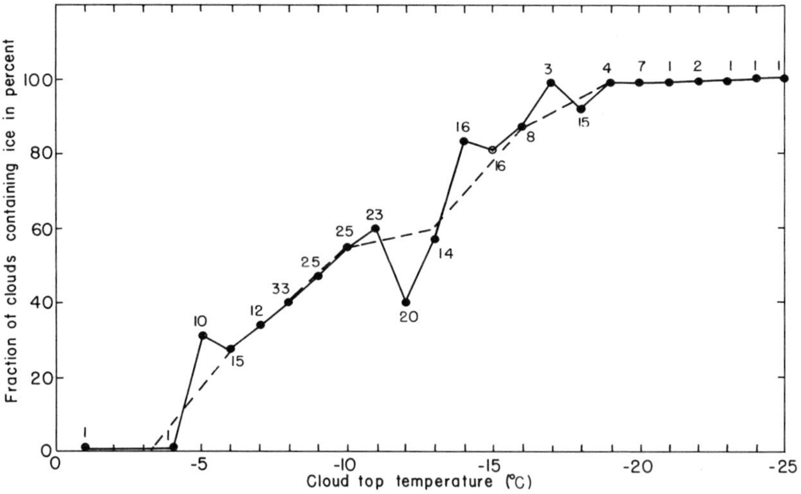


Figure 6.5 The fraction of clouds containing ice as a function of cloud top temperature, based on observations of 258 clouds by several investigators in different regions. The number above each point is the number of observations at that temperature. The dashed curve is the three-point running mean. After Hobbs et al. (1974).

Since the supersaturation in the real atmosphere seldom exceeds 1%, the only homogeneous nucleation process that can conceivably occur in the atmosphere is that of ice in water. There are only a few reports of supercooled clouds at temperatures in the vicinity of −40°C, but ice clouds typically are found at much warmer temperatures as shown in figure 6.5. It must be concluded that homogeneous nucleation is a rare phenomenon in the real atmosphere.

## 6.8 Heterogeneous Nucleation

Heterogeneous nucleation in the free atmosphere takes place on foreign particles in the mother phase. These particles, called nuclei, are a subset of the atmospheric aerosol. The previous discussion of homogeneous nucleation shows that the greatest barrier to nucleation is the small radii of molecular aggregates, which can persist only at large supersaturations. The primary function of a nucleus is to provide a relatively large radius *ab initio*. If condensation, for example, is to occur on an insoluble nucleus, the free energy of two boundary surfaces must be supplied: that between the nucleus and the liquid and that between the liquid and the vapor. This shows that the supersaturation required for condensation on an insoluble nucleus depends on not only the radius but also the specific surface free energy of the nucleus-water interface. In general, the properties of a substance at its surface differ from the properties within the substance. The surface may also exhibit fissures, pits, and other irregularities. When ice rather than water is nucleated, the crystal structure, bonding, steps, and dislocations become important. Thus heterogeneous nucleation is a very much more complex process than homogeneous nucleation.

## 6.9 Soluble Nuclei of Condensation

As discussed in section 1.13, a substantial fraction of the atmospheric aerosol consists of soluble substances, such as sulfates, chlorides and nitrates. These salts are hygroscopic and become aqueous solutions at saturation ratios less than unity. For example, a saturated solution of ammonium sulfate is in equilibrium at a relative humidity of about 80%, and the comparable value for sodium chloride is about 75%. These equilibrium values apply to droplets large enough so that the Kelvin radius of curvature effect is negligible. The case of smaller droplets is considered later in this section. Thus the hygroscopic particles are aqueous solutions well before condensation leading to the formation of cloud drops begins. Not only is there no second interfacial surface in this case, but the equilibrium vapor pressure over the droplet is less than that over pure water. This makes these hygroscopic particles exceptionally favorable sites for the formation of cloud drops. It can be argued that these particles should not be called condensation nuclei because the condensed phase is already in existence prior to the occurrence of cloudy condensation. Nevertheless, we shall follow the usual practice of calling them nuclei.

The process of condensation on hygroscopic particles is illustrated in figure 6.6. Each curve corresponds to a particle of the substance, in this case ammonium sulfate, of a selected dry mass. The curves give the equilibrium saturation ratios as a function of the particle radius. These are called Koehler (1926) curves because he was the first to compute them.

From the Kelvin equation (6.5), the effect of the solute is to reduce the equilibrium vapor pressure $e_{\mathrm{m}}$. The reduction of the equilibrium vapor pressure due to a dissolved electrolyte is given Raoult's law, modified by the van't Hoff factor as discussed by McDonald (1963):

$$\frac{e'_{\mathrm{m}} - e_{\mathrm{m}}}{e_{\mathrm{m}}} = -\frac{iM'}{iM' + M}, \tag{6.14}$$

where $e'_{\mathrm{m}}$ is the equilibrium vapor pressure over the solution, $e_{\mathrm{m}}$ is that over pure water, $M$ and $M'$ are, respectively, the moles of the water and of the solute, and $i$ is the van't Hoff factor. The van't Hoff factor varies with the chemical nature of the solute, and its concentration and is related to the dissociation of the solute. It may be evaluated from experimental measurements of the osmotic coefficient or of the equilibrium vapor pressure as a

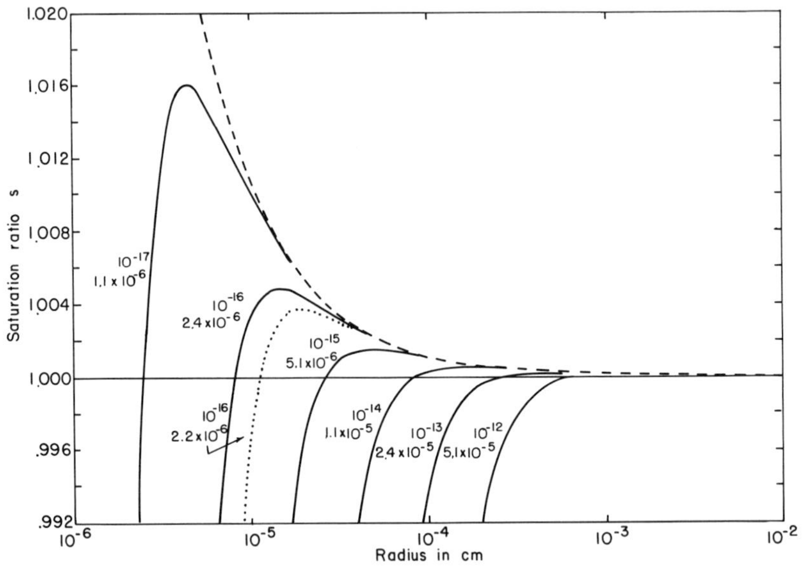


Figure 6.6 Koehler curves for ammonium sulfate nuclei. The dashed curve is a plot of the Kelvin equation for water. The dotted curve is for a sodium chloride nucleus for comparison. The upper number for each curve is the nucleus mass in grams, and the lower number is the radius of a dry salt nucleus of the same mass in cm. All data are at 20°C.

function of the molality of the solution. Since the latter is what is wanted in the Kelvin equation, it is not necessary to evaluate $i$ or to use (6.14) in computing the Koehler curves. Numerical values of $e'_m/e_m$ and the density and surface tension of solutions of common hygroscopic salts are given by Low (1969) and Kunkel (1970).

In figure 6.6 it is seen that the curves rise very steeply to a broad maximum. This corresponds to the rapid dilution of the solute, which leads to an increase of the equilibrium vapor pressure toward that of pure water. The curves are reversible to the left of the maximum and show how the particles may wax and wane with the ambient relative humidity. This has been used to explain observations that the visual range is inversely related to the relative humidity when the latter is fairly high. Note that the upper dashed curve in figure 6.6 is for pure water and is simply a plot of the Kelvin equation. It shows graphically the effect of the hygroscopic particles.

To the right of the maxima the curves have little meaning, since the radius increases with decreasing saturation ratio and the computed equilibrium cannot exist. When the particle reaches the maximum, the nucleus is said to be "activated," and its further behavior is governed by other processes, to be discussed in chapter 7. If one imagines a lifting process in

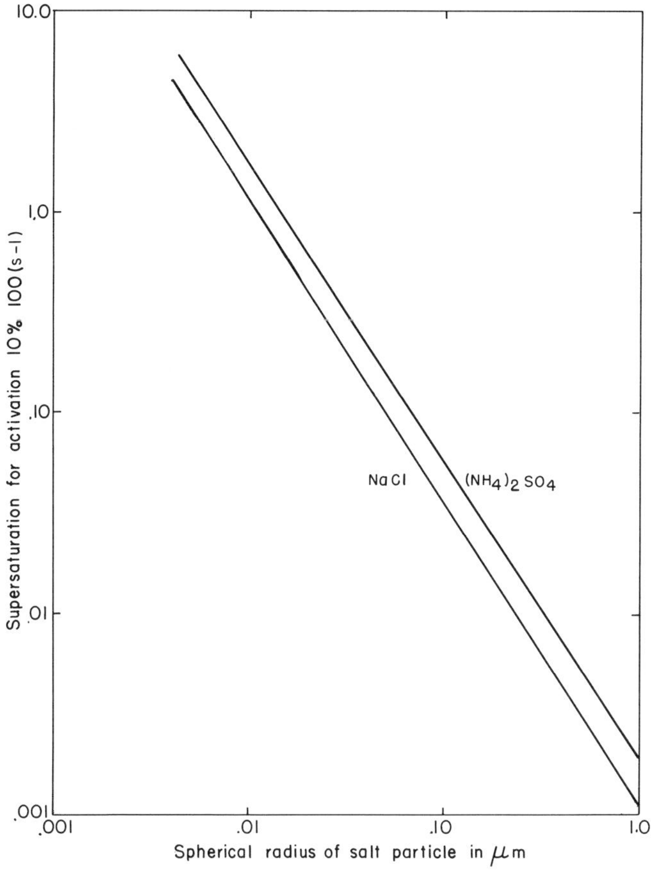


Figure 6.7 Percent supersaturation required to activate $(NH_4)_2SO_4$ and NaCl nuclei as a function of the spherical radius of the dry salt particle.

the atmosphere that causes the saturation ratio to increase steadily, particles of all the sizes present move along the corresponding Koehler curves until the largest particles reach their maximum. Their activation provides a sink for water vapor and slows the rate of increase of the saturation ratio. Activation of progressively smaller particles ultimately brings the rising saturation ratio to a halt and then reverses it. The smaller particles remain unactivated on the left-hand portion of their Koehler curves. This is a rather idealized representation of a complex process, but it does show how nature selects a portion of the nucleus population for activation. The number of nuclei activated depends on the size spectrum of the nuclei and, most important, on the rate at which the saturation ratio is forced up by the lifting process. In general, rapid lifting (or cooling) leads to a relatively large concentration of activated nuclei, and conversely.

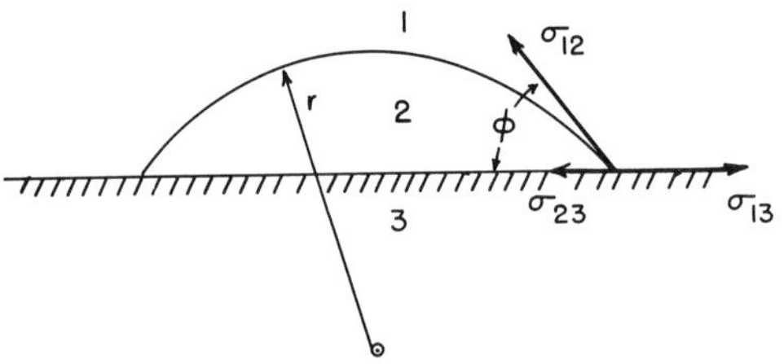


Figure 6.8 Spherical segment of liquid on a plane substrate. The numerals 1, 2, and 3 identify the vapor, liquid, and substrate, respectively; $\sigma_{12}$ is the specific surface free energy of the liquid-vapor interface, and $\sigma_{23}$ and $\sigma_{13}$ have analogous meanings.

The solid curves in figure 6.6 are for ammonium sulfate nuclei in view of the evidence presented in section 7.18 that this is the most common hygroscopic constituent of natural condensation nuclei. In comparison, the dotted curve is drawn for a sodium chloride nucleus as assumed by Koehler and others. Figure 6.7 shows the supersaturation required to activate the nuclei, the ordinates of the maxima of the Koehler curves, as a function of the equivalent spherical radii of the dry salt particles. For a given particle radius, the activation supersaturation for ammonium sulfate is about twice that for sodium chloride. Supersaturations in the atmosphere seldom exceed 1%. Together with the rapid decrease of particle concentration with increasing particle size (section 1.6), figure 6.7 shows that the important size range of hygroscopic condensation nuclei is about 0.01–0.1 $\mu$m radius.

Earlier attempts to confirm experimentally the theoretical predictions discussed here suggested that the activation supersaturation was some three times as large as predicted by the theory. However, careful experiments by Gerber et al. (1977) using sodium chloride and ammonium sulfate nuclei have given points that fall closely on the curves of figure 6.7. Twomey (1977) conducted similar experiments using nuclei formed by ultraviolet irradiation of natural air. Although the composition of these nuclei was not known, the experimental activation supersaturations clustered around the theoretical curve for ammonium sulfate. These investigations seem to provide an adequate confirmation of the theory. The acknowledged difficulties of simultaneous measurements of the activation supersaturation and of particle size in the 0.01–0.1-$\mu$m range are probably responsible for the deviations from theory found in earlier work.

## 6.10 Insoluble Condensation Nuclei

Some of the features of condensation on an insoluble surface are illustrated in figure 6.8. From the geometry of the figure

$$\sigma_{13} - \sigma_{23} = \sigma_{12} \cos \phi,$$

or

$$\cos\phi = m = \frac{\sigma_{13} - \sigma_{23}}{\sigma_{12}}. \tag{6.15}$$

The angle $\phi$ is the so-called contact angle, and its cosine, $m$, is a measure of the wettability of the surface. It is one for a perfectly wettable surface (contact angle zero) and minus one for a completely unwettable surface. The more wettable the surface, the smaller is the free energy of the interface. A perfectly clean glass surface is nearly completely wettable. Certain waxes and silicones are about as nonwettable as any real surface and may have a value of $m$ of around $-0.5$.

For the system described by figure 6.8, the counterpart to (6.4) is

$$\Delta G = V\rho R_w T \ln(s) + A_{12}\sigma_{12} + A_{23}(\sigma_{23} - \sigma_{13}), \tag{6.16}$$

where $V$ is the volume of the spherical cap, $A_{12}$ is the area of the cap, and $A_{23}$ is the area of the substrate covered by the liquid. Inserting the values of these areas, the volume, and (6.15) gives

$$\Delta G = \pi r^2(1-m)\left[\frac{(2+m)(1-m)}{3} r\rho_e R_w T \ \ln(s) + 2\sigma_{12} - (1+m)m\sigma_{12}\right]. \tag{6.17}$$

Setting the derivative to zero, the modified form of the Kelvin equation is

$$r_* = \frac{2\sigma_{12}}{\rho R_w T \ln(s)}, \tag{6.18}$$

which is the same as (6.5) except that $r_*$ is now the radius of the spherical cap, which is always larger than the radius of a sphere of equal volume as long as $\phi < 180°$.

Substitution of (6.18) in (6.17) gives the critical energy for nucleation:

$$\Delta G(r_*) = \frac{16\pi(\sigma_{12})^3}{3\rho_e^2 R_w^2 T^2 (\ln s)^2} \frac{(m+2)(m-1)^2}{4}. \tag{6.19}$$

This is the same as the barrier in the homogeneous nucleation case except for the factor involving $m$. For a perfectly wetted surface, $m = 1$, there is no energy barrier and nucleation would occur at $s = 1$. At the other extreme, $m = -1$, the factor is one and the substrate has no effect on the nucleation, as would be expected.

A more realistic geometry for condensation in the atmosphere is a sphere. This case has been treated by Fletcher (1958, 1959). The procedure

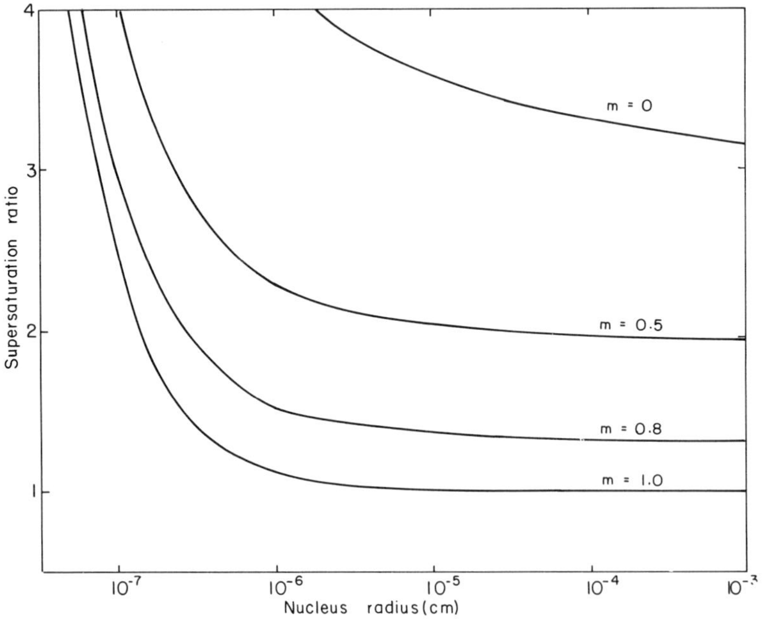


Figure 6.9 Critical supersaturation ratios for spherical insoluble nuclei as a function of radius for four values of the cosine of the contact angle $m$. After Fletcher (1958).

is the same as for the case of the plane substrate, but the spherical geometry introduces algebraic complexities. For this reason only Fletcher's results will be given here, in the form of the curves in figure 6.9. The bottom curve, $m = 1.0$, is for a perfectly wetted nucleus and is therefore simply a plot of the Kelvin equation. In comparison with figure 6.5, it is seen that all of the nucleation curves for hygroscopic nuclei lie below the Kelvin curve, while all of those for insoluble nuclei lie above. Even for the larger nuclei, for which the Kelvin effect is small, saturation ratios larger than those believed to exist in the atmosphere are seen to be required to activate insoluble nuclei unless they are nearly perfectly wetted. Thus unless there is a large disparity in nucleus size, the above theory predicts that most, if not all, atmospheric condensation takes place on hygroscopic particles. This conclusion is in accord with the early experimental work of Aitken (1923), who showed in the laboratory that condensation did not occur at modest supersaturations unless hygroscopic nuclei were present.

The predictions of (6.19) for plane insoluble surfaces have been partially confirmed by Mahata and Alofs (1975), who also discuss data of other investigators. They found reasonably close accord with the theory for surfaces with contact angles $<20°$. For larger contact angles, the observed

Table 6.3
Activation supersaturations of mixed nuclei consisting of salt and an insoluble wettable part

| Volume fraction of salt | 1.0 | 0.4 | 0.1 | 0.01 | 0.0001 | 0.0 |
|---|---|---|---|---|---|---|
| | Activation supersaturations (%) | | | | | |
| Dry radius $10^{-6}$ cm | 1.4 | 2.3 | 3.7 | 6.1 | 8.4 | 13.0 |
| Dry radius $10^{-5}$ cm | 0.04 | 0.06 | 0.13 | 0.34 | 0.65 | 1.2 |

supersaturation was found to be less than that predicted and to deviate more as the contact angle was increased. They found that adsorption of water did not occur and that artificial roughening of the surface had a relatively small effect in the anticipated sense. They conclude that it is unlikely that insoluble particles have a significant role as condensation nuclei.

On the other hand, Jiusto and Kocmond (1968) put silver iodide and microcrystalline wax nuclei into a thermal gradient diffusion chamber and obtained the activation supersaturations. The contact angle of water on silver iodide is fairly small and is large for the wax. Nevertheless, the observed activation supersaturations in both cases corresponded to that for completely wettable particles of the measured size. This experiment has apparently not been repeated. The results are in marked contrast with those for plane substrates and leave the efficacy of insoluble particles as condensation nuclei in doubt.

Many atmospheric particles are mixtures of soluble and insoluble substances. Junge and McLaren (1971) have studied theoretically the activation supersaturation of mixed nuclei, composed of soluble salt and a completely wettable insoluble substance, as a function of the fraction by volume of salt and the dry particle radius. They found that the activation supersaturation was not very different from that of an all-salt nucleus as long as the salt comprised 10% or more of the dry volume. This is illustrated in table 6.3, taken from the work of Junge and McLaren (1971). They did not consider the case of an insoluble portion that is not completely wettable. This would increase the activation supersaturation because of the need to supply the free energy of the water-solid interface.

Although of no importance as atmospheric condensation nuclei, mention is made of the important discovery by Wilson (1897, 1899) that condensation will occur on small ions at a saturation ratio somewhat smaller than that for homogeneous nucleation. This discovery was the basis for the Wilson cloud chamber, which has long been used for the detection of high-energy particles and radiations, which leave ion tracks that become visible when condensation occurs on them.

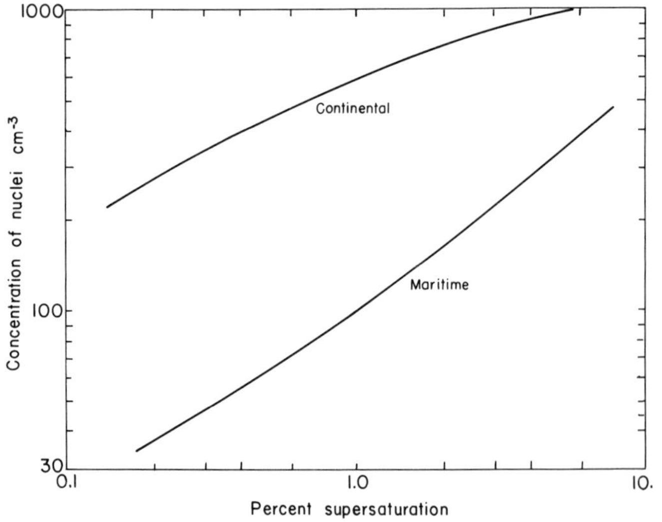


Figure 6.10 Median nucleus spectra over continental and maritime regions. After Twomey and Wojciechowski (1969).

## 6.11 Cloud Condensation Nuclei

Condensation nuclei that are activated at supersaturations of about 1% and less are called cloud condensation nuclei, abbreviated CCN. This distinguishes the nuclei on which cloud drops form from the much more numerous particles that are activated only at supersaturations greater than those that occur in the atmosphere.

Small expansional cloud chambers of the type first developed by Aitken produce large supersaturations and are useful only for measuring the total aerosol concentration. Refined expansional cloud chambers have been developed that can produce supersaturations of a few percent, but the thermal diffusion chamber has been found most useful in measuring CCN concentration. In this device gradients of temperature and water vapor are established between an upper warm, wet plate and lower cold plate by diffusion. Because of the nonlinearity of the vapor pressure-temperature relation, a region of supersaturation is developed, the magnitude of which may be controlled through the temperature difference between the plates. The diffusion chamber can maintain a supersaturation ranging from about 0.1 to several percent. This type of chamber was first used by Wieland (1956) to obtain a CCN spectrum in the atmosphere. A CCN spectrum is a plot of the concentration of CCN as a function of the supersaturations, as shown in figure 6.10, after Twomey and Wojciechowski (1969), Twomey (1959), using a similar chamber in Australia, showed that the concentration of CCN was about an order of magnitude larger in continental than

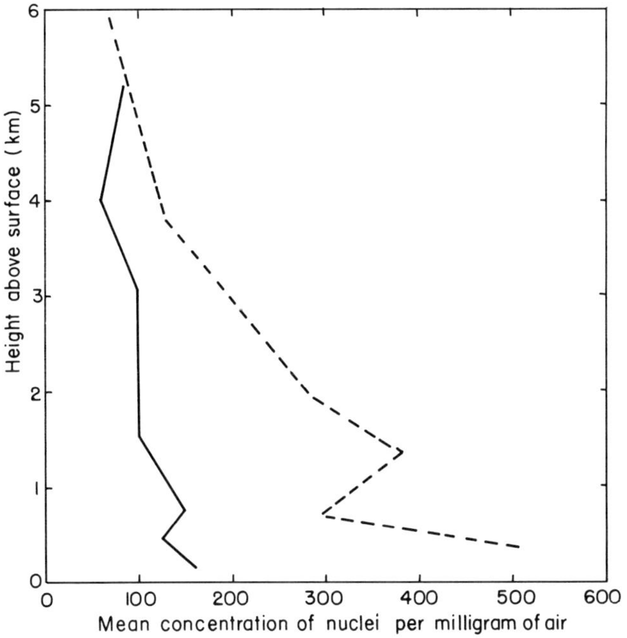


Figure 6.11 Mean concentration of cloud condensation nuclei active at 0.75% supersaturation as a function of height. The solid curve is from data taken over the Caribbean Sea, and the dashed curve from data over Colorado. After Squires and Twomey (1966).

in maritime air, leading to a correspondingly smaller concentration of larger cloud drops in maritime air. This has important consequences in the release of precipitation, as discussed in chapters 7 and 8.

The difference between the CCN spectra in maritime and continental air is shown in figure 6.10. These median spectra come from many hundreds of observations made from an airplane while covering some 100,000 miles in the midtroposphere. Individual spectra may differ from these median spectra by an order of magnitude or more in continental air, but by a lesser amount in maritime air. Since typical concentrations of cloud drops are 50–100 $cm^{-3}$ in maritime cumuli and 300–500 $cm^{-3}$ in continental cumuli, figure 6.10 shows that the activation supersaturation is generally less than 1%.

The variation of CCN concentration with height above the surface is illustrated in figure 6.11, which is taken from the observations of Squires and Twomey (1966) over the Caribbean and Colorado just to the east of the Rocky Mountains. The decrease with height of the Colorado profile suggests a surface source of the nuclei, while the uniform oceanic profile may be due to vertical mixing or to a source in the free atmosphere. These results are seen to accord with similar profiles of the aerosol discussed in sections 1.6, 1.7, and 1.9.

## 6.12 Heterogeneous Nucleation of Ice

Ice may be nucleated in the vapor and in the liquid on suitable particles, acting as heterogeneous nuclei. Wegener (1911) apparently was the first to postulate the existence of sublimation nuclei that would nucleate ice in the vapor, by analogy with condensation nuclei. Since sublimation is the evaporation of a solid, it now seems better to call these particles deposition nuclei. Freezing nuclei are particles that nucleate ice in the liquid. Since it is often difficult in practice to know whether ice has formed in the vapor or in the liquid, the term ice nuclei is often used to encompass both kinds of ice-forming nuclei.

Ice nuclei play an important role in the release of precipitation; an essential feature of the ice crystal precipitation process is that the concentration of ice particles, and hence of ice nuclei, is very much smaller than the typical concentration of cloud drops or CCN. The small concentration of ice nuclei, together with their usually small size, makes it difficult to identify, count, and size them.

Wegener and others assumed that the counterpart of hygroscopicity of CCN was a similarity of the crystal structure of ice nuclei to that of ice. It also seemed likely that ice nuclei would be highly insoluble. These assumptions were seemingly confirmed by Vonnegut's (1947) discovery that small silver iodide crystals acted as ice nuclei at temperatures only a few degrees below 0°C. Vonnegut selected silver iodide by scanning a tabulation of crystalline substances for low solubility and crystal lattice dimensions close to those of ice.

## 6.13 Substances That Act as Ice Nuclei

After Vonnegut's discovery of the nucleating ability of silver iodide, many experiments were carried out to find other ice-nucleating materials, both pure and natural substances. For this purpose, large expansion chambers and mixing chambers were used. In both cases a supercooled water cloud was formed. The material being tested was introduced into the chamber in a finely divided form; in the expansion chambers it was usually introduced prior to the expansion while it was dropped into the supercooled cloud in the mixing chambers. The presence of the cloud ensured water saturation, but it was not possible to know whether the ice was nucleated in the liquid or in the vapor. At first, ice crystals were detected and their concentration estimated by visual observations in a light beam in which the crystals scintillated. Later the crystals were allowed to fall out onto a tray of supercooled sucrose solution or a supercooled detergent film. The crystals grow rapidly on these media and are easily counted.

In other studies the finely divided nucleating material was introduced into carefully purified water and the temperature lowered until freezing occurred. It is clear that this test determines the temperature at which the substance acts as a freezing nucleus. In general, the freezing temperature was found to be nearly the same as the temperature at which ice appeared in the chambers for the same substance.

An abbreviated listing of the ice-forming temperatures of pure and natural substances appears in table 6.4. In each class, the substances are arranged by decreasing nucleation temperature. The similarity of the lattice dimensions of silver and lead iodide to those of ice is apparent and seems to confirm Vonnegut's hypothesis. However, an inspection of the rest of the table clearly shows that similarity of the crystal lattice dimensions is, at best, only one determinant of the nucleation temperature of ice. The temperatures in table 6.4 are not definitive, since various investigations have yielded somewhat different results. It is included here primarily to show that crystal lattice dimensions are only one aspect of ice nucleation and that many substances will nucleate ice at temperatures no lower than about $-15$°C.

## 6.14 Ice Nucleation on Large Crystals

Several investigators have observed the nucleation of ice on relatively large single crystals of known ice nucleants in chambers in which the supersaturation over ice could be carefully controlled. Ice crystals nucleated on single crystals of silver iodide, lead iodide, cadmium iodide, mica, and biotite are found to be uniformly aligned. This alignment with a crystal axis of the substrate is considered to be evidence of epitaxy or oriented overgrowth. This is the process visualized by Vonnegut as being responsible for the nucleating ability of substances with lattice dimensions similar to those of ice. The mismatch between the two lattices sets up a strain that increases the free energy barrier to nucleation. Accordingly, the greater the strain, the lower the temperature required for ice nucleation.

Mason and van den Heuvel (1959), using a large single crystal of silver iodide, found that ice crystals formed only at water saturation down to $-12$°C but that at lower temperatures ice crystals appeared at vapor pressures intermediate between ice and water saturation. They concluded that the silver iodide acted as a deposition nucleus at temperatures below $-12$°C and at supersaturations with respect to ice of at least 12%.

In experiments of the same sort Anderson and Hallett (1976) found that ice was nucleated on occasion at ice supersaturations as low as 2% at temperatures of $-5$–$-24$°C. The number of crystals nucleated increased

Table 6.4
Temperatures at which various substances nucleate ice

| Substance | Crystal lattice dimension a axis (Å) | c axis (Å) | Temperature to nucleate ice (°C) | Comments |
|---|---|---|---|---|
| **Pure substances** | | | | |
| Ice | 4.52 | 7.36 | 0 | — |
| AgI | 4.58 | 7.49 | −4 | Insoluble |
| $PbI_2$ | 4.54 | 6.86 | −6 | Slightly soluble |
| CuS | 3.80 | 16.43 | −7 | Insoluble |
| CuO | 4.65 | 5.11 | −7 | Insoluble |
| $HgI_2$ | 4.36 | 12.34 | −8 | Insoluble |
| $Ag_2S$ | 4.20 | 9.50 | −8 | Insoluble |
| $CdI_2$ | 4.24 | 6.84 | −12 | Soluble |
| $I_2$ | 4.78 | 9.77 | −12 | Soluble |
| **Minerals** | | | | |
| Vaterite | 4.12 | 8.56 | −7 | |
| Kaolinite | 5.16 | 7.38 | −9 | (Silicate) |
| Volcanic ash | — | — | −13 | |
| Halloysite | 5.16 | 10.1 | −13 | |
| Vermiculite | 5.34 | 28.9 | −15 | |
| Cinnabar | 4.14 | 9.49 | −16 | |
| **Organic materials** | | | | |
| Testosterone | 14.73 | 11.01 | −2 | |
| Chloresterol | 14.0 | 37.8 | −2 | |
| Metaldehyde | — | — | −5 | |
| $\beta$-Naphthol | 8.09 | 17.8 | −8.5 | |
| Phloroglucinol | — | — | −9.4 | |
| Bacterium *Pseudomonas Syringae* | — | — | −2.6 | (Bacteria in leaf mold) |

with ice supersaturation and increasing supercooling. This difference from the Mason and van den Heuvel (1959) results may be due to differences between the crystals used. In any event, it confirms the nucleation of ice by deposition on single crystals of silver iodide and copper sulfide. An important finding of both investigators was that the ice crystals reappeared at the same sites repeatedly when they were evaporated and then reformed. In many cases these active sites occurred at steps or at other identifiable features of the crystal surface.

These observations do not necessarily apply to small airborne nuclei of the same or other nucleating substances. The concentration of nucleation sites on the single crystals used by Anderson and Hallett was about

$10^5$ cm$^{-2}$. Others have found that silver iodide particles of about 0.1 $\mu$m radius can act as deposition nuclei. If such particles contain a minimum of one active site, the site concentration would be about $10^9$ cm$^{-2}$, or four orders of magnitude greater than on the single crystal. This discrepancy may be due to a large number of sites formed by the thermal vaporization-quenching process used to form the small particles or, more likely, the experimental evidence that only a very small fraction of the small particles are active nuclei. If only 1 in $10^4$ particles nucleated ice and contained a nucleation site, the discrepancy would be resolved.

## 6.15 Theoretical Speculations

One of the first attempts to develop a theory for the nucleation of ice by deposition and freezing on small particles was made by Fletcher (1962). He applied the Volmer theory, used earlier for condensation, to the nucleation of ice from the vapor and the liquid. He found that the size of deposition nuclei was not important for radii greater than about 0.1 $\mu$m but that the necessary supercooling increased rapidly for smaller particles. For nucleation from the liquid, the comparable radius was about 0.03 $\mu$m. In both cases the supercooling required for nucleation was found to increase rapidly with the contact angle between ice and the substrate, or with $m$ as defined in (6.15). Fletcher found that nuclei with $m = 0.95$ should act as deposition nuclei at a sufficiently large supercooling. This theory was found to lead to deposition or freezing nuclei spectra that were qualitatively similar to those observed. However, it is now believed that the Volmer theory as applied by Fletcher is an inadequate representation of the physics of ice nucleation. It is reasonable that there is a lower limit to the size of a nucleus, and the numerical values given earlier seem to be of the right order of magnitude.

The active nucleation sites observed on the large single crystals, as described earlier, has led to speculation that they are an important aspect of the physics of ice nucleation. Mere dimensional similarity of the crystal lattices of the substrate and ice is not adequate; it is essential that there be molecular bonding. This may occur at hydrophilic impurity centers. The existence of steps, screw dislocations, etch pits, and similar features at preferred sites may be due to the tendency for adsorbed molecules to become trapped there. The relative importance of these irregularities in the crystal and of impurities is not known. It is of interest that the silver iodide used in cloud seeding is formed by burning a mixture of silver iodide and sodium iodide in acetone. The resultant particles appear to be mixed crystals of silver and sodium iodide, the latter substance being soluble, but they are more active as ice nuclei than pure silver iodide particles.

Fletcher (1969) has used a simple model of a nucleus of silver iodide with an active site. The particle is assumed to be spherical with a conical cavity having its base at the surface. At a supercooling determined by the angle of the cone and the contact angle parameter $m$, ice nucleation occurs at the vertex of the cone and ice fills the cavity. For silver iodide, this occurs at a supercooling of about 4°C. The active site on which the analysis is based is the circular patch of ice at the surface of the particle. The nucleation of such a particle can be examined by a small extension of the Volmer theory for spherical particles. The analysis assumes that the probability of a particle containing an active site is proportional to its surface area. The activity spectrum of a typical silver iodide smoke size distribution is found to be similar in shape to observed activity spectra. This is a promising first attempt to examine nucleation on particles containing one type of active sites. Further developments along these lines must probably await more complete knowledge of the nature of active sites and the probability of their occurrence on small particles.

At the time of writing there are many clues to the physical processes involved in the heterogeneous nucleation of ice from the liquid and from the vapor. However, there is as yet no comprehensive theory that relates all or most of the known physiochemical properties of ice nuclei to their activity.

## 6.16 Contact Nucleation

As early as 1949, Rau (1949) found that water drops, supercooled by just a few degrees, froze when they were contacted by dry particles of various substances. This process is called contact nucleation and is a special case of freezing nucleation. In general, it is found that contact nucleation occurs at a significantly smaller supercooling than that required when the same particles are immersed in water. Thus, Gokhale and Goold (1968) reported that particles of clay, volcanic ash, CuS, and AgI froze water drops by contact nucleation at temperatures 5–10°C warmer than the same substances when immersed in the water; the most marked difference was exhibited by AgI.

Odencrantz (1969) found that 47 of 89 organic substances caused water drops to freeze by contact nucleation at a temperature of $-3$°C. This was not meant to be a comprehensive experiment, and only one temperature was used, but it shows that many organic substances act as contact nuclei at temperatures much warmer than those found in the various cloud chamber experiments.

Langer et al. (1978) have presented evidence that AgI particles larger than 0.01 $\mu$m diameter act as contact nuclei and that their activity increases with size to 0.03 $\mu$m; the activity of even larger particles seems to be independent of their size.

It appears that contact nucleation probably plays an important role in the atmosphere. It seems capable of explaining the reports of frozen drizzlesize drops and of graupel in convective clouds at temperatures as warm as $-5°C$. Contact between nuclei and supercooled water drops in the free atmosphere may occur by collisions in the gravitational field in the case of micron size nuclei or by thermophoresis, diffusiophoresis, and Brownian motion in the case of submicron nuclei.

There is as yet no viable theory of contact nucleation. It has been speculated that the oriented double layer at the surface of the water, a structure quite different from that of the water within the drop, plays an important role.

## 6.17 Activity Spectra of Ice Nuclei

It has been customary to plot the concentration of ice crystals formed from a polydisperse aerosol as a function of supercooling, and the result is termed an ice nucleus spectrum. It is a very useful plot because it may be used to predict the concentration of ice particles in an atmosphere, the temperature structure of which is known. Similarly, an ice nucleus spectrum of an artificial nucleant, such as silver iodide smoke, is invaluable in determining the seeding rate for precipitation modification.

Ice nuclei spectra are determined by means of one of the several types of cloud chambers, and the results are dependent on the operating conditions of the chambers. Under the best of current conditions, the concentrations of natural ice nuclei as measured by different types of cloud chambers agree to within one order of magnitude; curiously, the discrepancy may be several orders of magnitude for silver iodide smoke, which is often considered as a standard ice nucleus substance.

Some of the cloud chambers used are (a) the Bigg-Warner 10-liter expansion chamber, which is pressurized, humidified, and then expanded; (b) the isothermal, which is an enclosed "deep freeze" with a moisture source to produce a fog; and (c) membrane filters through which a large sample of air is passed—the filter is then placed in a diffusion chamber and "developed" by increasing the supersaturation over ice until crystals appear. In the first two a supercooled fog is formed, and nuclei may act by deposition, freezing, or contact. The operation of the Bigg-Warner chamber results in supersaturation with respect to water, and the particles may act first as condensation nuclei. Natural nuclei often are mixtures of soluble and insoluble substances, so that condensation on them is possible at very slight water supersaturations. A variety of observations suggests that there is a time delay in the activity of some ice nuclei ranging from a few seconds to minutes. Expansion chambers are not active long enough to activate such nuclei. The membrane filter technique has the advantage of close control of the ice supersaturation, and the ice nucleus spectrum can be obtained from a single sample, whereas a separate run for each temperature is required in the other chambers. However, the filter technique

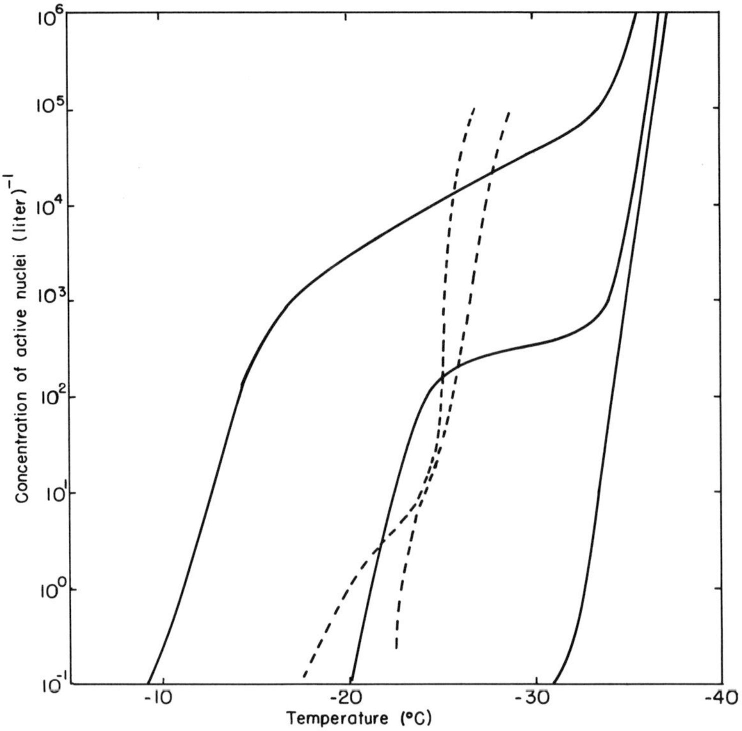


Figure 6.12 Sample ice nuclei activity spectra. The solid curves are after Smith and Heffernan (1954); the dashed curves are after Murgatroyd and Garrod (1957).

generally gives lower counts than the other instruments. This is attributed to condensation nuclei, or the filter itself, acting as water vapor sinks and a failure to detect small ice nuclei. This very brief account is intended only to alert the reader to the significant uncertainties in any current or past measurements of the concentration and activation temperatures of ice nuclei. This situation will certainly improve in the future because many or most of the causes are now understood at least in part.

Some samples of ice nuclei activity spectra are presented in figure 6.12. Other samples with somewhat different shapes are to be found in the literature, but it does not seem useful to discuss details in view of the uncertainties of such measurements. Note in figure 6.12 that the supercooling required to yield one ice crystal in 10 liter of air, roughly the detection limit, ranges from −9 to −31°C. The spectra tend toward very large concentrations as the homogeneous nucleation temperature is approached. About all that can be said about the shape of the curves is that

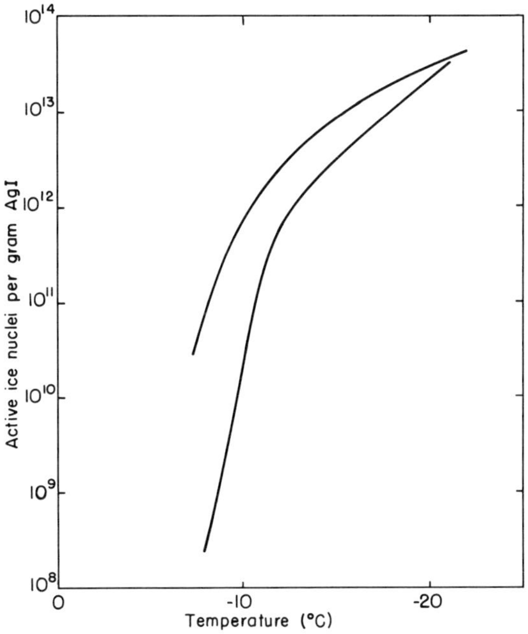


Figure 6.13 Activity spectra for silver iodide smokes produced by two different generators. After Fuquay and Wills (1957); the curves drawn from data in their table 1.

they are very roughly exponential with rather steep slopes. Use will be made of this in the discussion of stratiform precipitation in sections 8.23 and 8.24.

Two sample activity spectra of silver iodide smokes of the kind used in cloud seeding are given in figure 6.13. Note that the ordinates are concentrations of ice crystals per gram of silver iodide and their magnitudes cannot be compared with the ordinates of figure 6.12. However, the scales are the same, so the slopes of the curves may be compared, and it is evident that they are not very different, at least in the lower half of the curves. The decrease in slope of the curves of figure 6.13 at temperatures below about −10°C presumably reflects an approach to a limit set by the total number of silver iodide particles large enough to act as ice nuclei.

Huffman (1973) has shown that when the concentration of ice nuclei is plotted against supersaturation with respect to ice instead of supercooling, points taken at temperatures of −12, −16, and −20°C all fall on a single curve, as shown in figure 6.14. The data were taken by using a membrane filter that was developed in a thermal diffusion chamber in which the humidity did not exceed a slight supersaturation with respect to water.

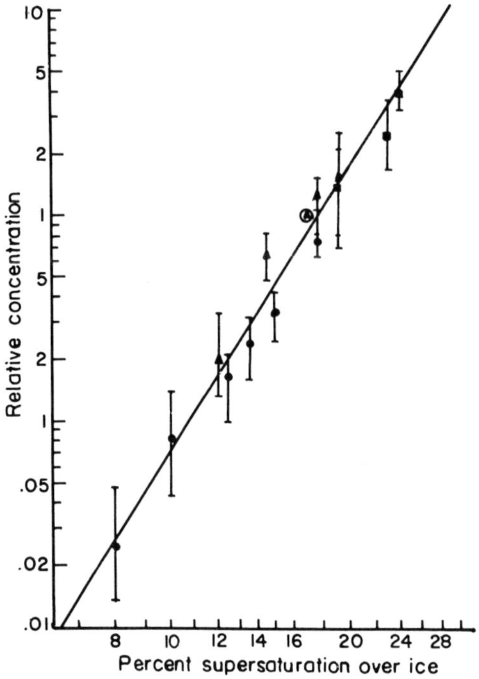


Figure 6.14 Concentration of ice nuclei versus the supersaturation over ice. Key: ●, −12°C; ▲, −16°C; ■, −20°C. After Huffman (1973).

Huffman believes that the particles acted as deposition nuclei. He obtained similar results on silver iodide smokes and natural nuclei at several locations.

In the development of (6.9), the alternative form of the Kelvin equation, (6.8), was derived from the Clausius-Clapeyron equation. This may be rewritten as

$$SC = 273 R_w \frac{T}{\bar{L}} \ln s, \tag{6.20}$$

where SC is the supercooling in °K, 273 is the nominal melting point of ice in °K, and the other quantities have been previously defined. Over the temperature range used by Huffman $T/\bar{L}$ is sensibly constant and (6.20) shows that supercooling and the supersaturation ratio are equivalent parameters.

## 6.18 Time Variations of Natural Ice Nuclei

The concentration of ice nuclei is at least as variable in time as that of CCN. The variability is explored by operating a cloud chamber at a single temperature, typically $-20$ or $-25°C$, over extended periods of time. On occasion the concentration of nuclei rises rapidly by one or more orders of magnitude and then declines as rapidly after some hours or days. Observations of high concentrations on the summit of Mount Washington, New Hampshire, were associated with air trajectories from the arid southwestern United States. The high concentration peaks observed in Australia did not correspond to any observable feature of the atmospheric circulation.

## 6.19 Composition of Natural Ice Nuclei

The observations that many natural soils, and particularly the clay minerals, act as ice nuclei at temperatures of $-15$–$-20°C$ and higher strongly suggests that these are a source of natural ice nuclei. This inference has been strongly supported by direct observations of the nuclei in natural snow crystals. The procedure, apparently first used by Kumai (1951), is to collect a natural snow crystal on an electron microscope grid, make a low-power photomicrograph of the snow crystal to establish its form and location on the grid, and then to place the grid in an electron microscope. In most cases a particle of about 1–10 $\mu$m diameter is found at what had been the center of the ice crystal before it sublimed. Some inference as to the composition of the central nucleus can be made from its morphology, but a definitive identification of crystalline substances may be made by focusing the electron beam so as to obtain the electron diffraction pattern. In addition to the central particle, most of the snow crystals contained many smaller particles scattered throughout the region occupied by the original snow crystal. Many of these are condensation nuclei, and some are crystalline substances of the same composition as the central particle. It is generally assumed that these small particles were captured by the snow crystal during its fall through the atmosphere. The results of the several investigations of the composition of the central nuclei by the method discussed earlier are summarized in table 6.5.

The preponderance of clay minerals, together with the known activity of clays as ice nuclei (see table 6.4) and their wide distribution in surface soils, is striking. It has been argued that there is only circumstantial evidence that the central particles are the actual ice nuclei, but the evidence that they are is about as compelling as is ever found in atmospheric phenomena. As table 6.5 shows, a significant fraction of the central nuclei are sea salt, which is not known to be an ice nucleus (but see later). There are also a few cases in which no central nucleus was found. There is other evidence that volcanic particles act as ice nuclei, and it has been suggested by Bowen (1956) that there is an extraterrestrial source of ice nuclei.

Some mention should be made of the nucleation of supercooled water clouds by solid carbon dioxide (dry ice). This was discovered by Schaefer

Table 6.5
Composition of central nuclei found in natural snow crystals

| Geographical location | Clay minerals (%) | Sea salt (%) | Unidentified other minerals (%) | No nucleus observed (%) | Reference |
|---|---|---|---|---|---|
| South Pole | 60 | 20 | 5 | 15 | Kumai (1976) |
| Hokkaido | 57 | 19 | 19 | 5 | Kumai (1951) |
| Honshu | 88 | 0 | 12 | 0 | Isono and Ono (1959) |
| Houghton, Michigan | 87 | 1 | 11 | 1 | Kumai (1961) |
| Greenland | 85 | 0.6 | 10.7 | 3.7 | Kumai and Francis (1962) |

(1946) and used by him and Langmuir in the first cloud-seeding experiments. Dry ice sublimes at $-78.5°C$ and has a latent heat of sublimation of 134 cal $g^{-1}$. The air in the immediate wake of the falling pellet is rapidly chilled, resulting in a very high supersaturation and homogeneous condensation, followed by homogeneous freezing. These processes produce very high concentrations of small ice crystals. Any other equivalent cold source would lead to a similar result. Thus dry ice is not an ice nucleus in the sense discussed here.

## 6.20 Biogenic Ice Nuclei

A new source of ice nuclei has been found by Schnell and Vali (1972) in decaying leaf litter. Subsequent studies by Schnell and Vali (1976) and by Vali et al. (1976) identified the bacterium *Pseudomonas Syringae* as the causitive agent and suggested that the bacteria themselves are ice nuclei. These biogenic nuclei act as freezing nuclei at temperatures as high as $-1.3°C$, and the concentration of nuclei active at $-10°C$ per gram of leaf mold ranges from $10^2$ for the least active molds to $10^9$ for the most active. Substantial fluxes of these biogenic nuclei into the atmosphere have been detected. They are found in all vegetated areas where tests have been made. The lowest concentrations were found in tropical regions, and the highest in midlatitude rainy climates with cold winters.

Investigations by Schnell (1977) have shown that biogenic nuclei are present in seawater, marine fog water, and marine air. Concentrations of the nuclei are highest in areas of phytoplankton blooms. The transfer of the biogenic nuclei from the water to the air is presumed to occur through the bubble-bursting mechanism discussed in section 1.15. Laboratory experiments have shown that organic material and bacteria are transferred to the air by this mechanism.

## 6.21 Comments on Ice Nuclei

It is apparent that many of the basic features of both artificial and natural ice nuclei are still poorly understood. On the theoretical side, there is no comprehensive model of heterogeneous ice nucleation that would permit, for example, the prediction of the activity of a given substance as an ice nucleus. Many known ice nuclei are crystalline and insoluble; others are soluble and/or amorphous. In many cases it is not known whether nucleation has resulted from deposition, freezing, or contact. Biogenic nuclei are a new entrant on an already confused scene. It is too early to tell whether this will prove to be an important discovery, but if it is, it will probably be necessary to revise some of the current concepts of the physics of nucleation. With each new discovery, there is a natural tendency to view it as a panacea. In view of the wide range of substances that have been found to nucleate ice and the diversity of the earth, its oceans, and its atmosphere, it is more reasonable to suppose that a variety of substances shares the burden of providing ice nuclei.

## Problems

1.
What acts as the nucleus in homogeneous nucleation?

2.
What is the most important property of a nucleus in heterogeneous nucleation?

3.
Derive the form of Kelvin's equation if the ice embryo is a thick hexagonal plate rather than a sphere.

4.
Show that (6.9) is an alternative form of (6.5).

5.
Why is it expected that nucleation of the liquid on a soluble nucleus will require less supersaturation than on an insoluble nucleus?

# 7

# Growth Processes of Water Drops and Ice Particles

## 7.1 Introduction

Once water drops or ice particles are nucleated in the vapor, they will continue to grow by condensation (vapor deposition) so long as the imposed lifting (cooling) process continues. This type of growth is governed by molecular diffusion, modified by forced ventilation as the particles become large enough to fall through the air at a few centimeters per second or more.

When the largest drops in a typical size spectrum fall at a speed of several centimeters per second or more, they begin to overtake and collide with the smaller drops. This growth process has been variously termed coalescence, accretion, or collisional growth and is responsible for a large fraction of the growth of particles to precipitation size. This process depends on the size of the largest particles, the drop size spectrum, and the concentration of the drops; it is modified by the aerodynamics of the flow around the falling drops, which causes small particles to be diverted from the larger particles.

These two growth processes are not mutually exclusive, but may act simultaneously under certain circumstances. Past failure to recognize this fact has led to underestimates of the growth rate of nascent precipitation particles.

## 7.2 Growth by Water Vapor Diffusion

Molecular diffusion is described by Fick's law, which may be written

$$\frac{d\rho_w}{dt} = D\nabla^2\rho_w, \tag{7.1}$$

where $\rho_w$ is the water vapor density, $D$ is the molecular diffusion coefficient of water vapor in air, and $\nabla^2$ is the Laplacian operator. Under atmospheric conditions the size of the particle changes so slowly with time that it is permissible to use the steady state form of (7.1). Many solutions of the Laplace equation have been found because of the many physical problems that it describes. An example of the use of one of these solutions is given in section 7.2.

For the case of a spherical droplet surrounded by a spherically symmetric vapor density field, one may write down an appropriate form of the equation on the basis of the definition of $D$, which is the mass flux by diffusion per unit time, per unit area, and per unit vapor density gradient. Thus

$$\frac{dm}{dt} = 4\pi R^2 \frac{d\rho_w}{dR}, \qquad (7.2)$$

where $m$ is the mass of the droplet and $R$ is the radial distance taken positive outward. Note that a positive vapor density gradient leads to a negative (inward) flux of water vapor, which yields a positive rate of mass increase. In a steady state the mass flux, and hence $dm/dt$, is independent of $R$.

Integrating (7.2) from the surface of the drop ($R = r$) to infinity gives

$$\frac{dm}{dt} = 4\pi r D(\rho_w - \rho_{0w}), \qquad (7.3)$$

where $\rho_w$ and $\rho_{0w}$ are, respectively, the vapor density in the environment at a large distance from the drop and the vapor density at the drop surface.[1]

Equation (7.3) is specialized for a spherical particle. A solution for nonspherical bodies, which might approximate the shapes of ice crystals, was presented by Houghton (1950), who used one of the several solutions of Laplace's equation. The steady state current flow from an isolated electrically charged body of electrical capacity $C$ in a medium of uniform conductivity $\sigma$ is given by

$$-\frac{dQ}{dt} = 4\pi C\sigma(V_\infty - V_0), \qquad (7.4)$$

where $Q$ is electric charge and $V_\infty$ and $V_0$ are the potentials at a great distance and at the surface of the body, respectively. The negative sign is due to a different choice of the sign of the radial direction. Otherwise (7.4) is a direct analog of (7.3), which is made even clearer by the knowledge that

1. Equation (7.3) and all subsequent equations for drop growth apply equally to drop evaporation with $\rho_{0w} > \rho_w$ and $dm/dt$ negative.

$C = r$ for a sphere. In the subsequent treatment of the growth of ice crystals by vapor deposition, $C$ will be used is place of $r$ in (7.3).

## 7.3 The Temperature of a Growing Particle

When a particle is growing by condensation or deposition, the latent heat is released to the particle and raises its temperature. This heat is conducted away to the environment. It is readily shown that heat transfer by radiation or convection is negligible. The molecular conduction of heat is also described by LaPlace's equation, and one may write by inspection of (7.3)

$$\frac{dq}{dt} = 4\pi CK(T_0 - T), \tag{7.5}$$

where $dq/dt$ is the heat flow away from the particle, $K$ is the molecular coefficient of heat conductivity, $C$ is the electrical capacity of the particle in centimeters, and $T_0$ and $T$ are, respectively, the temperatures at the surface of the particle and of the environment. Noting that $dq/dt$ is $L\,dm/dt$, where $L$ is latent heat of condensation, (7.5) and (7.3) may be combined to give

$$(T_0 - T) = (DL/K)(\rho_{\mathrm{w}} - \rho_{0\mathrm{w}}). \tag{7.6}$$

If the particle has a radius of curvature greater than about 1 $\mu$m, the Kelvin effect is negligible. The effect of dissolved salt nuclei of typical radii (0.01–0.1 $\mu$m) becomes negligible when the drop radii exceed about 5 $\mu$m. For these and larger drops $\rho_{0\mathrm{w}}$ is the vapor density over pure water. Equation (7.6) may then be solved by introducing the relation between $\rho_{0\mathrm{w}}$ and $T_0$ through the Clausius-Clapeyron equation. This has been done by Mason (1971), who found for spherical particles, after some simplifications, that

$$\frac{dm}{dt} = 4\pi r(s - 1)\left(\frac{L^2}{KR_{\mathrm{w}}T^2} + \frac{R_{\mathrm{w}}T}{De_{\mathrm{m}}}\right)^{-1}, \tag{7.7}$$

where $s$ is the saturation ratio of the environment, $e/e_{\mathrm{m}}$.

In most cases $T_0 - T$, which is nearly equal to the wet bulb depression, is small, typically less than a 1°C. For occasional computations from (7.6), a linear interpolation in a table of $\rho_{0\mathrm{w}}$ versus $T_0$ is a simpler approach than (7.7) and is also more accurate.

In some cases the Kelvin effect and/or that due to the solute is not negligible. Examples are the computation of Koehler curves shown in figure 6.7 and the growth of large salt particles of either natural origin or those introduced to modify warm clouds. Fletcher (1962) has developed an approximate expression, including the Kelvin effect, the effect of dissolved nuclei, and diffusional growth, that yields quite accurate results and has been used by many workers:

$$\frac{dm}{dt} = \frac{4\pi\rho_L D r}{1 + \rho_w D L^2/R_w K T^2}[(s-1) - a/r + b/r^3], \tag{7.8}$$

where

$$a = \frac{2\sigma}{\rho_L R_w T} \approx 3.3 \times 10^{-5}/T,$$

$$b = \frac{imM_0}{(4/3)\pi\rho_L M} \approx 4.3im/M,$$

$M$ and $M_0$ are the gram molecular weights of the solute and water, respectively, $i$ is the van't Hoff factor, $m$ is the mass of solute in the drop, and $\sigma$ is the surface tension and $\rho_L$ is the density of the solution. The principal approximations in the derivation of (7.8) are the use of the first few terms of a power series to represent the exponential in Kelvin's equation for dilute solutions and the smallness of the temperature difference between the drop surface and the environment.

The three terms in square brackets in (7.8) represent, in order, diffusional growth, the Kelvin effect, and the modification due to the dissolved substance. This permits one to determine the orders of magnitude of these terms for a given set of circumstances to provide a basis for the neglect of one or both of the last two terms.

## 7.4 The Effects of Knudsen Flow

The equations considered earlier for the growth of water drops or ice particles by molecular diffusion in the steady state are based on the equation developed by Maxwell (1890) in his study of the wet bulb thermometer. This classical approach assumes that molecular diffusion applies all the way to the drop surface, a valid assumption when the drop is very large compared with the mean free path of the water molecules. However, for small drops, cognizance must be taken of the region of free molecular or Knudsen flow within about a mean free path from the drop surface. In the spherically symmetric steady state case the diffusion field may be thought of as an assembly of conical regions with apices at the center of the drop. The vapor flux is uniform in each cone, independent of position along the axis. It is easily seen that, for a sufficiently small drop, maintenance of the flux prescribed by molecular diffusion will finally be impossible because there is an insufficient number of molecules to carry the flux. This is accentuated by the fact that the fraction of the molecules reaching the drop that stick, called the condensation coefficient, is small compared with unity. Since the flux along the conical axis must be constant to maintain a steady state, it must fall below the flux defined by molecular diffusion. In other words, the growth rate of small particles is always less than that predicted by the diffusion equation.

This problem has been recognized for some time, but the several treatments have led to different quantitative results. In part, at least, these differences have been due to somewhat different formulations, to a failure to recognize that the transfer of heat is also affected, and by the lack of definitive values of the condensation coefficient and its thermal counterpart.

One of the more complete treatments, and the one most often referenced, is that of Fukuta and Walter (1970). In the steady state, the mass flow due to diffusion must equal that due to Knudsen flow very close to the drop surface. They used the following expression for Knudsen flow:

$$\left(\frac{dm}{dt}\right)_K = \frac{\beta(e - e_m)}{(2\pi R_w T)^{1/2}}, \qquad (7.9)$$

where $\beta$ is the condensation coefficient, the fraction of the molecules striking the drop that condense or stick, and $R_w$ is the gas constant of water vapor. Equating (7.3) and (7.9) and making some minor simplifications, they obtain

$$\frac{dm}{dt} = 4\pi r f_\beta D(\rho_w - \rho_{0w}), \qquad (7.10)$$

where $f_\beta$ is a "normalization factor,"[2] given by

$$f_\beta = \frac{r}{r + l_\beta}, \qquad l_\beta = \frac{D}{\beta}\left(\frac{2\pi}{R_w T}\right)^{1/2}, \qquad (7.11)$$

where $l_\beta$ is a mean path.

A similar procedure for the transfer of heat leads to

$$\frac{dg}{dt} = 4\pi r f_\alpha K(T_0 = T), \qquad (7.12)$$

where $f_\alpha$ is the normalization factor for heat transfer:

---

2. Others have chosen to incorporate $f_\beta$ in a modified $D$.

$$f_\alpha = \frac{r}{r + l_\alpha}, \qquad l_\alpha = \frac{K(2\pi M_a RT)^{1/2}}{\alpha p(c_v + R/2)}, \tag{7.13}$$

where $R$ is the universal gas constant, $M_a$ is the molecular weight of air, $p$ is the total pressure and $c_v$ is the molar specific heat of air at constant volume. Equations (7.10) and (7.12) may be combined with the aid of the Clausius-Clapeyron relation and some minor approximations to give (7.14), the counterpart of (7.7) for Knudsen-limited condensation:

$$\frac{dm}{dt} = 4\pi r(s - 1)\left(\frac{L^2}{KR_w T^2 f_\alpha} + \frac{R_w T}{e_m D f_\beta}\right)^{-1}. \tag{7.14}$$

Equation (7.14) reduces to (7.7) when $f_\alpha = f_\beta = 1$.

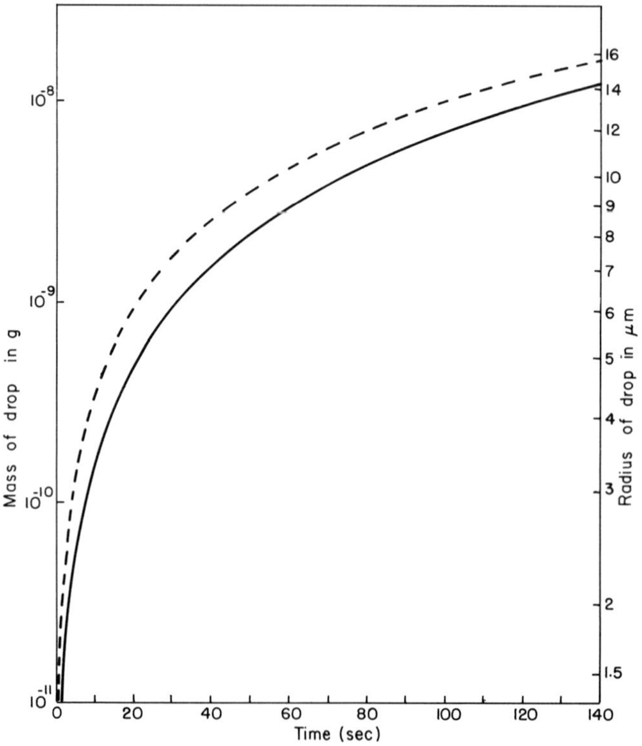


Figure 7.1 Diffusional growth of a pure water drop of initial radius 1.0 $\mu$m. The solid curve was computed from (7.14), while the dashed curve is based on (7.7). The supersaturation ratio is 1.01, $p = 1000$ mbar, $T = 283$°K, and $\beta = 0.04$. Based on Fukuta and Walter (1970).

Table 7.1
Time required for a drop to grow from selected initial to final radii by Knudsen-limited flow and Maxwellian flow[a]

| | | Elapsed time | | |
|---|---|---|---|---|
| Initial radius ($\mu$m) | Final radius ($\mu$m) | Knudsen-limited (sec) | Maxwellian (sec) | Ratio |
| 1 | 5 | 21.1 | 13.4 | 0.635 |
| 1 | 10 | 72.8 | 55.5 | 0.762 |
| 1 | 20 | 261 | 224 | 0.858 |
| 1 | 50 | 1,497 | 1,400 | 0.935 |
| 5 | 20 | 240 | 210 | 0.875 |
| 10 | 20 | 188 | 168 | 0.894 |
| 10 | 50 | 1,424 | 1,345 | 0.945 |
| 20 | 50 | 1,236 | 1,177 | 0.952 |
| 40 | 50 | 524 | 504 | 0.962 |

a. Conditions as in figure 7.1

The quantitative effects of Knudsen-limited condensation depend on the values of $\alpha$ and $\beta$, which are poorly known. It is generally believed that $\alpha$ is near unity while the most recent values of $\beta$ range from 0.02 to 0.04. Figure 7.1 shows the effect graphically for a case chosen by Fukuta and Walter (1970) in which $\alpha = 1$, $\beta = 0.0415$, $p = 1000$ mbar, $T = 283$°K, and $s = 1.01$. For comparison, a growth curve has been computed from (7.7) for the same conditions. Table 7.1, based on the computations for figure 7.1, shows the time required for a drop to grow from selected initial radius to final radius. It is evident that the ratios of the times required for Knusden-limited growth to the Maxwellian are lowest for the small drops and that the differences are still apparent for the largest drops listed.

The results of the numerical evaluations are dependent on the ambient conditions selected and the value of $\beta$. Fitzgerald (1978) has derived a more complete equation than (7.14), and he finds that his numerical results are within a few percent of those from (7.14) for droplet radii $\geq 1$ $\mu$m. The splicing of the diffusion and Knudsen flows at a small distance from the drop is an approximation that may introduce some error. Further experimental work on the growth rates of small drops and the determination of the condensation coefficients is needed.

There is no doubt that all of this applies as well to the growth of ice crystals by vapor deposition, but the results given earlier assume spherical geometry and cannot be applied quantitatively to ice crystals.

## 7.5 Preliminary Inferences

Equation (7.3) may be written

$$r\frac{dr}{dt} = \frac{D}{\rho_L}(\rho_w = \rho_{0w}). \tag{7.15}$$

This shows that growth will occur only when $\rho_w > \rho_{0w}$. Except in the early stages of growth on hygroscopic nuclei, the environment must be somewhat supersaturated.

Cloud drops do not grow indefinitely in size by condensation. It is sometimes argued that the limit to growth is set by the duration of the condensation process. If a "reasonable" supersaturation is selected, perhaps 0.1%, it may be shown by integration of (7.15) that it would take hours to grow a drizzle drop ($r \simeq 0.1$ mm) and days to grow a raindrop ($r \simeq 1$ mm) by condensation. Convincing as this may seem, the argument is faulty because the supersaturation cannot be prescribed but, rather, is set by the condensation process itself. Condensation results from an imposed lifting, or cooling, and this is the forcing function. The supersaturation attains whatever value is required to produce condensation at the prescribed rate. The supersaturation depends primarily on the number and size of the drops in a unit volume of air. The limit on the size of the drops is set by the total amount of water vapor available for condensation and the number of drops that share the condensate. For example, if a sample of air has an initial water vapor mixing ratio of 10 g $kg^{-1}$ and contains 100 drops $cm^{-3}$, the average drop radius after all of the vapor has condensed is about 30 $\mu$m. Any other reasonable combination of the mixing ratio and drop concentration will lead to a similar result. Therefore the basic limitation on the size attained by cloud drops by condensation is the relatively large concentration of the drops that must share the condensate. In contrast, the concentration of ice crystals may be five orders of magnitude smaller than that of liquid cloud drops. For this and other reasons, ice crystals attain larger masses than cloud drops, and their growth may be limited by the time they spend in an environment in which they can grow by vapor deposition (see section 7.20). Even in the case of ice crystals, the supersaturation is determined by the imposed rate of deposition.

To compare the growth rates of drops of different sizes in the same environment, (7.15) may be integrated with the right-hand side held constant to give

$$(r_2^2 - r_1^2) = \frac{2D}{\rho_L}(\rho_w - \rho_{0w})\Delta t, \tag{7.16}$$

where $\Delta t$ is the time required for a drop of radius $r_1$ to grow to $r_2$ in the

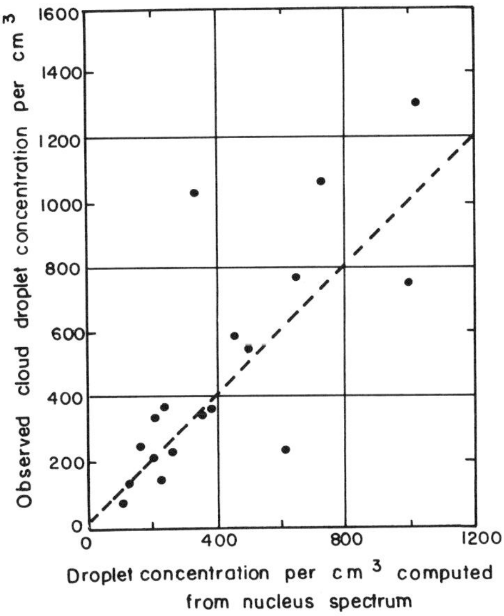


Figure 7.2 Comparison of mean droplet concentration observed in cloud with the concentration computed from the observed spectra of cloud nuclei for an updraft of 3 m sec$^{-1}$. The dashed line represents exact agreement between observed and computed values. After Twomey and Warner (1967).

specified environment. This parabolic relation shows that an initially broad spectrum of droplets will narrow as condensation proceeds if, as usual, the drop size is specified by the radius. For example, if two drops are of initial radii 1 and 10 units, when the smaller grows to 10 units the larger will have a radius of only about 14 units.

## 7.6 Cloud Drop Concentration

If the supersaturation spectrum[3] of the CCN is known, the concentration of the activated nuclei, and hence of the cloud drops, may be computed from (7.8), or its equivalent, for a given vertical air velocity and ambient conditions. In fact, an extension of this computation leads to a prediction of the cloud drop spectrum as a function of time. Twomey and Warner (1967) measured the supersaturation below the bases of a number of nonprecipitating cumuli, the vertical air velocity, and the cloud drop concentration just above cloud base. Figure 7.2 shows the comparison between the computed and measured cloud drop concentrations. The

3. The supersaturation spectrum is the concentration of the nuclei activated as a function of the supersaturation.

Table 7.2
Cloud drop concentrations in continental and maritime cumuli[a]

| Period | Maritime cumuli ($cm^{-3}$) | Continental cumuli ($cm^{-3}$) |
|---|---|---|
| First quartile | 19 | 150 |
| Median | 42 | 290 |
| Third quartile | 66 | 470 |
| Maximum observed | 400 | 2,810 |

a. After Squires (1958).

dashed line represents exact agreement, and the correlation coefficient is 0.9.

The number concentration of cloud drops is inversely related to an average drop size. As will appear in more detail later, growth by collision is more rapid when the drops are large, even though, by consequence, their concentration is smaller. The number concentration of the cloud drops increases with the concentration and size of the CCN and the vertical air velocity at the cloud base. It also increases with decreasing cloud base temperature. This is because the rate of growth of the activated nuclei decreases with temperature, resulting in higher supersaturations and activation of more CCN.

Squires (1958) made the important discovery that the concentration of cloud drops is nearly an order of magnitude larger in continental than in similar maritime clouds. Some of his results are shown in table 7.2. The total mass of cloud water in a unit volume of air (the liquid water content) was about the same in both kinds of clouds, which means that the drops in the maritime clouds are roughly twice as large as those in the continental clouds. These results have been confirmed and extended by others and have proved to be of critical importance in understanding rain formation in clouds warmer than 0°C. The basic cause of the lower drop concentrations in maritime clouds is the lower concentration of CCN in maritime air, as discussed in section 6.11.

## 7.7 Observed Cloud Drop Spectra

Many examples of observed cloud drop size spectra are to be found in the literature. Most of the methods of measurement have involved the collection of a droplet sample by impaction on a surface treated to preserve the droplets or to leave an image proportional to the drop size. The aerodynamics of the impaction process discriminates against the smaller drops, and impaction at airplane speeds often disrupts large drops. Corrections for small-drop discrimination are now commonly made, but this was not done for earlier observations. More recently other means of measurement

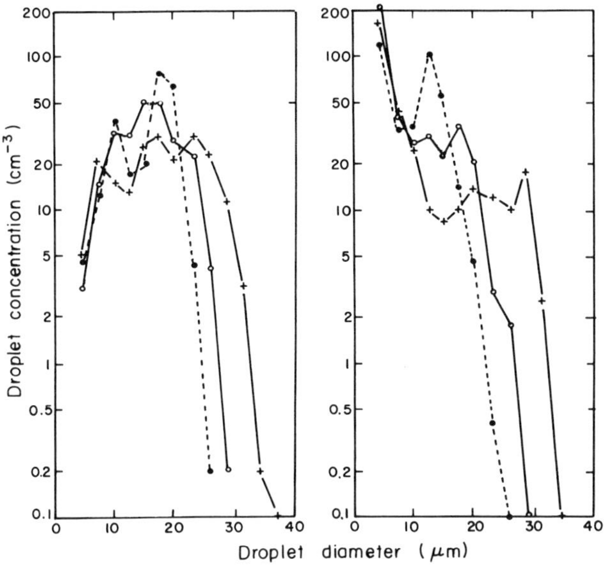


Figure 7.3 The variation with height of the droplet spectrum in two clouds, showing also the height above cloud base at which the samples were taken, the average total drop concentration at that height, and the number of slides exposed at each height: left, 10 December 1967; right, 20 March 1966. After Warner (1969).

| Date | 10 December 1967 | | | 20 March 1966 | | |
|---|---|---|---|---|---|---|
| Symbol | ⊙ | ○ | + | ⊙ | ○ | + |
| Height above base (m) | 600 | 1,100 | 1,800 | 200 | 650 | 1,100 |
| $\bar{n}$ (cm$^{-3}$) | 238 | 235 | 202 | 364 | 388 | 329 |
| Slides | 4 | 9 | 12 | 13 | 5 | 10 |

of drop size have been developed, including electrically charged probes, direct photography, shadow casting, and optical scattering. Each of these methods has its own limitations, but when carefully used and calibrated, they are reasonably satisfactory.

Samples of cloud drop spectra in nonraining convective clouds near the eastern Australia coast as observed by Warner (1969) are shown in figures 7.3 and 7.4. Corrections have been applied to rectify, at least partly, the observational bias against small drops. Figure 7.4 is an example of a markedly bimodal spectrum, which Warner found in about 20 of 300 cases.

## 7.8 Computations of Cloud Drop Spectra

Integrations of (7.8), or its equivalent, over time to yield cloud drop spectra were first performed by Howell (1949). A sample of his results is shown in figure 7.5, which is for a relatively low concentration of CCN and a moderate vertical air velocity. It is apparent that the initially broad CCN spectrum is rapidly converted to a narrow cloud drop spectrum, confirming the deduction made earlier on the basis of (7.16). Also note the absence of drops much smaller than the mode radius as condensation proceeds.

Mordy (1959) and Neiburger and Chien (1960) carried out similar but more elaborate calculations on electronic computers. In general, these studies simply confirmed and extended Howell's work.

Based on a more complete version of (7.8), Fitzgerald (1972, 1978) has derived cloud drop spectra that correspond reasonably well to observed spectra just above cloud base. He also had considerable success in explaining observed drop spectra in sea fogs, as illustrated by figure 7.6. These recent results seem to have reduced the discrepancies between observed and computed spectra, but some questions remain in both the small and large ends of the spectra.

## 7.9 Comments on Cloud Drop Spectra

It would seem that the only means for maintaining the observed relatively large concentration of drops of radius $\leq 5\ \mu m$ against the strong trend toward the rapid growth of such drops is evaporation followed by recondensation. It is well known that convective clouds entrain substantial amounts of unsaturated environmental air. It is usually assumed that this entrained air is nearly instantaneously mixed into the cloud, resulting in a decrease in the rate of condensation. It is more realistic to expect that the entrained air enters in parcels that are slowly eroded and mixed with the cloud. Airplane penetrations usually show rapid spatial changes in liquid water content, as shown in figure 7.7, and sometimes cloud-free holes. It is plausible that nearly complete evaporation[4] can occur in these incompletely mixed parcels; continuation of the lifting process may lead to recondensation. Evaporation and recondensation are the most probable explanation for bimodal drop spectra such as that in figure 7.4.

Baker and Latham (1979) have studied the effects of inhomogeneous mixing of entrained environmental air on the evolution of the droplet spectrum. A simple model, based on observations in the laboratory, was used in which the droplet evaporation time constant is small compared with that for turbulent mixing. The resultant cloud drop spectrum is much more realistic than that derived for rapid homogeneous mixing, containing both more small and more large drops. The enhanced growth rate of the

4. For example all drops of less than 10 $\mu m$ radius will evaporate in about 5 sec at a relative humidity of 90%.

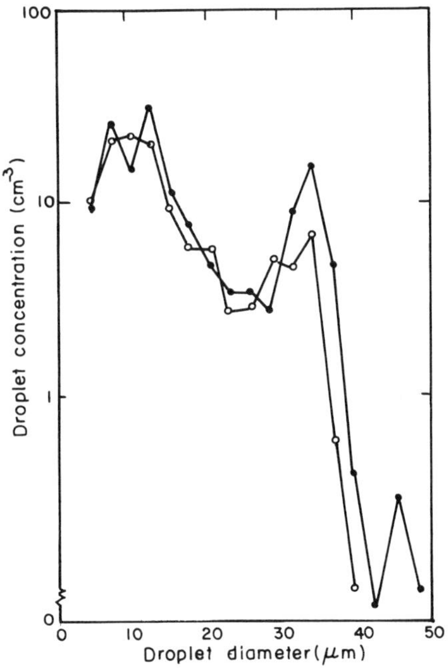


Figure 7.4 Adjacent samples taken 100 m apart near the top of a cloud 1,400 m deep. Both show a strongly bimodal distribution. After Warner (1969).

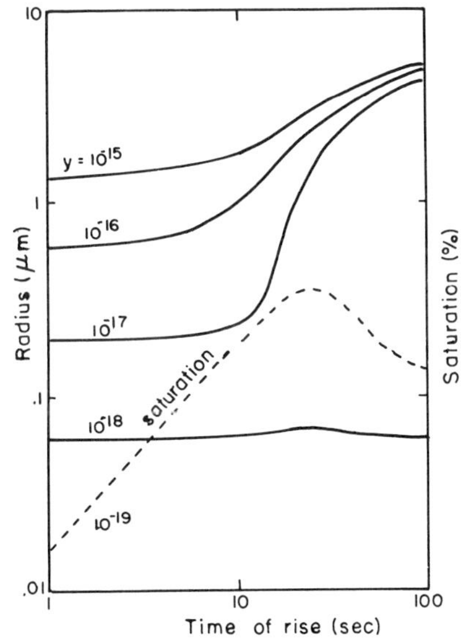


Figure 7.5 Growth of nuclei of the indicated molar masses to cloud drops as a function of time. The concentration of nuclei is 500 cm$^{-3}$, and the vertical air speed is 4 m sec$^{-1}$. After Howell (1949).

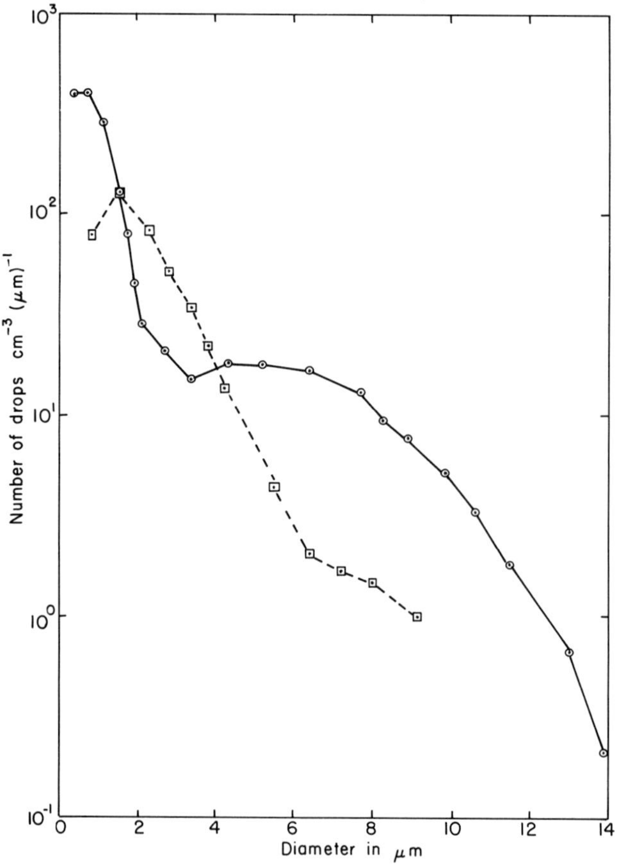


Figure 7.6 Fog droplet size spectra in a marine fog. The dashed curve is the observed spectrum, and the solid curve is the computed spectrum. After Fitzgerald (1978).

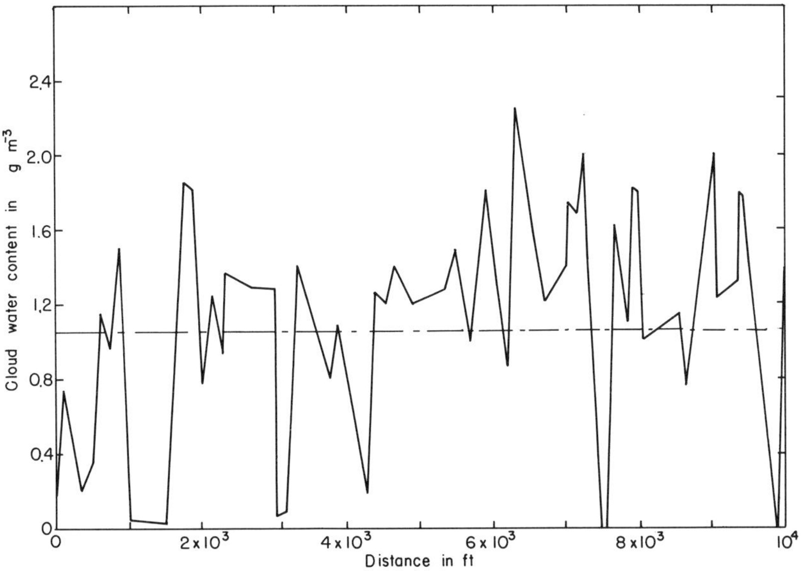


Figure 7.7 Cloud liquid water content recorded during a penetration of a cumulus cloud at a height of 4,300 m over the central United States. The cloud top was at 5,500 m, and the base was at 2,740 m. The mean value of the liquid water content was about 1.1 gm $m^{-3}$. After Draginis (1958).

larger drops is due to an increased supersaturation. This results from the complete evaporation of the drops in an unmixed parcel of environmental air and their replacement by newly activated droplets, which demand a higher supersaturation. It should be understood that this is a preliminary study that is not supported by observational information on the mixing of entrained air into the cloud. It further supports the idea that complete evaporation and recondensation plays an important part in the evolution of the cloud drop spectrum.

The largest drops in the drop spectrum are of particular importance because they are the ones that start the accretion process of precipitation. The typical concentration of large drops required to initiate precipitation is one per liter of air, or only $10^{-5}$–$10^{-6}$ of the total concentration of cloud drops. These large drops presumably form on giant nuclei, which are again the one-in-a-million nuclei. Unless special measuring devices are used that count only these rare giant nuclei or large drops, they will be lost among the vastly greater numbers of smaller particles. This may partly explain some of the difficulties encountered in explaining rain from clouds that seem not to have drops large enough to initiate precipitation.

## 7.10 Ventilation

When a cloud droplet becomes large enough to have a free-fall speed of a few centimeters per second or more, the spherically symmetric vapor field around the drop is distorted. The vapor density gradient is increased ahead of the drop, and the rates of heat and mass transfer are also increased.

The increased mass and heat transfers resulting from ventilation are introduced in (7.3) and (7.5) by multiplying the right-hand sides by the ventilation coefficient $c_v$, which is dimensionless and has a value of unity for a stationary drop. In principle, the value of $c_v$ for mass transfer may differ from that for heat transfer, but it will be assumed here that they are equal. Theory predicts that the ventilation coefficient for mass transfer depends on the Reynolds number and the Schmidt number; for heat transfer the Prandtl number replaces the Schmidt number. Since both the Schmidt and the Prandtl numbers are constant in the atmosphere, the Reynolds number is the important factor.

An apparently successful theory of the effect of ventilation has been developed by Woo and Hamielic (1971), but for the present, reliance will be placed on the experimental results of Beard (1976). His experiments were performed in a vertical wind tunnel in which the drops could be maintained in a fixed position as they evaporated in a known environment. He determined the velocity coefficient for mass transfer, which will be assumed to be also valid for heat transfer. He summarized his results in two empirical expressions (Re = Reynolds number):

$$\text{for Re} > 2.5, \qquad c_v = 0.78 + 0.275\text{Re}^{1/2}; \tag{7.17}$$

$$\text{for Re} < 2.5, \qquad c_v = 1.00 + 0.086\text{Re}. \tag{7.18}$$

## 7.11 Fall Speeds of Cloud Drops

It has been customary to compute the fall speeds of cloud drops from Stoke's law, which assumes viscous flow and no slip:

$$v = \frac{2g}{9\eta}(\rho_L - \rho_A)r^2, \tag{7.19}$$

where $v$ is the terminal fall speed in cm sec$^{-1}$, $\eta$ is the dynamic viscosity of air, $g$ is the gravitational constant, and $\rho_L$ and $\rho_A$ are the densities of the liquid and of the air, respectively. Beard (1976) showed that (7.19) underestimates the fall speeds of drops smaller than about 4 $\mu$m radius. Equation (7.19) also overestimates the fall speeds of drops larger than about 30 $\mu$m radius. Thus its useful range just about covers cloud drops formed by condensation. The fall speeds of large water drops will be considered in section 8.28. Beard gives formulas from which the terminal speeds of drops ranging from 0.25 to 3.5 $\mu$m radius for arbitrary atmospheric pressure and temperature can be obtained.

From (7.18) and (7.19) it is readily found that the ventilation coefficient for drops of 15 $\mu$m radius is about 1.006 and reaches 1.10 at a radius of about 40 $\mu$m. Thus it is frequently adequate to neglect the ventilation effect on the growth of cloud drops of typical size. We shall return to the effects of ventilation on the growth of ice particles in section 7.15.

## 7.12 Forms of Snow Crystals

The most striking feature of snow crystals is their intricate geometry. It has been said that no two ice crystals are exactly the same, but even if this is true, they can be classified into a small number of groups, which may be further subdivided. Examples of the principal habits or shapes of ice crystals are shown in figure 7.8. The three basic crystal habits are ice needles, planar crystals, and hexagonal columns. Ice needles are subdivided into individual needles or sheaths, combinations of these forms, and long, slender columns. The planar crystals embrace the greatest number of subspecies, most of which show hexagonal symmetry, although a few are trigonal or duodenal. They range in detail form from simple hexagonal plates to crystals with broad branches, stellar forms, and lacy dendritic crystals. Columns may be solid or hollow, shaped like bullets, cups, or pyramids. Various combinations of planar and columnar forms are often found.

The first comprehensive classification of snow crystals was due to Nakaya (1954). His book contains over a thousand photographs of natural and artificial snow crystals. An amended classification has been given by Magono and Lee (1966), and this is the most commonly used. This paper contains some 60 microphotographs illustrating the classification scheme. Mention is made here of a large collection of snow crystal photographs of natural snow crystals by Bentley and Humphreys (1931). The several thousand photographs seem to have been selected in large part because of their symmetric shapes and their beauty. It should be pointed out that many natural snow crystals are nonsymmetric or malformed, and many apparently broken fragments are found.

## 7.13 Dependence of Crystal Habit on Ambient Conditions

In a classical series of experiments, Nakaya (1954) demonstrated that the crystal habit was dependent on the ambient temperature and supersaturation. His results have since been confirmed and refined by others. Table 7.3 is a somewhat subjective averaging of the results of seven investigators. The temperature boundaries between habits are only 1–2°C

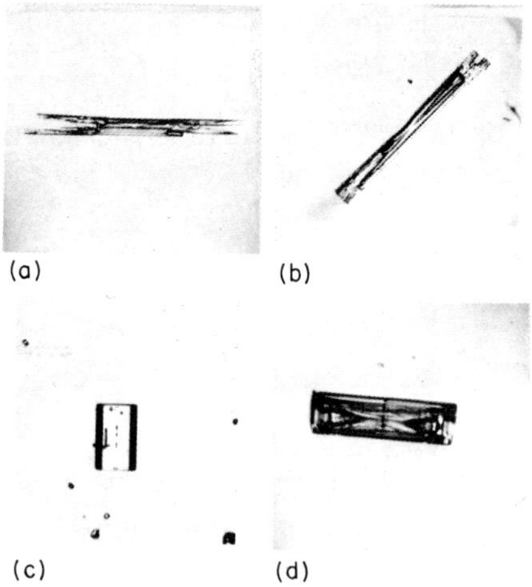


Figure 7.8 Photomicrographs of ten snow crystals of different habits, collected and classified by Magono and Lee (1966): (a) elementary needle, ×20.6; (b) elementary sheath, ×16.4; (c) solid column, ×34; (d) hollow column, ×22; (e) solid thick plate, ×44.5; (f) hexagonal plate, ×14.5; (g) crystal with sectorlike branches, ×10; (h) crystal with broad branches, ×18.3; (i) stellar crystal, ×18.3; (j) ordinary dendritic crystal, ×9.7. My sincere thanks to Professor C. Magono for supplying me with these photomicrographs.

Table 7.3
Ice crystal habit as a function of ambient conditions

| At or above water saturation (°C) | Between ice and water saturation (°C) |
|---|---|
| 0 to −4: plates | 0 to −4: plates |
| −4 to −6: needles | −4 to −10: columns |
| −6 to −10: columns | −10 to −20: thick plates |
| −9 to −13: plates, sector plates | −16 to −40: columns |
| −13 to −17: stellars, dendrites | |
| −14 to −21: thick and sector plates | |
| −21 to −40: columns, plates, bullets | |

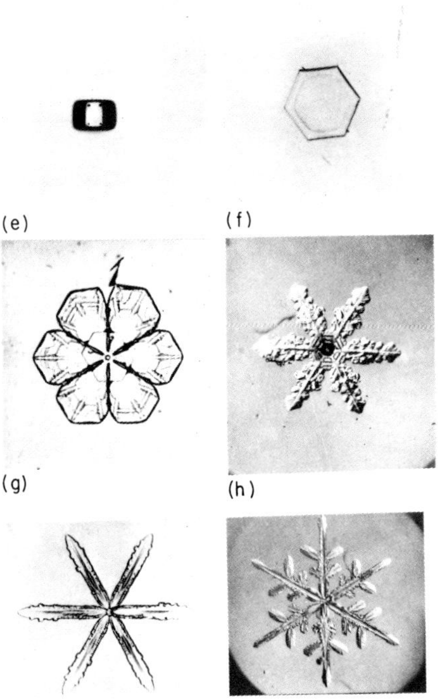

(e) (f) (g) (h) (i) (j)

wide, but the boundary temperatures are still somewhat uncertain. The indicated overlapping is due partly to the simultaneous occurrence of two habits and partly to the uncertainty of the habit boundaries. All of the more delicate crystal forms occur only at or above water saturation. Note the tendency, both at or above water saturation and below water saturation, for the change of habit with decreasing temperature to go from plates to columns. Hallett and Mason (1958) showed the temperature dependence of the crystal habit in a striking experiment in a diffusive cloud chamber in which the temperature varied with height. Crystals were grown on a vertical fiber and assumed the habits corresponding to the ambient temperatures. The fiber was then lowered to bring all of the crystals to a lower temperature. The new growth started at the outer portions of the crystals with habits appropriate to the new temperature. For example, ice needles formed at $-5$°C and transferred to $-14$°C grew stellar crystals on the tips of the needles.

As yet there is no complete explanation for the marked dependence of crystal habit on temperature. It is almost certain that this is due to the properties of the crystal surface on a molecular scale. Growth of a crystal surface results from the incorporation of vapor molecules at a step that, in consequence, moves across the crystal surface. The molecules migrate to

Table 7.4
The electrical capacity of bodies of simple shapes

| Shape | Capacity |
| --- | --- |
| Sphere | $r$ |
| Thin circular disk | $2r/\pi$ |
| Prolate spheroid[a] | $2ae/\ln[(1+e)/(1-e)]$ |
| Oblate spheroid | $ae/\sin^{-1} e$ |

a. For a long, slender prolate spheroid $C = a/\ln(2a/b)$, $r$ is radius, $a$ is semimajor axis, $b$ is semiminor axis, and $e$ is eccentricity.

the step from an adsorbed layer or layers by surface diffusion. All of these quantities are temperature sensitive, but they fail to explain the nearly discontinuous change of the growth rates of the basal and prism faces found experimentally. Lamb and Scott (1974) show that the observed behavior can be explained by the rapid increase in adsorption with temperature and the assumption of spiral growth steps, but their analysis involves a number of assumptions and speculations.

## 7.14 Growth of Ice Crystals from the Vapor

Diffusive growth of ice crystals follows the same law as the growth of water drops by condensation; the difference lies in the complex nonspherical geometry of ice crystals. The electrical analog, as given by (7.4), provides a means for dealing with simple nonspherical shapes. The electrical capacities of some pertinent shapes are given in table 7.4.

Common ice crystal shapes include hexagonal plates, columns, stellars, and dendrites, as illustrated in figure 7.8. The substitution of simple shapes for the much more complex crystal shapes was first suggested by Houghton (1950). He argued from the electrical analog that it was known experimentally that the distortion of the electric field due to fine structure on a charged body was evident only within a distance of a few radii from the body. Applying this to the growth of ice crystals, the total mass or heat flux to or from a crystal is almost the same as that to a simple shape of the same gross dimensions. This conclusion was largely confirmed by McDonald (1963), who measured the capacities of a number of brass models having shapes typical of natural snow crystals. Table 7.5 gives suggested values of the capacity for various ice crystal habits based in part on McDonald's results. It is evident that some approximation is introduced in substituting simple shapes for the real shapes. Further, there are other crystal habits for which there are no obvious simple substitute shapes.

In order to compute the growth rate of a crystal it is necessary to know the relation between the capacity and the mass of the natural crystal. For

Table 7.5
Suggested expressions for the capacities of some crystal habits

| Crystal habit | Substitute shape | Capacity |
|---|---|---|
| Thin hexagonal plate | Circular disk | $2r/\pi$ |
| Thick hexagonal plate | Oblate spheriod | $ae/\sin^{-1}e$ |
| Broad arm stellar | Circular disk | $1.77r/\pi$[a] |
| Narrow arm stellar | Circular disk | $1.46r/\pi$[a] |
| Plane dendrite | Circular disk | $1.66r/\pi$[a] |
| Needle | Slender prolate spheroid | $a/\ln(2a/b)$ |

a. Numerical factor based on McDonald (1963).

Table 7.6
Empirical relations between principal dimensions of ice crystals[a]

| Crystal habit | Empirical relation[b] |
|---|---|
| Thin hexagonal plate | $c = 2.02a^{0.449}$ |
| Thick hexagonal plate | $c = 0.402a^{1.018}$ |
| Stellar | $c = 2.028a^{0.431}$ |
| Plane dendrite | $c = 2.801a^{0.377}$ |
| Needle | $a = 1.099c^{0.611}$ |
| Column | $a = -8.48 + 1.002c - 0.00234c^2$ |

a. After Aver and Veal (1970).
b. $a$ and $c$ are in micrometers.

most snow crystals we must rely on measurements of the linear dimensions and masses of natural crystals. Because of the variations of shape of crystals of the same habit, only average relations can be obtained.

Nakaya and Terada (1935) first measured the size and mass of a number of snow crystals. More recently Ono (1970) and Auer and Veal (1970) measured the dimensions of hundreds of natural snow crystals. Their measurements were made on plastic replicas of the crystals, and it was not possible for them to measure the masses of the crystals.

Auer and Veal (1970) derived the empirical relations given in table 7.6 from their data. In table 7.6 $a$ is the crystal diameter in the basal plane and $c$ is its length in the direction normal to the basal plane. Thus $c$ is the thickness of planar crystals or the length of columnar crystals and $a$ is their diameter.

In order to obtain the mass of a crystal from its dimensions it is necessary to know its bulk density, which is defined as the crystal mass divided by the volume of the hexagonal column or plate that just encloses the crystal. Ono (1970) made painstaking microscopic examinations of many replicas of

Table 7.7
Bulk densities of columnar ice crystals[a]

| Long ($c$) axis ($\mu$m) | Solid column ($g\,cm^{-3}$) | Hollow column ($g\,cm^{-3}$) | Sheath ($g\,cm^{-3}$) | Needle ($g\,cm^{-3}$) |
|---|---|---|---|---|
| 50 | 0.90 | 0.87 | 0.89 | 0.86 |
| 100 | 0.89 | 0.83 | 0.87 | 0.81 |
| 200 | 0.88 | 0.77 | 0.83 | 0.75 |
| 400 | 0.87 | 0.70 | 0.80 | 0.66 |
| 600 | 0.86 | 0.65 | 0.77 | 0.61 |
| 800 | 0.85 | 0.62 | 0.75 | 0.56 |
| 1,000 | 0.85 | 0.59 | 0.73 | 0.52 |

a. After Ono (1970).

columnar crystals, from which he determined the bulk densities, some of which are given in table 7.7. The bulk densities are smaller than the density of ice because of cavities and irregularities in the external crystal shapes. A few other determinations have given bulk densities of hollow columns of 0.55–0.68, sheaths 0.4–0.6, and needles of 0.36, suggesting that Ono's values are somewhat too large. The bulk density of thin hexagonal plates is close to that of ice, but thick plates often show internal cavities and nonplanar surfaces, which will lead to a smaller bulk density. The bulk density of stellar crystals has been estimated as 0.25 for narrow arm types and 0.6 for broad arm crystals.

Although there have been only a relatively few direct measurements of crystal mass as a function of habit and principal dimensions, they obviate reliance on the poorly known bulk densities. Data on the relation between crystal mass and the principal dimensions are given in table 7.8 in the form of empirical expressions. The differences between the equations for a given habit are believed to be real and to represent differences in the crystal populations sampled.

## 7.15 Ventilation Coefficients for Ice Crystals

The ventilation coefficients of ice crystals are poorly known. The only direct observations appear to be those of Thorpe and Mason (1966), who used thin hexagonal plates cut from bulk ice. They also gave approximate values for a few large natural dendrites. For Reynolds numbers from 10 to 200, they express their results by

$$c_v = 0.65 + 0.39\,\mathrm{Re}^{1/2}. \qquad (7.20)$$

Table 7.8
Empirical equations relating ice crystal mass to the principal dimensions[a]

| Crystal habit | Empirical equation[b] | Reference |
|---|---|---|
| Needle | $m = 2.9 \times 10^{-5}c$ | Nakaya and Terada (1935) |
| | $m = 2 \times 10^{-4}c^2$ | Bashkirova and Pershina (1964) |
| Thin hexagonal plate | $m = 1.9 \times 10^{-2}a^3$ | Bashkirova and Pershina (1964) |
| Stellar crystal | $m = 9.4 \times 10^{-4}a^2$ | Higuchi (1956) |
| Planar dendrite | $m = 3.8 \times 10^{-4}a^2$ | Nakaya and Terada (1935) |
| | $m = 5.9 \times 10^{-4}a^2$ | Higuchi (1956) |
| | $m = 2.5 \times 10^{-3}a^{5/2}$ | Bashkirova and Pershina (1964) |
| Spatial dendrite | $m = 1.0 \times 10^{-3}a^2$ | Nakaya and Terada (1935) |
| Column (and bullet) | $m = 0.325\,a^2c$ | Higuchi (1956) |
| | $m = 4.6 \times 10^{-2}c^3$ | Bashkirova and Pershina (1964) |

a. The equations have been fitted to the curves drawn by Bashkirova and Pershina (1964).
b. $a$ and $c$ are, respectively, the crystal diameter in the basal plane and the crystal length normal to the basal plane.

Equation (7.20) does not satisfy the requirement that $c_v \to 1.0$ as $\mathrm{Re} \to 0$ and is not useful in the important Reynolds number range of 1–10.

Skelland and Cornish (1963) and Pasternak and Gauvin (1960) have measured the effects of ventilation on the mass and heat transfers of bodies of simple shapes in moving airstreams at Reynolds numbers ranging from about one hundred to several thousand. The particle shapes included spheres, spheriods, hemispheres, circular disks, cubes, and rectangular prisms; some shapes were tested at several orientations to the airstream. Their most important finding was that the heat and mass transfers for all of the shapes could be represented by a single expression if the characteristic length in the Reynolds number was taken as the surface area of the particle divided by its perimeter in a plane normal to the air flow. As shown by Koenig (1971), the ventilation coefficient can be written as

$$c_v = \left(1 + \frac{\kappa A}{4\pi C l_*}\mathrm{Re}_*^n\right), \tag{7.21}$$

where $l_*$ is the characteristic length defined earlier, $\mathrm{Re}_*$ is the corresponding Reynolds number, $A$ is the surface area of the particle, and $\kappa$ and $n$ are constants. The constant $n$ is approximately one-half for moderate and large Reynolds numbers, but approaches one for very small Reynolds numbers. the numerical values of $\kappa$ obtained at high Reynolds numbers are of doubtful value for the low fall speeds of ice crystals. It is readily found that for spheres (7.21) reduces to

$$c_v = 1 + (\kappa/2)\,\mathrm{Re}^n. \tag{7.22}$$

Numerical values of $\kappa$ may be found by equating (7.21) to (7.17) or (7.18) as appropriate. This numerical value of $\kappa$ may then be used in (7.21) with $A$, $C$, and $l_*$ of the selected simple shape. This procedure amounts to using (7.21) as a means of adapting (7.17) and (7.18) to nonspherical shapes. The validity of this procedure was tested by comparing the results with those measured by Thorpe and Mason (1966), and they were found to agree within 5%. This procedure is expected to be fairly satisfactory for plates and columns, but not for the more delicate habits, such as dendrites and stellars. For the latter we have only the sparse observations of Thorpe and Mason (1966) already discussed.

## 7.16 Fall Speeds of Ice Crystals

When the Reynolds number is less than one, ice crystals may fall in any orientation; at larger Reynolds numbers ice crystals fall with their long axes horizontal. Ice crystals tend to oscillate or tumble at Reynolds numbers larger than 100. The most satisfactory way of getting crystal fall speeds is by direct observation. Data of this type have been presented by Nakaya and Terada (1935) and Kajikawa (1972). It is also possible to compute the fall speeds of simple shapes, such as circular disks, of known size and mass from information on the drag coefficients. This procedure is of greatest value in converting observed fall speeds to other ambient conditions. The fall speeds are usually measured near the ground, whereas the crystals are formed and grow aloft, where the fall speed of a given crystal is greater.

It will suffice here to show the fall speeds of a number of crystal habits in figure 7.9. Except for the needles and columns, these data are taken from Kajikawa (1972). The data points found on the original figure are not reproduced here, but their scatter is considerable, as would be expected from the variations in shape of crystals of a given habit.

## 7.17 Growth Rates of Ice Crystals

Following Houghton (1972), the equation for the growth of ice crystals by vapor deposition may be written as the product of two factors:

$$\frac{dm}{dt} = [4\pi C c_v]\,[D(\rho_w - \rho_i)]. \tag{7.23}$$

The second term in square brackets depends almost entirely on the ambient conditions. The first square-bracketed term depends primarily on the size

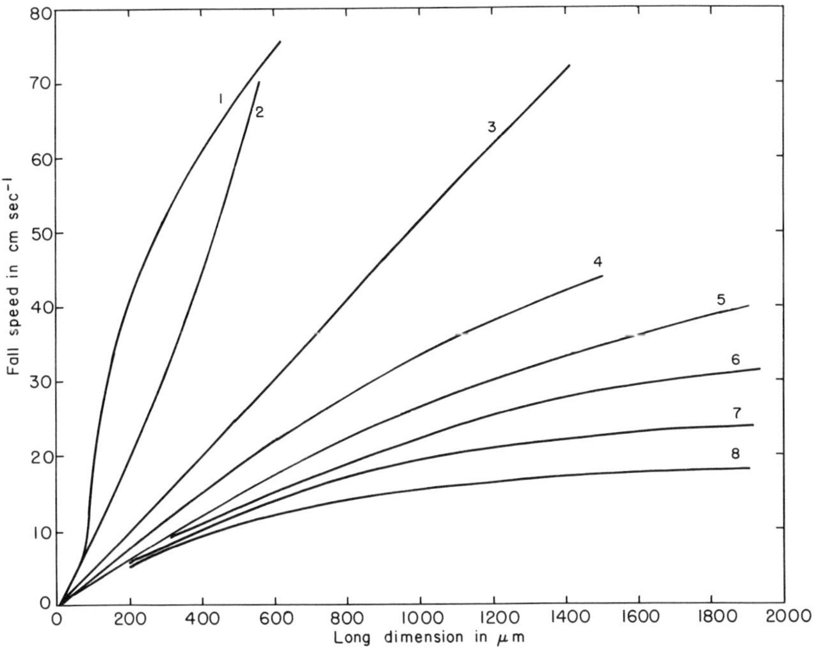


Figure 7.9 Fall speeds of snow crystals of different habits as a function of the long dimension of the crystals: 1, columns; 2, thick hexagonal plates; 3, thin hexagonal plates; 4, sector plates; 5, broad arm crystals; 6, needles; 7, dendrites; 8, stellar crystals. After Kajikawa (1972).

and habit of the crystal. There is a very weak dependence on the ambient conditions through the crystal fall speed and hence the ventilation coefficient.

The environmental factor in (7.23) is plotted in figure 7.10 as a function of pressure and temperature in a water-saturated atmosphere. This factor varies very nearly linearly with the supersaturation over ice, so that growth rates may be readily found for ambient conditions between water and ice saturation.

The factor $[4\pi C c_v]$ for a few ice crystal habits are plotted in figure 7.11. These are taken from Houghton (1972) and are subject to some modifications on the basis of more recent information, but the general course of the curves is believed to be correct. The curve for columns lies only slightly below that for hexagonal plates and was not plotted to avoid confusion.

The growth rate of ice crystals is readily found from figures 7.10 and 7.11 after the environmental conditions and the habits and masses of the crystals are selected. In figure 7.12, $dm/dt$ is given for five crystal habits. For each habit an environmental temperature was selected at which the habit forms (see table 7.3), and the corresponding pressure was taken from

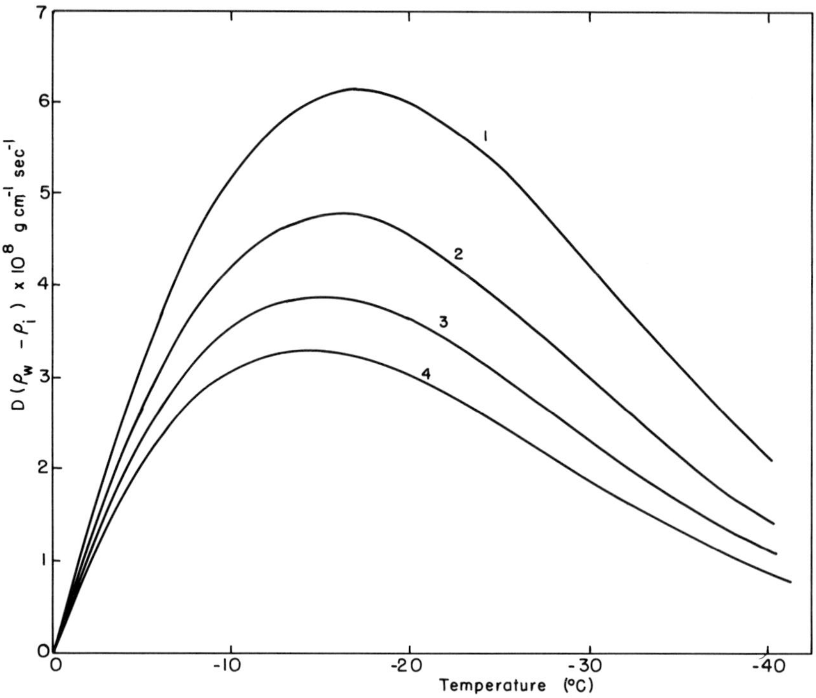


Figure 7.10 The environmental factor, $D(\rho_w - \rho_i)$, in the ice crystal growth equation (7.23) as a function of temperature and pressure at water saturation: 1, 400 mbar; 2, 600 mbar; 3, 800 mbar; 4, 100 mbar.

the Supplemental Standard Atmosphere for January at latitude 45°N. The molecular diffusion coefficient varies inversely with the pressure and therefore the environmental factor in (7.23) does likewise. In figure 7.13 the curves of figure 7.12 have been numerically integrated with an arbitrary lower limit of $10^{-8}$ g. After 400 sec of growth the mass of the delicate dendrite is more than two orders of magnitude larger than that of the hexagonal plate at $-3$°C. This is due in large part to a near maximum of both factors in (7.23) for dendritic crystals. Figure 7.13 is only an example, although curve 1 comes close to being an upper limit.

## 7.18 Comparison with Observed Ice Crystal Growth

A number of investigators have reported that observed growth rates agreed rather well with those computed from the equivalent of (7.23) and certainly within a factor of two. However, Isono et al. (1956) and Fukuta (1969) found observed crystal masses only one-third to one-half of those computed, discrepancies that, they felt, were much larger than experimental

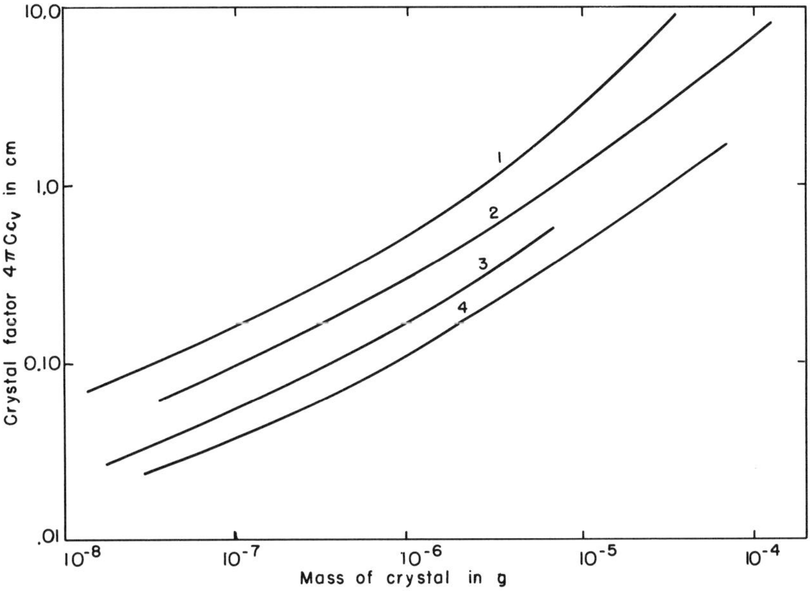


Figure 7.11 The factor $[4\pi Cc_v]$ in the ice crystal growth equation (7.23) as a function of crystal mass: 1, delicate planar dendrites; 2, dense planar dendrites; 3, needles; 4, hexagonal plates.

errors. In both cases, Isono et al. in a natural fog and Fukuta in a laboratory chamber, time was counted from nucleation and the crystals were quite small when they were collected for measurement. These results immediately suggest that the discrepancies are due to Knusden flow, as discussed earlier in connection with the growth of small water drops. The most reliable values of the condensation or deposition coefficient of ice appear to be those of Choularton and Latham (1977), who found a nearly linear dependence on temperature ranging from $\beta = 10^{-3}$ at $-37$°C to $3.3 \times 10^{-3}$ at $-19$°C. The nonspherical shape of ice crystals precludes a rigorous computation, but it is apparent from figure 7.1 for water drops with $\beta = 0.04$ that a factor of two to three in growth time can be explained if $\beta$ for ice is an order of magnitude smaller than that for water drops.

## 7.19 Comments on Growth of Ice Crystals

Very small ice crystals have simple solid habits, such as plates and prisms; elaborate habits such as dendrites are not found with diameters smaller than about 200 $\mu$m. Computations have shown that the growth rates of small crystals of simple form are much the same as the growth rates of solid ice spheres of the same mass. It is suggested that integrations of the growth rate from nucleation to mass of $10^{-8}$–$10^{-7}$ g be made for ice spheres for

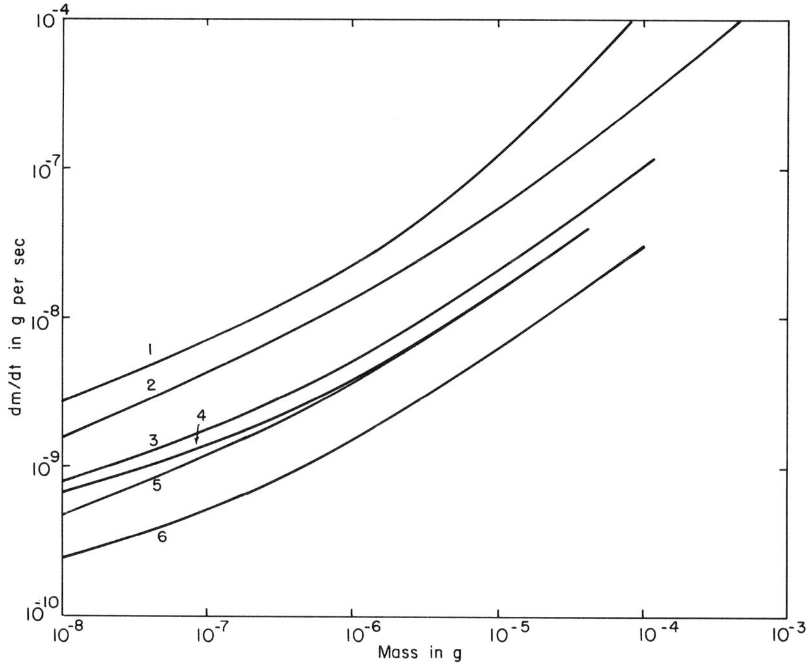


Figure 7.12 Mass growth rates of ice crystals of several habits versus crystal masses in various environments, but all at water saturation:1, delicate planar dendrites at −15°C and 645 mbar; 2, denser planar dendrites at −15°C and 645 mbar; 3 hexagonal plates at −20°C and 575 mbar; 4, hexagonal columns at −10°C and 735 mbar; 5, ice needles at −5°C and 900 mbar; 6, hexagonal plates at −3°C and 950 mbar.

which the factor $4\pi Cc_v$ reduces to $4\pi r$, thus permitting a simple analytic solution of the integral. Further growth may be computed by the methods discussed earlier. In view of the Knusden-limited flow for small crystals, the time of growth from nucleation to a long dimension of around 100 $\mu$m should be multiplied by a factor of two to three.

Certain crystal habits form only at or near water saturation, including needles, stellars, and dendrites. The simpler and more solid habits, such as simple plates and columns, may grow below water saturation (but above ice saturation). Water saturation generally occurs only in the presence of supercooled water drops. As with dropwise condensation, the forcing function is the rate of lift of the air that, with the ambient conditions, determines the condensation or depositional, rate. If the applied condensation rate exceeds the depositional rate of all of the ice crystals in a unit volume of air, supercooled cloud drops will coexist with the ice crystals, and the latter will grow at water saturation. Conversely, if the condensation

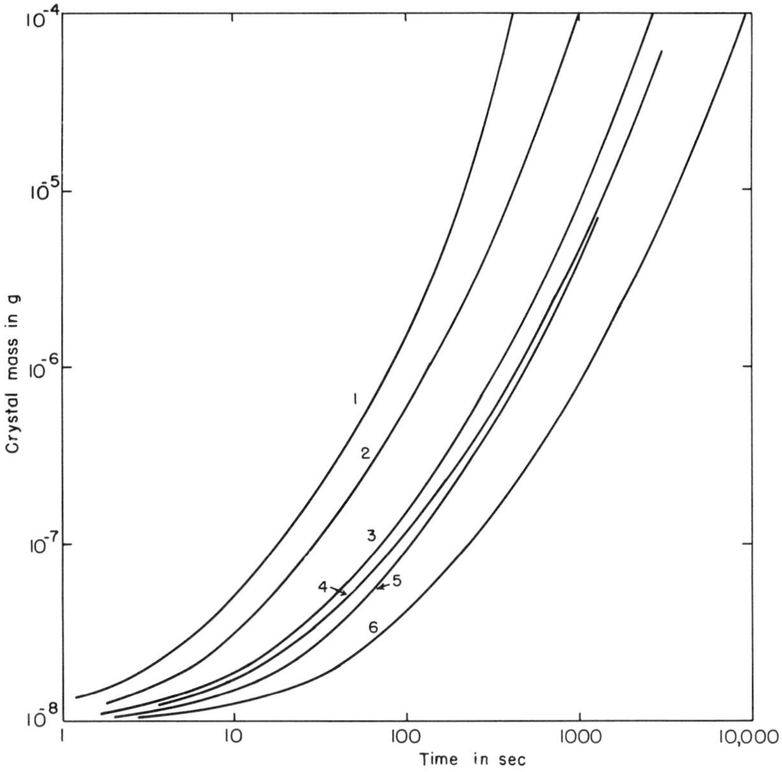


Figure 7.13 Time required for ice crystals of several habits and environmental conditions to grow from an initial mass of $10^{-8}$ to an arbitrary larger mass (water saturation is assumed): 1, delicate planar dendrites at −15°C and 645 mbar; 2, denser planar dendrites at −15°C and 645 mbar; 3, hexagonal plates at −20°C and 575 mbar; 4, hexagonal columns at −10°C and 735 mbar; 5, ice needles at −5°C and 900 mbar; 6, hexagonal plates at −3°C and 950 mbar.

rate is smaller than the sum of the crystal growth rates, the vapor pressure will be less that water saturation, and any water drops will evaporate. The vapor pressure will adjust itself so that the condensation rate and the sum of the growth rates of the ice crystals are equal.

Only the growth of ice crystals by vapor deposition has been considered so far. Ice particles may also be introduced by the freezing of supercooled liquid drops. If ice crystals and supercooled cloud drops coexist, collisions between the crystals and the drops leads to rimed ice crystals (see section 7.24), in which both the ice crystal and the frozen cloud drops can be recognized. Further collisions of supercooled drops with ice crystals or with frozen drops forms graupel (that is, granular snow pellets) in which the original ice particle can only be found, if at all, by dissection.

## 7.20 Observed Sizes and Masses of Snow Crystals

The observed sizes and masses of snow crystals are realistic measures of depositional growth as it operates in the atmosphere. The available information is not as extensive as one would like, but it is sufficient to establish masses within an order of magnitude and linear dimensions within a factor of two. Table 7.9 contains the average and range of the long axes of several crystal habits, the mass of a crystal of average size, and the radius of a water drop of the same mass. The data on crystal dimensions were taken from Nakaya and Sekido (1930) and Hobbs et al. (1977). The former are from Japan, which has a modified continental climate, and the latter from the Washington State, which has a maritime climate. The two sets of data were averaged in a somewhat subjective fashion. The crystal masses were computed from the list of empirical relations in table 7.8. The relatively large mass of a dendrite of average size is due to a near maximum of both factors in (7.23). The radii of melted crystals correspond to drizzle drops, except for the dendrites, which may lead to small raindrops. Additional growth by accretion is responsible for raindrops of typical size.

## 7.21 Accretional Growth of Liquid Drops

Larger drops fall faster than smaller drops, resulting in the larger drops overtaking smaller drops and colliding with them. This process is growth by accretion, sometimes called coalescence, coagulation, or collisional growth. Here, accretion will be used to designate the overall process by which a small drop is incorporated into the larger drop. Physically, this process is comprised of a collision, followed by coalescence.

The collision process is illustrated in figure 7.14 for viscous (Stokes) flow around a sphere of radius $R$. The small drops of radius $r$ are assumed to have no effect on the flow; that is, they are considered to behave as mass points. The momentum of the small drops tends to cause them to follow a linear trajectory. As soon as they leave their initial streamlines they are acted upon by a viscous force that tends to move them toward the initial streamline. The result is a droplet trajectory that is intermediate between a straight line and a streamline. Since the inertia is proportional to the mass of the droplet, while the viscous force depends on its radius, the larger the mass, the more nearly straight is its trajectory. The trajectories depend on the ratio of the radii of the small and large drops.

In figure 7.14, $a$ designates a grazing trajectory, while $b$ is a sample of a collision trajectory. The distance from the axis to the grazing trajectory in the undisturbed upstream region is designated $y$. The collision efficiency is given by $[y/(R + r)]^2$. It is the cross section containing all the droplets that will collide with the large drop divided by the geometric cross section. Some workers define a so-called linear collision efficiency by $y/R$.

Table 7.9
Observed size and estimated mass of natural snow crystals

| Crystal habit | Long dimension (cm) | | Mass of average crystal (g) | Radius of drop of same mass (cm) |
|---|---|---|---|---|
| | Average | Range | | |
| Thin hexagonal plate | 0.05 | 0.01–0.15 | $1.6 \times 10^{-6}$ | 0.007 |
| Stellar | 0.135 | 0.04–0.4 | 15 | 0.015 |
| Column | 0.05 | 0.02–0.2 | 8 | 0.013 |
| Needle | 0.1 | 0.02–0.3 | 2.5 | 0.008 |
| Plane dendrite | 0.34 | 0.07–0.65 | 90 | 0.028 |

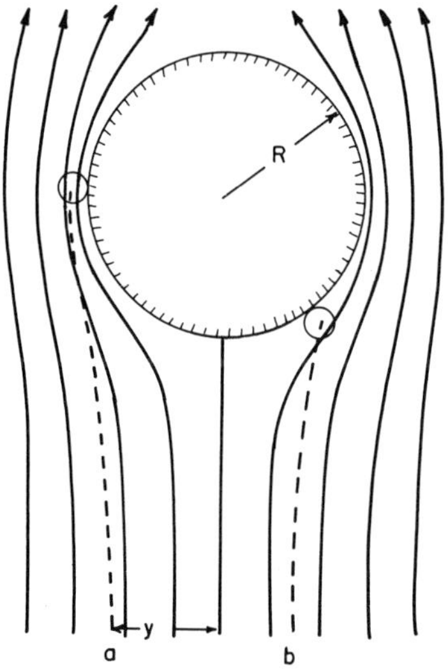


Figure 7.14 The solid curves are streamlines of viscous flow around a spherical drop of radius $R$. The dashed curves are the trajectories of small drops considered as mass points. Trajectory $a$ is a grazing trajectory, while $b$ is a collision trajectory.

Viscous flow, as assumed in figure 7.14, obtains only when the Reynolds number is less than about 0.5–1.0, or for drop radii less than 30–40 $\mu$m. The flow around larger spheres is rather well known, and the equation of motion of a mass point in such a field is readily written down, but numerical solutions are required. When $r/R$ is no longer small, it is necessary to consider the flow around each sphere by linear superposition. In order to follow the rapid changes as the two drops begin to approach each other, it is necessary to take very small time increments in the numerical solution if large errors are to be avoided. This regime is pertinent to the collisions between cloud drops of nearly the same size, the process that may initiate accretion. The results of computations of the collision efficiency are still not in satisfactory agreement in all cases. It has been suggested that the remaining differences are due to different drag force expressions, failure to minimize the error in the trial-and-error solution required to find the grazing trajectory, and computational instabilities. Figures 7.15 and 7.16 exhibit graphically the collision efficiencies of liquid drops as a function of the radii of the small and large drops. Figure 7.16 shows only the region where the efficiencies approach unity, where the logarithmic plot of figure 7.15 is not clear.

Perhaps the most interesting feature of figure 7.15 and 7.16 is the rapid increase in collision efficiency as the two drops approach equal size. This

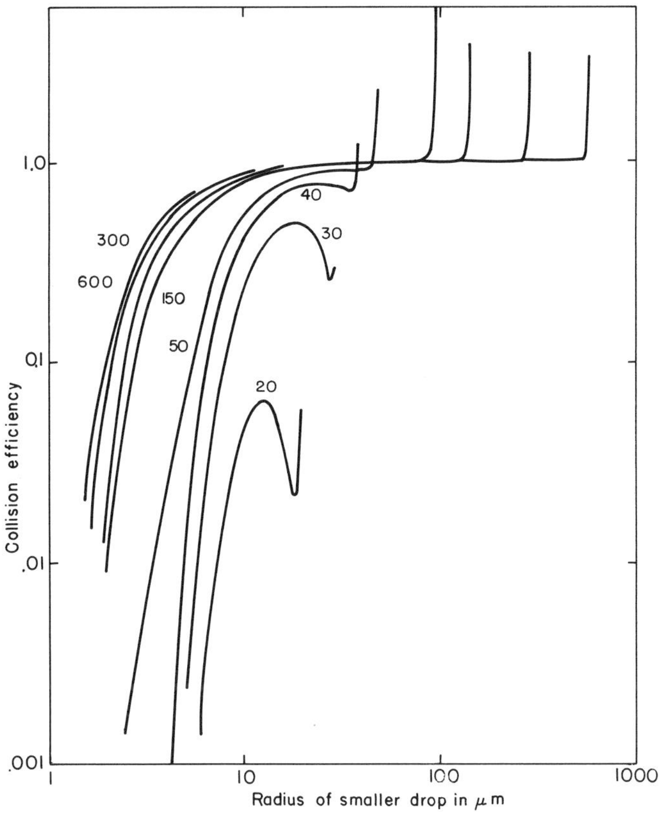


Figure 7.15 Collision efficiencies of pairs of water drops of different sizes. Numbers on the curves are radii of the larger drops in micrometers. Based on Shafrir and Gal-Chen (1971), Beard and Grover (1974), and Almeida (1977).

is due to inertial effects, which accelerate both drops, but more so for the upper drop, resulting in a finite closing velocity even for drops of equal size. Experiments have confirmed collision efficiencies greater than unity for drops of nearly equal size. This phenomenon is not of much importance in the atmosphere because of the long time required to achieve a collision and the relatively small probability that drops of nearly the same size will be situated so as to favor this process.

Figure 7.15 shows that the maximum collision efficiency of a 20 $\mu$m collector drop is about 0.65. A 10 $\mu$m collector drop has a collision efficiency of less than $10^{-3}$ for all sizes of the collected drops. This shows that accretional growth cannot be of much importance until the condensation process forms drops of 20–25 $\mu$m radius.

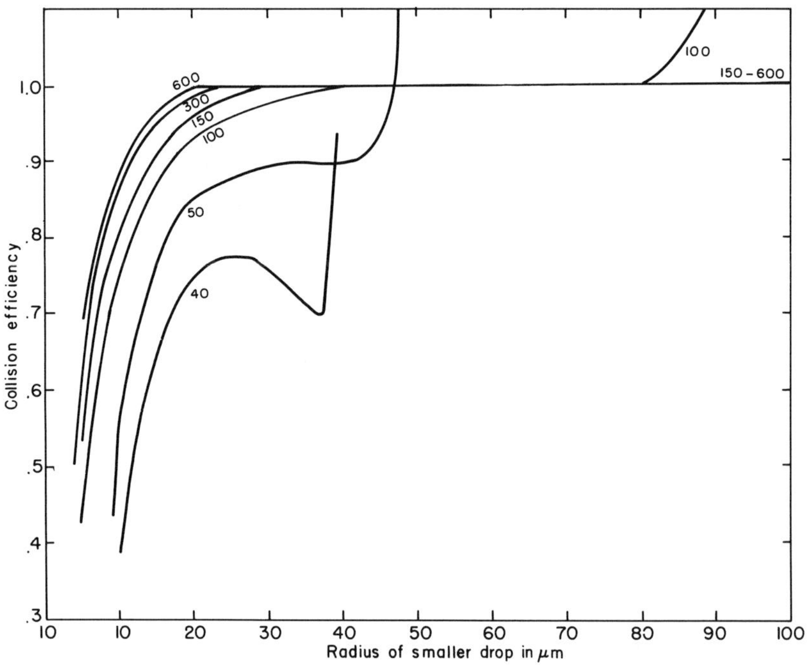


Figure 7.16 Collision efficiencies of pairs of water drops of different sizes. Similar to figure 7.15 except for the linear scale, which provides more detail in the region of large efficiencies. Numbers on the curves are the radii of the larger drops in micrometers. Data are from figure 6 of Lin and Lee (1975).

## 7.22 Coalescense of Liquid Drops

A number of experiments have shown that some drops on collision trajectories do not coalesce but bounce off. If the drops actually come into contact, coalescence will occur, since the surface energy of the combination is less than the sum of the surface energies of the two drops. It is generally agreed that the principal barrier to coalescence is the cushion of air between the two drops that must drain before they can come into contact.

Whelpdale and List (1971) observed that droplets on collision trajectories near the limb of the large drop tended to bounce off. Three collector drop radii of 0.5, 0.85, and 1.75 mm and droplets of 35 $\mu$m radius were used in their experiments. They reported that their results can be given by

$$E^1 = \left(\frac{R}{R+r}\right)^2, \qquad (7.24)$$

where $E^1$ is the coalescence efficiency and $R$ and $r$ are the radii of the large and small drops, respectively. Equation (7.24) is an upper limit of $E^1$, but gives results only a few percent larger than the experimental values.

Levin and Machnes (1977) repeated the procedures of Whelpdale and List for a wider range of large drops (0.01–1.25 mm radius) and for a greater range of $r/R$ (from 0.01 to about 1.0). They found consistently smaller coalescence efficiencies than (7.24), but were able to show a dependence of the coalescence efficiency on the size of the collector drop as well as on the ratio $r/R$. They offer no explanation for their substantially smaller coalescence efficiencies than those of Whelpdale and List.

Arbel and Levin (1977) developed a numerical model of coalescence based on the drainage of the air between the approaching drops assuming that the drops are distorted so that they are flat in the region of contact as they approach. This model predicts a decrease of coalescence efficiency as the ratio $r/R$ increases, but one that is much more rapid than the experimental data. This theory must be considered as an important first attempt, but is not useful for predicting numerical values. Until more experimental data are available, it is suggested that (7.24) be used. This has the virtue of simplicity and gives numerical results that seem in general accord with indirect evidence.

Some question remains as to the possible effects of adsorbed surface layers on coalescence. It has been found that external electric fields and/or electric charge on the drops increases the coalescence efficiency.

## 7.23 Effects of Electric Fields and Charges

As noted in section 10.1, the atmosphere has a fair weather electric field of about 1 V $cm^{-1}$, which may increase to some thousands in thunderclouds. Cloud drops carry electric charges, usually not exceeding a few tens of elementary charges. Charges of both signs are present, and the net charge on a cloud volume is typically small except in thunderclouds. These charges and fields produce Coulomb forces that affect the motion and collision efficiency of cloud drops.

Because of the difficulties in the way of experimental work, most studies of the effects of electrical forces on collision efficiency have been based on numerical models. The most complete set of such computations is due to Schlamp et al. (1979). There are many combinations of electric field directions and magnitudes and of the signs and magnitudes of the drop charges, and not all of them have been completely examined. Data on droplet charges and fields within real clouds are scanty, and it is not yet possible to examine the net result of electric forces on accretional growth in a macroscopic cloud.

Some of the conclusions reached by Schlamp et al. may be briefly summarized as follows: (a) external fields with uncharged drops invariably enhance the collision efficiency, being most pronounced for small collector drops; (b) when the two drops carry charges of opposite sign and there is no field, the collision efficiency is increased, particularly for small collector drops; (c) the above effects are negligible for collector drops of radius greater than about 70 $\mu$m, even for the largest fields and drop charges observed in thunderclouds; (d) when both a field and charged drops are present, the situation becomes very complex, and a decrease in collision efficiency may even result.

Freire and List (1979) have examined the case of a vertical field in which the drops carry opposite charges, so that collector drops are acted on by a force opposite to that of gravity. A critical field exists at which the larger and smaller drops have the same fall speed. In the neighborhood of this critical field the collision efficiency is very much larger than unity. This is attributed to the enhanced duration of the two-droplet interaction. This phenomenon may occur in relatively small fields, and it is also suggested that it occurs with relatively large, strongly charged drops.

It is tentatively concluded that electric forces may significantly affect individual droplet collisions under conditions that obtain in clouds but that the macroscopic effects on accretional growth are problematical. Although experimental and numerical results are scanty, it is intuitively likely that electric forces are important in promoting the coalescence of colliding drops.

## 7.24 Collision Efficiencies of Droplets on Ice Crystals

Inspection of snow crystals shows that many of them are more or less densely covered with frozen cloud drops. These are called rimed snow crystals. The collision efficiencies of cloud drops on platelike and columnar ice crystals have been computed with results as shown in figures 7.17 and 7.18. Columnar crystals fall with their long axes horizontal, and the collision efficiency is defined as $y/(R + r)$ instead of $[y/(R + r)]^2$, which is used for spheres and plates. The lengths and diameters of the columnar crystals in figure 7.18 were chosen to have the mean dimensional ratios found by Auer and Veal (1970). Other ratios occur, and figure 7.18 is only approximate for such crystals.

The collision efficiency drops to zero for all droplet radii for plates smaller than about 150 $\mu$m radius and for columns smaller than about 50 $\mu$m diameter and 70 $\mu$m length. From figures 7.17 and 7.18 it is seen that, in general, cloud drops smaller than 5–10 $\mu$m radius have zero or small collection efficiencies. The upper cutoffs, where the collision efficiency again drops to zero, occur at drop radii that are near or above the largest cloud drops usually observed. The above conclusions are qualitatively confirmed by the observations of Nakaya (1954) and Ono (1969). Nakaya

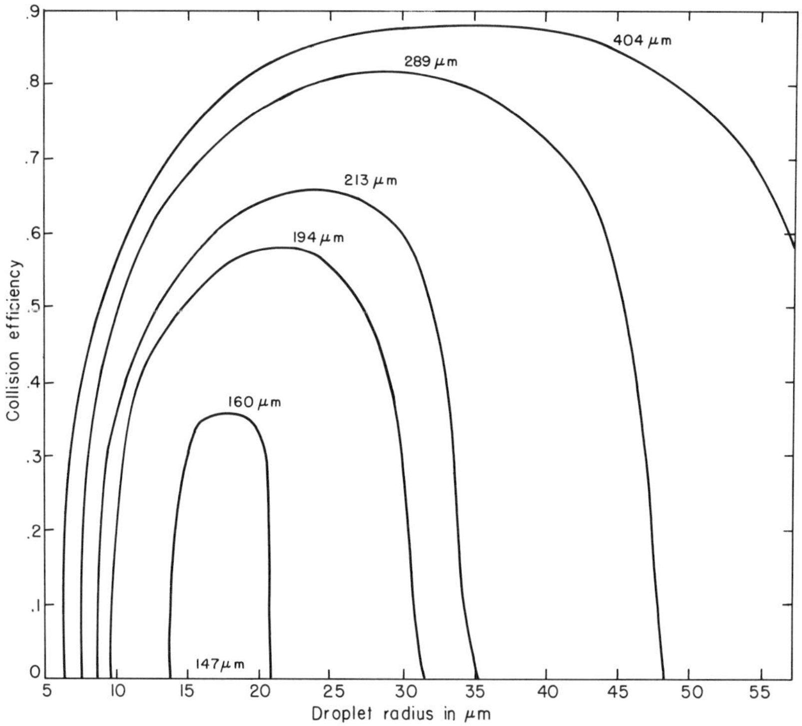


Figure 7.17 Collision efficiencies of water drops on thin ice plates. Numbers on the curves are radii of the thin plate. Ice plates are approximated by oblate spheroids of axis ratio 0.05 and density 0.92 gm cm$^{-3}$. The pressure is 700 mbar and the temperature is $-10$°C. Collision efficiency falls to zero for all droplet sizes at an ice plate radius of 147 μm. After Pitter (1977).

measured the frozen droplet size spectra on natural crystals and found a mode of about 15 μm radius and a range from about 7–40 μm. Ono found modes of about 10 μm radius on a rimed dendrite and about 20 μm on a rimed cylinder. The largest drops were about 32 μm radius and the smallest about 9 μm. The two droplet spectra were from crystals collected on different days, and the differences are not necessarily due to the collection efficiencies. Ono also found that the onset of riming on columnar crystals occurs at a diameter between 50 and 90 μm. He found riming on planar crystals $>200$ μm radius, but no riming for a radius $<150$ μm.

Moderately to heavily rimed ice crystals evidently have larger masses and greater fall velocities than their unrimed counterparts of the same principal dimension. The greater fall speeds and the changes in structure of rimed crystals can also be expected to modify the collision efficiencies. Since the degree of riming is continuous and generally unknown, it is not possible to quantify the effects of riming noted earlier. As a gross approxi-

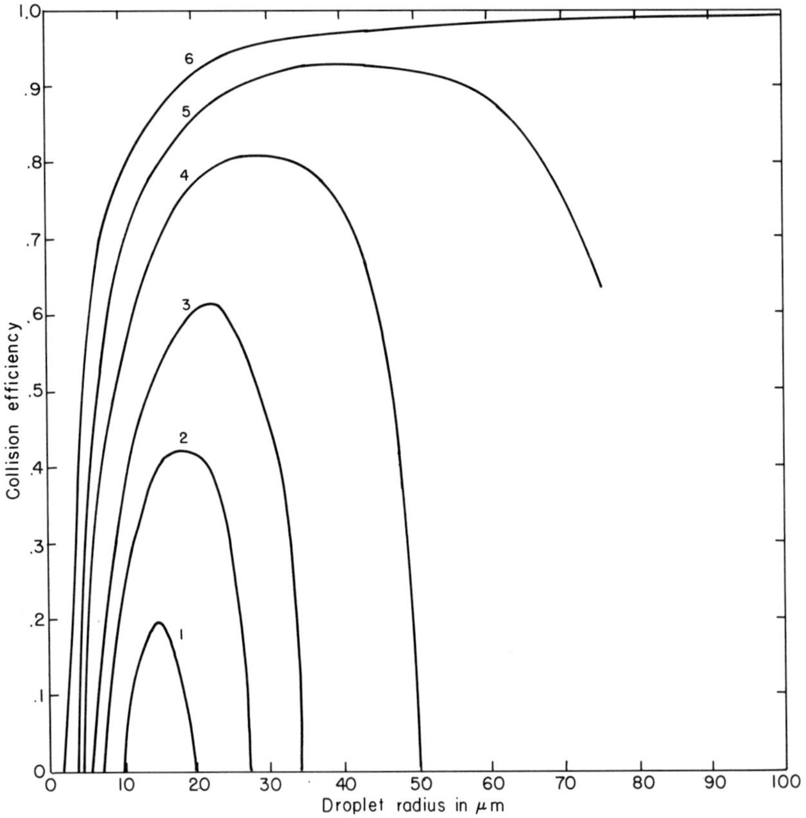


Figure 7.18 Collision efficiency of droplets on cylinders, used to approximate columnar ice crystals, for several cylinder diameters and lengths. The pressure is 800 mbar, and the temperature is −8°C. After Schlamp et al. (1975).

| Curve | Diameter (μm) | Length (μm) |
|---|---|---|
| 1 | 65.4 | 93.3 |
| 2 | 73.2 | 112.6 |
| 3 | 83 | 138.3 |
| 4 | 106.8 | 237.4 |
| 5 | 154.9 | 514.9 |
| 6 | 292.8 | 2,440 |

mation, the full speeds of moderately rimed crystals may be taken as double those of an unrimed crystal of the same long axis.

Continuation of riming leads to the obscuration of the crystal habit and the formation of graupel. Graupel may also form on frozen drizzle or raindrops. Consideration of graupel and of hail will be deferred to chapter 8.

### 7.25 Growth by Accretion

The accretional growth equation may be written down from the definition of the collision efficiency:

$$\frac{dm(R)}{dt} = \pi E(R + r)^2 w_r(V_R - V_r), \tag{7.25}$$

where $m(R)$ is the mass of the larger particle, $E$ is the accretion efficiency (collision efficiency times the coalescence efficiency), $w_r$ is the liquid water content of the small drops in grams per unit volume of air, and $V_R$ and $V_r$ are, respectively, the fall speeds of the larger and smaller particles. Note that for a cylinder or hexagonal column, $\pi(R + r)^2$ in (7.25) is to be replaced by $2(R + r)L$, where $L$ is the length, in view of the necessary difference in the definition of the collision efficiency. As written, (7.25) applies to a single $R$ and $r$. In the usual case, there is a size spectrum of both the large and small particles. Since all of the terms in (7.25) depend on $R$ or $r$ or both, numerical solutions of (7.25) must be obtained for a suitable range of $R$ and $r$ defined by the assumed size spectra. Equation (7.25) may be simplified when $R$ is very large compared with $r$ by neglecting $r$ and $V_r$. A further simplification may be made by selecting an average value for $E$. Inspection of figure 7.14 shows that a reasonable average $E$ may be selected when the collector drop is fairly large and most of the liquid water is in drops larger than 10 $\mu$m radius. With these approximations, (7.25) becomes

$$\frac{dm(R)}{dt} = \pi R^2 w_r \bar{E} V_r \tag{7.26}$$

or by a simple transformation, for spherical drops

$$\frac{dR}{dt} = \frac{w_r \bar{E}}{4P_L} V_r, \quad \text{or} \quad dR = \frac{w_r \bar{E}}{4P_L} dZ, \tag{7.27}$$

where $P_L$ is the density of the drop and $Z$ is vertical distance with respect to the air and is positive down.

## 7.26 Stochastic Accretion

The accretional growth equations discussed earlier assume that the small drops are uniformly distributed throughout space, much like a gas, so that each air volume, no matter how small, contains its proportional share of the liquid water content. As first clearly stated by Telford (1955), each collision is a discrete event; either the large particle collects a drop or it does not. Equation (7.25) is now called the continuous accretion equation to distinguish it from the more realistic discontinuous accretion process, which is called stochastic accretion for reasons to emerge later.

If it be reasonably assumed that the cloud droplets are randomly distributed throughout the ambient atmosphere and if a number of larger drops of equal initial size fall through the cloud, it is clear that they will not all have the same collision success and hence that they will become of different size. The continuous accretion equation would predict that all of the drops would grow equally. Raindrops have a typical radius of 1 mm, and a typical cloud drop has a radius of 10 $\mu$m, or some $10^6$ cloud drops are required to make one raindrop. The concentrations of cloud and raindrops differ by about five orders of magnitude. Thus only 1 cloud drop out of $10^5$–$10^6$ grows to raindrop size. This process is indicated schematically in

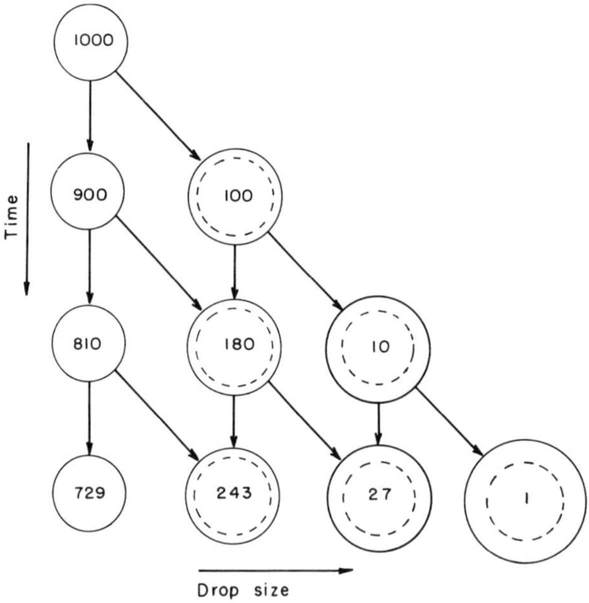


Figure 7.19 Schematic representation of stochastic accretion. Dashed circles are for reference, to indicate the size of the initial 1,000 drops. After Berry (1967).

figure 7.19. It is assumed that only 1 of 10 of the large drops will collect a small drop during one time interval; the other 9 do not collect a small drop. In figure 7.19 this process is carried out for three time steps. The result is a distribution of large drops ranging from drops that underwent no collisions to the one drop out of the thousand that collected a small drop in each time step. The continuous accretion model would predict that each of the 1,000 drops would collide with 0.3 small drops during the three time steps and that all of the drops would be of the same size. It is clear that the stochastic process will result in the favored few becoming substantially larger than the uniform size from continuous accretion. The stochastic process shows to best advantage when the collector and collected drops are of similar size, as in a cloud drop size spectrum. When the collector drops are much larger than the collected, there is relatively little difference in the predictions of the two accretion models because the mass increment per collision is small and the large air volume swept out in unit time tends to equalize the collisions experienced by the large drops.

Telford (1955) computed the growth of large drops by stochastic accretion using a simple model in which the small droplets were assumed to be of uniform size and concentration during the coalescence process. In a numerical example he assumed a cloud in which 90% of the drops had a radius of 10 $\mu$m, 10% of the drops had double the mass of the 10-$\mu$m drops, and the liquid water content was 1 g m$^{-3}$. He found that the time taken for 100 drops to experience their first 10 coalescences was 5.11 min, compared with 33.0 min for continuous accretion.

Twomey (1966), using a more complete model in which the droplets' radii were distributed according to a truncated Gaussian spectrum, carried out stochastic accretion computations for three model clouds each of which had a liquid water content of 1 g m$^{-3}$. Cloud I had 50 drops cm$^{-3}$ and a dispersion (standard deviation divided by mean radius) of 0.15; cloud II had 200 drops cm$^{-3}$ and a dispersion of 0.50. In a separate test he found that the dispersion had little effect in the range used. The results of the computations are presented in table 7.10. Cloud I is intended to be rep-

Table 7.10
Time required for growth by stochastic accretion in three clouds[a]

| Cloud | Droplet concentration (cm$^{-3}$) | Time for 100 drops m$^{-3}$ to appear (min) | |
|---|---|---|---|
| | | 100 $\mu$m radius | 400 $\mu$m radius |
| I | 50 | 14 | 23 |
| II | 200 | 39 | 51 |
| III | 800 | 63 | 77 |

a. After Twomey (1966).

resentative of maritime cumuli, while III is for continental cumuli and II is intermediate. It is seen that the maritime cloud that contains the fewest but largest drops yields large drops in the shortest time. Also note, particularly in clouds II and III, that most of the time required to form drops of 400 $\mu$m radius is taken up in forming 100-$\mu$m drops.

## 7.27 The Stochastic Accretion Equation

The initial condition is a cloud drop spectrum, based on observations, that has arisen from condensation and perhaps other processes. From the accretion equation one can write

$$K(m, m') = \pi(r + r')^2 E(r, r')[v(r) - v(r')]. \tag{7.28}$$

$K(m, m')$ is the accretion parameter or the collection kernel; primes refer to the collected drop, and the unprimed quantities to the collector drop. It may be seen that $K(m, m')$ is the effective air volume swept per unit time for collisions between drops of mass $m$ and those of mass $m'$. That is, it is the geometric volume swept per unit time multiplied by the accretion efficiency. The probability $P(m, m')$ that a particular drop of mass $m$ will accrete a drop of mass $m'$ in time $\delta t$ is

$$P(m, m') = K(m, m')\, f(m')\, dm'\, \delta t, \tag{7.29}$$

where $f(m')\, dm'$ represents the drop spectrum and is the mean number of drops of mass $m'$ to $m' + dm'$ as a function of $m'$. Equation (7.29) is the product of the effective volume swept in the time interval $\delta t$ and the number concentration of drops of mass $m'$. The mean number of drops of mass $m$ that will collect one drop of mass $m'$ in the time $\delta t$ is the probability $P(m, m')$ multiplied by the concentration of drops of mass $m$, or $P(m, m')\, f(m)\, dm$. The time rate of change of the number concentration of drops of mass $m$ is given by the difference between their rate of formation by accreting drop pairs of mass $m'$ and $m - m'$ and the rate of loss by accretions of drops of mass $m$ with drops of any other mass. Thus

$$\begin{aligned} \frac{d[f(m,t)]}{dt} = \frac{1}{2} \int_{m=m_0}^{m'=m} & K[(m - m'), m']\, f(m', t) f[(m - m'), t]\, dm' \\ - \int_{m'=m_0}^{\infty} & K(m - m')\, f(m', t)\, f(m, t)\, dm', \end{aligned} \tag{7.30}$$

where $m_0$ is the mass of the smallest drop in the assumed drop spectrum. The factor 1/2 in the first integral of (7.30) compensates for counting each accretion event twice. As a simple example, note that a mass of 10 (arbitrary units) can result either from the combination of $m - m' = 9$ with

$m' = 1$ or of $m' - m = 1$ and $m' = 9$. These are the same physical event, since the larger drop always overtakes the smaller.

Of a number of computations of the changes in size spectra due to stochastic accretion, two due to Berry and Reinhardt (1974) have been chosen for figure 7.20. This shows the time evolution of the size spectra at 5-min intervals. The ordinates are the droplet mass per unit of $\ln r$ and show how the liquid water is distributed with drop size. Area under the curves gives the mass of liquid water contributed by drops in any selected radius interval. The two cases depicted in figure 7.20 differ only in the initial mean droplet radii, which are 12 and 14 $\mu$m. At first the spectrum broadens on the large radius side, but then a second peak occurs at an ever increasing radius as time passes. It is this large drop portion of the spectrum that leads to precipitation.

## 7.28 Simultaneous Accretion and Condensation

Since the condensational process seldom leads to a drop radius of much more than 20 $\mu$m except on giant nuclei and the accretion process is very slow when the largest drops are smaller than about 25 $\mu$m radius, growth computations have often tacitly assumed a transition from growth by vapor diffusion to growth by accretion at this radius as though the processes were mutually exclusive. If all of the drops are liquid, the two processes cannot only coexist but are synergistic because the rates of growth of both processes are dependent on the droplet size.

This is shown in figure 7.21, taken from Leighton and Rogers (1974), which gives the change in the drop spectrum in 7 min by accretion alone, condensation alone, and both acting together. Accretion alone, acting on the rather narrow initial spectrum with a mass mean radius of 8 $\mu$m, leads to small changes in the spectrum: a slight increase in the number of somewhat larger drops that hardly shows in the plot and a small decrease in drops of the mean size. Condensation acting alone narrows the spectrum and shifts the mass mean drop radius to about 15 $\mu$m. When the two processes act together, the spectrum is radically changed to the right of the maximum, where a second maximum appears in the small raindrop region. The sharp maximum of the condensation only spectrum is retained in the joint process, as is the entire curve to the left of the maximum. This overlap is so close that this portion of the accretion-condensation spectrum has been omitted.

The striking results of this numerical experiment are due in part to the choice of the initial conditions. They are a cloud consisting of 600 drops per milliliter of air in the initial spectrum shown in figure 7.21, which leads to a liquid water content of 1.3 g/m$^3$, a cloud base temperature of 10°C, a vertical air speed of 7.5 m/sec, ascent following a moist adiabat, no fall out of drops, and cloud base at 2,300 m. Changes in these conditions will yield

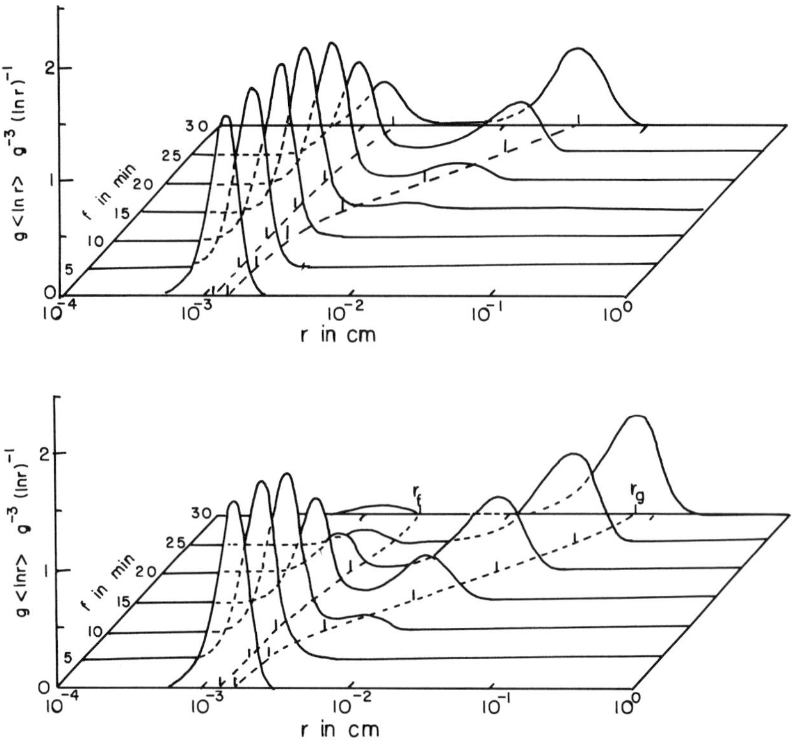


Figure 7.20 Time evolution of initial droplet spectra as a result of stochastic accretion. Upper diagram is for an initial mean drop radius of 12 $\mu$m, while for the lower diagram it is 14 $\mu$m. After Berry and Reinhardt (1974).

different numerical results, but it can hardly be doubted that the combined effects of condensation and accretion will lead to the appearance of large drops more rapidly than either process acting alone.

## 7.29 Some Other Aspects of Growth by Collection

Twomey (1976a) has shown that the stochastic collision equation is nonlinearly dependent on the liquid water content, particularly for the large drops in the "tail" of the droplet spectrum. Observations of the liquid water content in convective clouds show that it is variable along a path through the cloud, often approaching the adiabatic value, although the mean value is well below the adiabatic. Unfortunately, information on the three-dimensional size and the duration of these pockets of high liquid water content is so limited that a quantitative evaluation is not possible. If a few regions exist in 1% of the volume of the cloud in which the liquid water content is, say, 10 times the average value and they last for a few

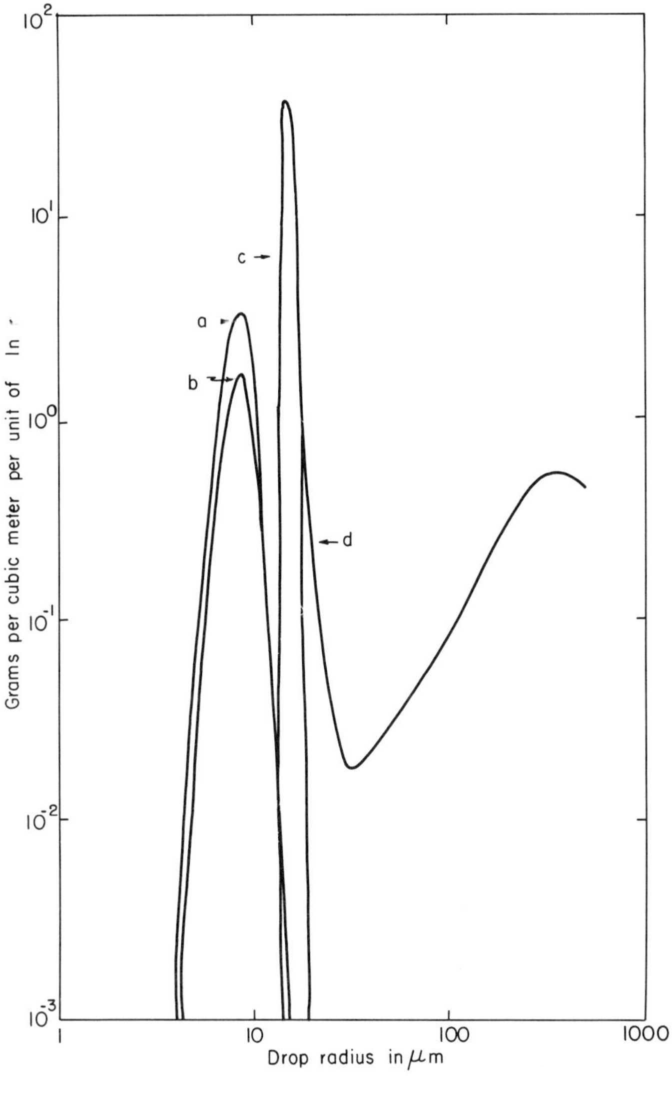


Figure 7.21 Evolution of droplet spectra due to diffusion and accretion acting for 14 min. Vertical air speed is 7.5 m $sec^{-1}$, and liquid water content is 1.3 g $m^{-3}$. Key: a, initial spectrum; b, growth by accretion only; c, growth by condensation only; d, growth by accretion and condensation. After Leighton and Rogers (1974).

minutes, Twomey shows that the concentration of large drops formed in these regions is orders of magnitude larger than in the bulk of the cloud. This process may be thought of as a second type of stochastic growth and deserves further study.

## 7.30 Effects of Turbulence

For many years it has been speculated that microscale turbulence enhances the collision efficiency of cloud drops, but adequate theoretical and experimental evidence has been lacking. A novel approach to this problem has been taken by Almeida (1979). He has developed a stochastic collision kernel that incorporates the effects of statistical turbulence in the microscale. His representation of turbulence assumes a locally homogeneous and isotropic velocity field in an incompressible fluid in the inertial subrange. Some of Almeida's results are shown in figure 7.22, in which collision efficiencies in a turbulent cloud are compared with the conventional collision efficiencies in a nonturbulent atmosphere. It is immediately apparent that the turbulence markedly increases the collision efficiency for collector drops of 15–20 $\mu$m radius. If confirmed, these results predict that accretion becomes important for collector drops larger than about 15 $\mu$m radius rather than about 25 $\mu$m without turbulence. Production of 15-$\mu$m drops by condensation is rapid, but the additional time to reach 25 $\mu$m is uncomfortably long.

The Almeida curves in figure 7.22 were computed for a turbulence dissipation rate of 10 $cm^2\,sec^{-3}$, which is a very modest turbulence intensity, particularly in convective clouds where values to $10^3$ $cm^2\,sec^{-3}$ have been suggested. This is an important matter because the collision efficiencies with turbulence decrease as the turbulent intensity increases beyond 10 $cm^2\,sec^{-3}$. This is due to the decrease in the turbulence time scale compared with the relaxation time of the drops. Almeida's approach is promising and deserves further exploration. Measurements of the turbulent dissipation rate, particularly in the region just above cloud base, would be very useful.

Jonas and Goldsmith (1972) observed drop collisions in a strong linear shear region of about 1 cm thickness in a special wind tunnel. These experimental results are similar to Almeida's in that the shear layer markedly enhanced the collision efficiences of collector drops of 10–20 $\mu$m radius. Although the results are quantitatively similar, the differences in the structure of turbulence in the dissipative region call for caution in comparing them. Nevertheless, it seems that microscale turbulence plays a role, and perhaps a very important one, in the difficult transitional drop growth process from condensation to accretion.

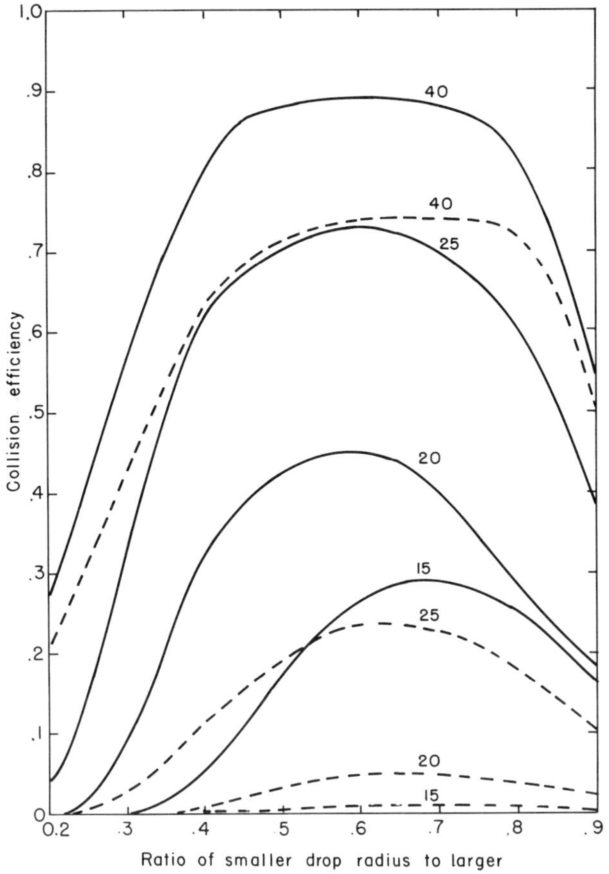


Figure 7.22 Collision efficiencies in a turbulent cloud (solid curves) according to Almeida (1979). The dashed curves are the same in a nonturbulent cloud. The numbers on the curves are the radii of the larger drops.

## 7.31 Accretion in the Ice Phase

Ice is present in the atmosphere as snow crystals, graupel, hail, and frozen drops. Collisions will occur, but there is a paucity of data on both collisions and "sticking" after collision. Latham and Saunders (1970) found an accretion efficiency of about 0.3 for thick plates and columns of mass median diameter of 4–5 $\mu$m on an ice sphere of 2 mm diameter at temperatures of $-7$°C to $-27$°C. They also found that electric fields of 500–1,500 V $\text{cm}^{-1}$ increased the accretion efficiency by 15–80%. This is a somewhat artificial model of ice-ice accretion in the real atmosphere, but it does show that a significant fraction of the ice crystals was captured. Other experiments have shown that when two ice spheres touch, they adhere even at rather low temperatures. This is usually attributed to sintering, which builds an ice bridge around the point of contact.

The most important type of ice-ice accretion is the aggregation of snow crystals to form what are often called snowflakes. These aggregates may consist of tens, hundreds, or even thousands of individual ice crystals, as can be readily seen under low magnification. Up to half or possibly more of the total snow fall in midlatitudes is in the form of aggregates. Snow crystals aggregates range up to several centimeters in maximum size, have masses ranging from around $10^{-4}$ to $10^{-2}$ g, and have fall speeds of 0.5–2 m $\text{sec}^{-1}$. Their melted radii range from 0.3 to 1.3 mm and are therefore of typical raindrop mass.

## Problems

1.
Radiosondes often report relative humidities below 100% while in a water cloud. As one test of such reports, estimate the time required to evaporate drops of 10 $\mu$m radius in an environment of temperature 10°C and a relative humidity of 95%.

2.
What factors determine the concentration of cloud drops in number per unit volume of air?

3.
A graupel pellet falling in a supercooled cloud may grow by vapor deposition and accretion of cloud drops. The accretion of cloud drops releases the latent heat of fusion, warms the pellet, and decreases growth by deposition. At a sufficiently large liquid water content, depositional growth will be suppressed. Develop an analytical expression for the critical liquid water content that just stops growth by deposition.

4.
At the base of a warm cloud the pressure is 900 mbar and the temperature is 10°C. There are 100 drops of radius 1 $\mu$m in a cubic centimeter of air. If the updraft is 5 m/sec, what is the supersaturation ratio?

5.
A raindrop of 1 mm diameter falls 1 km to the surface. What will be the size of the drop at the surface? Take the temperature at cloud base as 10°C and as 17°C at the surface.

# 8

# Precipitation Processes

## 8.1 Introduction

The microphysical growth processes discussed in chapter 7, accretion and condensation-deposition, lead to the formation of precipitation particles. The dynamics of the lifting processes act as the forcing function on the growth of precipitation particles. In turn, the latent heat released during the growth of precipitation and the mass of the precipitation modify the dynamics of the system. Thus the dynamics and the microphysics form a coupled system that cannot as yet be fully modeled. The emphasis in what follows is on the growth processes of the precipitation particles in the context of the dynamics of the clouds.

The lifting processes that lead to clouds and precipitation may be subdivided into convective and stratiform systems. The latter refers to large-scale lifting typified by the classical upglide motion along the warm front of an extratropical cyclone. Convective cells are often embedded in such stratified systems, but this does not negate the advantages gained by treating these two lifting processes separately.

Convective clouds typically are of small lateral and large vertical dimensions, have vertical air velocities of some meters per second or more, and have active lifetimes on the order of an hour. Stratiform clouds are of large lateral extent, typically consist of several layers, have vertical velocities of a few tens of centimeters per second, and have lifetimes on the order of a day.

## 8.2 Comparison of Microphysical Growth Processes

In chapter 7, the precipitation growth processes were identified as accretion and vapor deposition on ice crystals. These processes often act in concert, but it is important to understand their relative effectiveness in

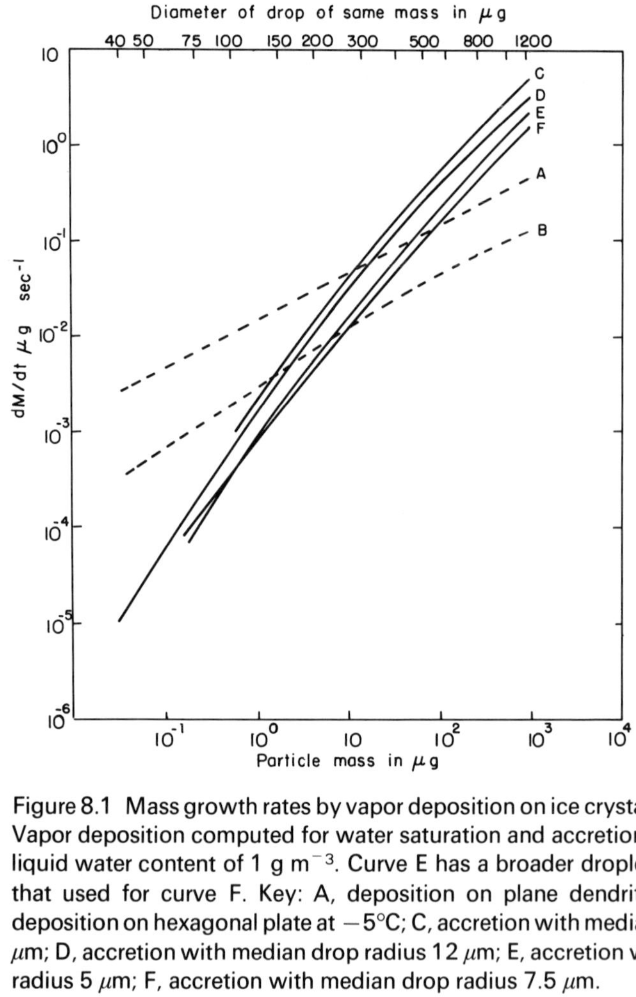


Figure 8.1 Mass growth rates by vapor deposition on ice crystals and accretion. Vapor deposition computed for water saturation and accretion computed for a liquid water content of 1 g $m^{-3}$. Curve E has a broader droplet spectrum than that used for curve F. Key: A, deposition on plane dendrite at −15°C; B, deposition on hexagonal plate at −5°C; C, accretion with median drop radius 25 μm; D, accretion with median drop radius 12 μm; E, accretion with median drop radius 5 μm; F, accretion with median drop radius 7.5 μm.

producing precipitation particles. Inspection of (7.3) and (7.23) shows that the rate of mass growth by deposition increases linearly with the radius, while that for accretion increases with the cube of the radius, if the fall speed is conservatively taken to be linearly dependent on the radius. This shows that the accretion process will yield a faster growth rate than deposition for a sufficiently large radius.

Some light on how large is sufficiently large is shed by figure 8.1, taken from Houghton (1950). Figure 8.1 shows the growth rate against particle mass for four cases of accretion and two of deposition. The two cases of deposition assume water saturation and were selected to yield a reasonable range of growth rates for the deposition process. The four cases of accretion are for a liquid water content of 1 $g\,m^{-3}$. Four cloud drop size spectra were selected with median radii of 5, 7.5, 12, and 25 $\mu$m, and the accretional growth of the largest drops was computed from the continuous accretion equation. Growth by stochastic accretion would be faster, but on the other hand, the drop size spectra used are relatively broad, thus favoring rapid growth. It is evident that other choices could be made and therefore that figure 8.1 is not quantitatively definitive. However, the steeper slope of the accretion curves and the presence of a region of intersection between the two sets of curves are reliable features. In figure 8.1 the region of intersection is centered around a particle mass of about 10 $\mu$g, the mass of a water drop of about 130 $\mu$m radius. Other indirect evidence suggests that this is a reasonable value. Thus for particles of the mass of cloud drops, depositional growth is much faster than accretion, while for particles of the mass of drizzle or small raindrops or more, accretion is more rapid than deposition.

## 8.3 Types of Convective Precipitation

The ambient temperature is of considerable importance, since depositional growth of ice crystals occurs only at temperatures significantly colder than 0°C. It has become customary to speak of warm, cool, and cold convective clouds. A warm cloud is one in which the temperature is above 0°C at all levels. Since ice practically never forms in the atmosphere at temperatures warmer than about −3°C, the definition of a warm cloud can be relaxed to this temperature. Cool convective clouds are those that intersect the 0°C isothermal surface at some level between the cloud base and the cloud top. Cold clouds are colder throughout than 0°C and consist primarily of ice crystals, although supercooled cloud drops may often exist in the early stages of cloud growth and precipitation initiation.

## 8.4 Observations of Warm Shower Clouds

For some years after the enunciation of the ice crystal theory of precipitation by Bergeron (1935), it was widely believed that all significant precipitation was initiated by ice crystals, even though Bergeron himself stated that precipitation could form in the tropics by liquid accretion. The nearly complete reliance on the ice crystal theory of precipitation was finally abandoned in the face of repeated observations of rain showers from clouds entirely warmer than 0°C.

Warm shower clouds are common in the subtropics, usually in maritime air. Vertical air velocities range from less than one to several meters per second. Clouds as shallow as 1.5 km may release rain; deeper clouds rain more heavily, and rainfall rates to 10 $mm\,hr^{-1}$ have been observed. The largest raindrops range up to about 2.5 mm in diameter. The time from the

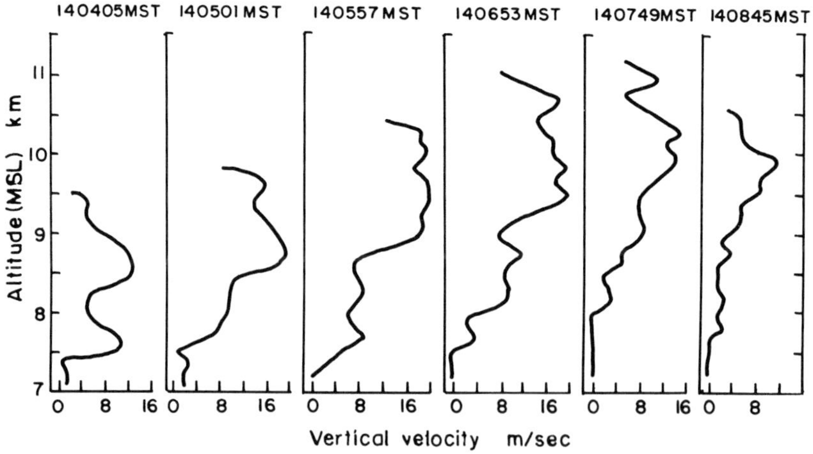


Figure 8.2 Vertical profiles of the upward air velocity in a thunderstorm in southern Arizona taken at intervals of 56 sec. Vertical air speeds were obtained from a vertical-pointing Doppler radar. After Battan and Theiss (1966).

initiation of the cloud to the release of rain is about 15–25 min, and the rain lasts another 10–30 min. Warm shower clouds have been observed in many places, but have been most intensively studied in Hawaii by Takahashi (1977) and in the Caribbean by, among others, Brown and Braham (1959) and Saunders (1965). They are also found in midlatitudes in summer over the continents, and these clouds are typically more vigorous and deeper than those in the maritime subtropics. A warm shower cloud usually consists of a single convective cell. These cells are seen on radar as approximately vertical echoes that reach the surface and then decay as the precipitation falls out. In examining such radar echoes it must be remembered that the echo does not appear until drops of drizzle size are present and the precipitation process is well advanced.

## 8.5 Vertical Air Velocity

The vertical air velocity in a convective cloud is important in the precipitation process because it largely determines the trajectories of the growing precipitation particles and the time they spend in the cloud. It is difficult to measure the vertical air velocity profile of a convective cloud because of turbulence and the rapid change of the profile with time. Repetitive penetrations at several levels by an instrumented airplane consume a substantial portion of the cloud's lifetime, and the first and last traverses correspond to different stages of evolution of the cloud. Doppler radar has been used by Battan and Theiss (1966) to observe the vertical velocity in a thundercloud. The Doppler spectrum is the net result of the rising air velocity and the fall

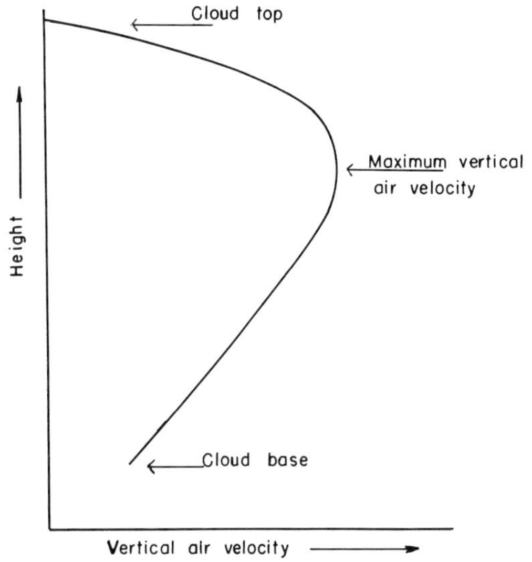


Figure 8.3 Schematic vertical air velocity profile in a convective cloud.

speeds of the precipitation particles. By assuming that the smallest radar detectable particles have a fall speed of about 1 m $sec^{-1}$ and knowing the fall speeds of larger drops, it is possible to infer the vertical air velocity within about $\pm 1$ m $sec^{-1}$; results are given in figure 8.2. The profiles at 140653 MST and 140749 MST (Mountain Standard Time) are similar to the mean profile for the mature stage of a thunderstorm found by Byers and Braham (1949). These profiles for thunderstorms are not directly applicable to shower clouds, but some qualitative similarities can be expected.

An idealized vertical velocity profile is given in figure 8.3 as an aid in the examination of the effects of vertical velocity on the precipitation process in convective clouds. Similar profiles have often been assumed to represent a steady state. It is important to repeat that convective clouds, and particularly shower clouds, are transients, as shown, for example, in figure 8.2.

## 8.6 Effects of Horizontal Wind Shear

The horizontal wind commonly changes in speed and direction with elevation. This wind shear is typically $1\text{–}5 \times 10^{-3}$ $sec^{-1}$, and it leads to the commonly observed sloping convective clouds. According to Malkus (1952) the action of the wind on a cloud is due to the horizontal momentum of the entrained air and the form drag of the updraft. The slope of the cloud is smaller than the slope of the vertical wind profile.

In general, sheared convective clouds are less vigorous and do not attain the height of vertical clouds in otherwise similar environments, due in part to the fact that buoyancy acts in the vertical. With strong wind shear aloft the cloud tops may be blown off and decay. In a cloud with a marked slope, the growing precipitation particles may fall outside the cloud. As will be seen in section 8.9, a major portion of the mass of a precipitation particle is acquired during its downward course through the cloud. It is generally concluded that strong shear is inimical to the development of most convective clouds and the release of precipitation.

An important exception to the last statement appears to be the giant squall line thunderstorms, which usually exhibit strong shear. The depth and vigor of these storms is such that the precipitation process is complete in the updraft. The shear may actually protect the updraft by causing the precipitation to fall out outside the updraft.

## 8.7 Convective Cloud Lifetimes

Convective clouds have an active lifetime of an hour or less, and this imposes constraints on the growth of precipitation. A warm shower cloud exhibits a single cell on a radar. Clouds of this type were observed by Saunders (1965) by photography and radar. He reported that, starting from fractocumulus, the clouds showed radar echoes in 25–35 min and that the radar rain echo reached the surface in another 6–10 min and decayed for 5 min or more as the rain fell out. Braham (1964) found that the active lifetime of convective clouds in Missouri was about the same as the time required for the precipitation process to produce echo-sized particles (about 400 $\mu$m diameter). Byers and Braham (1949) found that the individual convective cells in thunderstorms had a draft duration of about 20 min. Single-cell storms were observed, but there were many multicell storms. In the latter, cells appeared at irregular time intervals, so that the active life of the thunderstorm as a whole could be much longer than that of a single cell. The 20-min average duration of a cell may be an underestimate because of the observational difficulties, but it probably is no more than half an hour.

From these similar observations it may be concluded that the active lifetimes of unicellular convective clouds are typically one-third to two-thirds of an hour. By active lifetime is meant the time during which precipitation is forming and growing; alternatively, it is the time from the beginning of the principal convective process to the fallout of the bulk of the precipitation at cloud base.

## 8.8 Entrainment

The vertical growth of a convective cloud is due to the buoyancy resulting from an initial positive temperature increment and the latent heat released by condensation. As the cloud rises it entrains air from the resting and drier environment due to turbulent mixing and the inflow required to satisfy continuity in the upward-accelerating cloud. This entrainment is substantial, amounting to a doubling of the mass of the rising air in a pressure-height interval of 300–500 mbar. The entrained air is cooler than the cloud, and further cooling results from the evaporation of cloud water to saturate the incoming drier air. This cooling reduces the buoyancy, acceleration, and vertical momentum of the cloud. The net results are a temperature lapse rate intermediate between the dry and moist adiabatic lapse rates and a smaller vertical air velocity and liquid water content than if there were no entrainment. Since precipitation feeds on cloud water, entrainment reduces the growth rate of precipitation.

Byers and Braham (1948), Using balloons as tracers, have confirmed the convergence of environmental air in the region surrounding convective clouds. It has also been found that the lapse rate in convective clouds is nearly as large as the dry adiabatic, as predicted. Most important, the mean liquid water content in convective clouds increases only slowly, if at all, with height and is only about one-quarter to one-third of that computed for an adiabatic ascent from cloud base. Reference is made to figure 7.7, which shows the variation of liquid water along a horizontal path through a cloud. In this example the liquid water varies from zero to more than twice the mean, and larger variations have been observed. These fluctuations, which may play a role in the growth of precipitation particles, are probably due to incomplete mixing of the entrained air into the cloud. Heymsfield et al. (1979) have found protected cores in convective clouds within which the lapse rate is nearly moist adiabatic and the liquid water content approaches the adiabatic. This is further evidence of incomplete mixing of the entrained air across the updraft.

## 8.9 Computed Growth of Warm Rain

Rain from warm shower clouds can only be formed by the liquid accretion process. An example of early computations of the formation of raindrops by this process, due to Bowen (1950), is given in figure 8.4A. The computations were made from the continuous accretion equation using the collision efficiencies of Langmuir and Blodgett. The process was started on drops of 25.2 $\mu$m diameter in a cloud of uniform drops of 20 $\mu$m diameter and a liquid water content of 1 g m$^{-3}$. Figure 8.4A shows the trajectories of the growing drops for several uniform vertical air velocities starting from cloud base. In figure 8.4B the same results are plotted against drop diameter. These idealized steady state calculations showed that raindrops of typical size could form by accretion but that the time required was two to three times that observed. Also note from figure 8.4B that some 85% of the final drop diameter, or nearly all of the mass, is acquired on the downward portions of the trajectories.

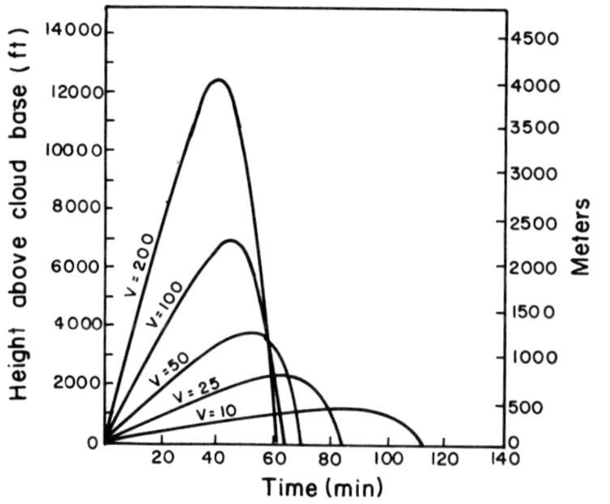


Figure 8.4A Time-height trajectories of drops growing by accretion in an idealized convective cloud for different values of the vertical air velocity $v$ (in cm $sec^{-1}$). Vertical air velocity and the liquid water content are assumed invariant with height and time. After Bowen (1950).

Much effort has been devoted to seeking an explanation for the observation that growth by liquid accretion is more rapid than that computed from the simple continuous accretion equation. These developments have been discussed in chapter 7. The most important of these are considered to be stochastic accretion and simultaneous growth by accretion and condensation. Taken together, these two processes reduce the time required for the growth of raindrops to an approximation of that observed. However, the feeling persists that the agreement is somewhat forced by favorable choices of the initial and ambient conditions.

Most computations of accretional growth begin with a cloud drop size spectrum tacitly assumed to result from condensation. In order to start accretion the cloud spectrum must include a few drops of radius greater than 20–25 $\mu$m, and the larger these drops, the more rapid the accretion. The size of the drops formed by condensation depends critically on the drop concentration. Large drop concentrations lead to small drops simply because the available water is distributed among a large number of drops. On the basis of such simple considerations, maritime clouds with drop concentrations less than about 100 $cm^{-3}$ may contain drops of radii greater than 20–25 $\mu$m, but this becomes increasingly unlikely at larger drop concentrations. Thus there appears to be a gap in the growth theory between the drop size formed by condensation and that required for significant growth by accretion. This gap, not apparent in nature, must be bridged to yield a satisfactory theory.

One of the most important features of the formation of precipitation in convective clouds by liquid accretion is that the time required to grow

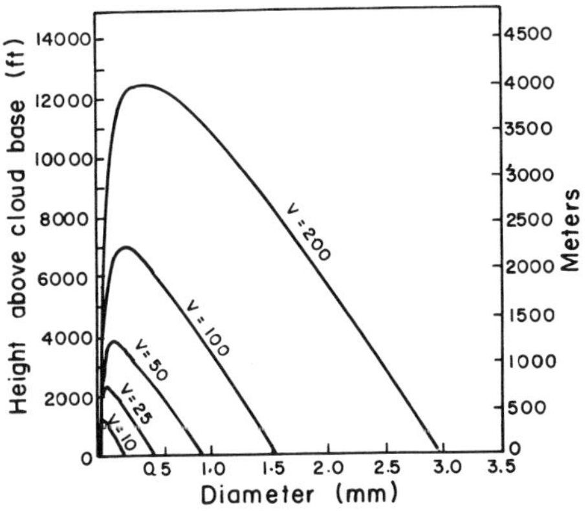


Figure 8.4B The computations that led to figure 8.4A are transformed to yield the diameters of the growing drops versus height. As in figure 8.4A, $v$ is in cm sec$^{-1}$.

raindrops of typical size is not very different from the lifetime of a convective cell. In consequence, the precipitation process must begin very early in the life of the cell and in the lower portion of the cell. The precipitation often initiates the downdrafts that signal the demise of the cell.

## 8.10 Accumulation Zone

The concept of an accumulation zone was introduced by Sulakvelidze et al. (1967) to explain the rapid growth of hailstones. The simple steady state vertical velocity profile in figure 8.3 may be used to show what is meant by such an accumulation zone. If the fall speeds of the precipitation particles are smaller than the vertical air velocity, they will rise above the level of maximum air velocity until their fall speeds equal the air velocity. As they continue to grow they will fall slowly with respect to cloud base. Thus the precipitation particles will remain in this region above the air velocity maximum for some time. This region is the accumulation zone, which is characterized by a high concentration of water in which the hailstones grow rapidly by accretion.

An accumulation zone as just described cannot exist because it ignores the effect of the mass of water added to the zone on the air velocity profile. A simple calculation shows that the addition of 1 g of water (or ice or anything) to 1 kg of air will produce a downward acceleration of about 1 cm sec$^{-2}$. For reference, the buoyancy acceleration due to a temperature increment of 1°C is about 3.5 cm sec$^{-2}$. The accumulation of some tens of grams of water per cubic meter of air as initially proposed would strongly modify and probably reverse the vertical air velocity.

A more realistic form of accumulation will occur wherever the vertical air velocity is larger than the fall speeds of the smaller precipitation particles and less than that of the larger particles. Figure 8.4 shows trajectories relative to cloud base of drops growing by accretion as calculated by Bowen (1950) for a steady state and uniform vertical velocities. In the region around the maximum of a trajectory, the velocity of the particles relative to cloud base is small and accumulation results. An analysis of accumulation can be made only with a cloud model in which the interactions of the cloud dynamics with the cloud microphysics is simulated. Nevertheless, accumulation occurs whenever the vertical air velocity and the fall speeds of the particles are similar, a common situation.

## 8.11 Precipitation Processes in Cool Clouds

Cool clouds, those that extend from below to above the melting level, are very common over both the oceans and the continents. Observations of such clouds with microwave radar over the central United States showed that half or more of the radar first echoes were entirely warmer than 0°C and that most of the others straddled the freezing level. It was generally concluded that the warm accretion process was responsible for the growth of the drops producing the echoes in most cool clouds in several different geographical areas. This conclusion represented a marked change from the earlier contention that clouds with cold tops precipitated by means of the ice crystal process.

More complete information on the precipitation process in cool clouds is provided by penetrations of the clouds by instrumented airplanes. An excellent example of this approach was provided by Braham (1964) in Project Whitetop, a study of shower clouds in southern Missouri. These clouds form in nearly unmodified maritime air. Cloud bases are at 1–1.5 km MSL, tops at 5–7.5 km. Airplane penetrations just below the cloud top frequently found precipitation-size liquid drops; subsequent penetrations of the same cloud found ice pellets of about the same size and concentration and no large liquid drops. The ice pellets were in the form of nearly clear frozen drops and graupel.[1] The observed concentrations of ice pellets were several orders of magnitude larger than that of the ice nuclei at cloud top temperature.

Similar results were found by Mossop et al. (1970) in cool maritime clouds off the coast of Tasmania. The dimensions and summit temperatures of these clouds were much the same as those of Project Whitetop. Drizzle-size water and ice particles were observed in these clouds, including frozen drops, graupel and ice crystals. The concentration of ice particles

---

1 Graupel are opaque lumps of rime, formed by accretion of supercooled cloud drops.

was found to be three orders of magnitude greater than that of the ice nuclei at cloud top temperature, suggesting that an ice multiplication process was involved. In many cases it was found that the liquid condensation-accretion process was able to produce large liquid drops, while ice particles of similar size were absent.

It appears that maritime cool clouds with summit temperatures warmer than $-10$ to $-15$°C precipitate by the liquid condensation-accretion process, much the same as warm maritime clouds. The observed ice pellets are almost certainly formed by the freezing of the large water drops. In view of the low concentration of ice nuclei at the lowest temperatures in these clouds, the freezing of the drops most likely results from contact with ice splinters produced by an ice multiplication mechanism.

In accordance with this, the ice pellets are not involved in the initiation of precipitation. However, the introduction of ice has important consequences through the release of the latent heat of fusion. This will evidently enhance the buoyancy and stimulate some further growth of the cloud. If a cloud region is completely converted to ice, the temperature will be increased by about 0.5–1.0°C, depending on the water content and ambient conditions.

Cool continental clouds in Israel have been probed by Gagin (1975). The typical cloud top temperature was $-20$°C. The concentration of ice particles near the cloud tops was found to be much the same as that of the ice nuclei, and only one-tenth of the ice particles were graupel, most of them being snow crystals. No large water drops were found. Knight et al. (1974) found that precipitation particles near the melting level in continental convective clouds in Colorado were graupel and rimed snow crystals.

In spite of the belief that the cloud drops in continental air are too small to initiate accretion, Mossop et al. (1972) and Isaac and Schemenauer (1979) found drizzle-size liquid drops as well as ice particles in continental clouds.

It is difficult to deduce the precipitation process acting in cool continental clouds from the incomplete and somewhat conflicting observational evidence outlined. Precipitation may be initiated by either cloud drop accretion or deposition on ice nuclei. The former is thought to require drops larger than 20–25 $\mu$m radius, which are rare in continental clouds. The concentration of ice nuclei is too small to develop significant precipitation at temperatures warmer than $-15$ to $-20$°C. As will be seen in section 8.12, ice multiplication accompanies riming of ice particles at temperatures of $-5 \pm 2$°C by cloud drops of more than 12 $\mu$m radius. However, riming of ice crystals does not begin until the larger crystal dimension exceeds a few hundred micrometers. These barriers to precipitation seem difficult to surmount, but nature does so with great regularity.

The approximate agreement between the ice particle concentration and that of the ice nuclei in the clouds probed by Gagin suggests that the precipitation (ice) particles have been formed by deposition on ice nuclei. However, these clouds were penetrated within 300 m of the cloud tops, and the results may not be representative of conditions farther down in the clouds. In other papers Gagin gives further evidence that the precipitation is initiated by the ice crystal process but that most of the subsequent growth of the particles is due to riming accretion. The shallower continental clouds referred to earlier were found to contain liquid drizzle drops as well as ice particles. It would seem that the liquid drops must have formed by accretion.

The finding of drizzle-size liquid drops in cool convective clouds in continental air may be due to the presence of cloud drops of radii greater than 20–25 $\mu$m or processes that enhance the collision efficiency of collector drops in the 15–20 $\mu$m radius range. The former could result from a small number of giant nuclei on which large drops could grow even though the total droplet concentration were typical of continental air. The latter could result from microturbulence or electrical forces, both of which may act to increase the collision efficiency of small drops as discussed in sections 7.23 and 7.30.

The presence of drizzle drops is evidence for the operation of a warm rain accretion process, as already concluded from studies of the radar first echoes. This process must start in the lower region of the cloud because of the high vertical velocities and the time constraints. Many, if not most, of the precipitating cool convective clouds are probably multicellular, making it possible for some immature precipitation particles to be recycled in a subsequent cell.

Ice seems to be a common, and possibly universal, component of the cool clouds. If the cloud top is cold enough, perhaps −20°C or lower, the precipitation may be initiated in the classical way by deposition on ice nuclei. In warmer clouds the precipitation seems to be initiated by liquid accretion, followed by an ice multiplication process that freezes the large water drops and also initiates ice crystals. Subsequent growth then results from the accretion of supercooled cloud drops on the ice particles to form graupel. Finally, liquid accretion adds further to the mass of the precipitation below the melting level.

## 8.12 Ice Multiplication

It has already been pointed out that the observed concentrations of ice crystals at temperatures of −5 to −15°C are orders of magnitude larger than those of ice nuclei at these temperatures. Both Hobbs (1969) and Auer et al. (1969) found that the ratio of ice particles to ice nuclei concentrations increased exponentially with temperature, ranging from near unity at −25 to −35°C to $10^3$ to $10^4$ at −5°C. Mossop (1970) has given a summary of the observational evidence and of the suggested explanations, none of

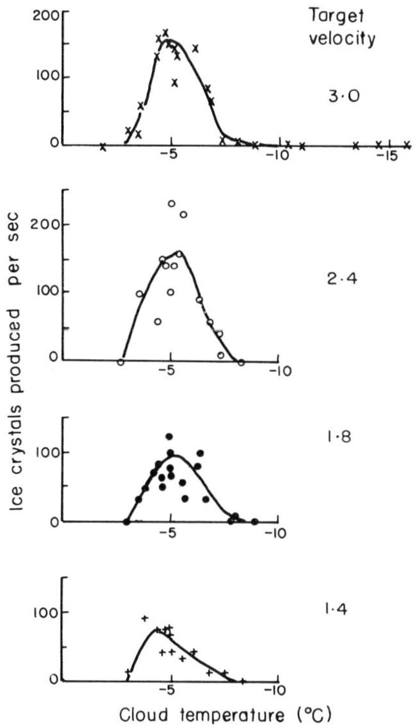


Figure 8.5 The rate of production of secondary ice crystals during growth of rime upon a moving rod 30 cm long and 1.8 mm diameter, as a function of cloud temperature and rod velocity. The target velocities are in m $sec^{-1}$. After Mossop (1976).

which seemed adequate at the time. The larger ratios at warm temperatures make it certain that this so-called ice multiplication is a real phenomenon and not a result of the admitted uncertainties of the measurements of ice nuclei.

Following up a number of suggestions that ice multiplication accompanied the growth of graupel in a supercooled cloud, Hallett and Mossop (1974) carried out laboratory studies of riming in which the temperature and the size of the supercooled drops could be controlled. Shortly thereafter, Mossop (1976) repeated the experiments to refine the results, and that paper is the basis for the following discussion.

The most important results of these experiments are that riming releases secondary ice particles, but only in the ambient temperature range of $-3$ to $-8$°C and only when drops larger than 24 $\mu$m diameter are accreted. These rather stringent limits probably explain the failure of earlier riming experiments to detect ice multiplication.

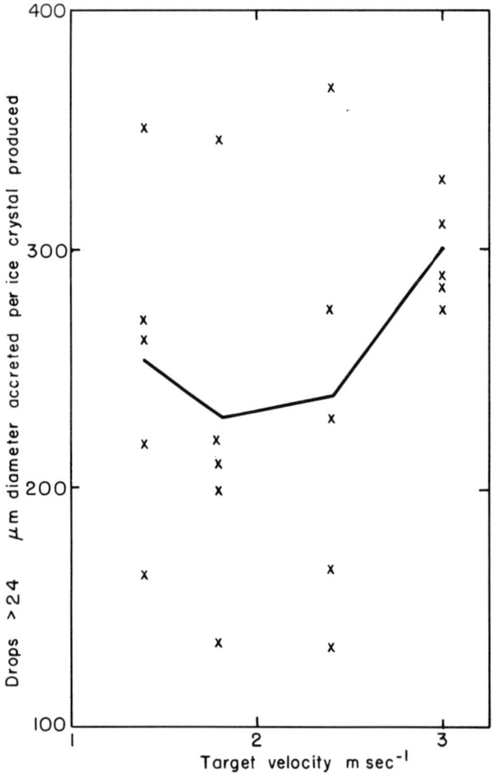


Figure 8.6 The number of drops of diameter greater than 24 $\mu$m accreted per secondary ice crystal produced as a function of target velocity. After Mossop (1976).

Some of the experimental results are shown in figure 8.5. These data were obtained with a rotating rod in a supercooled cloud. The four sets of data in figure 8.5 correspond to different rotation rates of the rod or to different target velocities, but the apparent differences in the number of ice particles is accidental. Figure 8.5 is intended to show only the sharp upper and lower bounds of temperature and that these limits are not sensitive to the target velocity. Figure 8.6 shows the number of drops larger than 24 $\mu$m diameter accreted per secondary ice particle formed. The scatter of the data, probably due to differences in the size spectra of the cloud drops, does not suggest any relation to target velocity. The average is about 250 drops of diameter $>24\ \mu$m per secondary ice particle formed.

These experimental results seem able to explain many of the observations of ice multiplication of ice particles in natural clouds. However, these results cannot be fully accepted until the physical mechanism of splintering during riming is understood. There is a need to conduct experiments on free-falling ice pellets of various sizes instead of a rotating rod. Mossop and Hallett feel that their results can be applied to graupel particles of 0.5–2.0 mm diameter. Under typical conditions the graupel particles will be subliming, and this may be a factor in the splintering mechanism.

## 8.13 Graupel

In the discussion of cool clouds it has been brought out that graupel is a characteristic type of precipitation in these clouds. Graupel particles are usually white and friable, but some approximate solid ice. Graupel pellets occur as irregular lumps and in roughly conical and hexagonal shapes. The sizes of graupel pellets, given by their average diameters, range from a few tenths to five or more millimeters. The bulk density of graupel pellets ranges widely from less than 0.1 to near 0.9. The density increases with the fall speed and the size of the accreted drops and decreases with decreasing temperature.

The mass and fall speed of graupel pellets must be obtained from direct measurements and are not uniquely related to the average diameter because of the variations of shape and bulk density. Of the several sets of such data, that of Locatelli and Hobbs (1974) has been chosen as an example. The curves in figures 8.7 and 8.8 were obtained by averaging the empirical expressions fitted by Locatelli and Hobbs to their extensive observations for lump and conical graupel. These curves should be considered only as rough approximations because of the great natural variability of graupel pellets.

Graupel pellets form by the accretion of supercooled cloud drops on frozen drops or on ice crystals. As has already been seen, frozen drops appear primarily at relatively warm temperatures by the freezing of drizzle-size drops formed by the warm rain process. These large frozen drops rime rapidly because of their size and large fall speed. Ice crystals do not begin to accrete cloud drops until their long dimensions are several hundred micrometers, and this crystal growth requires some time. Graupel pellets are found in convective clouds because the large vertical air velocity maintains a supercooled cloud against the depleting effects of the growth of precipitation.

## 8.14 Growth of Graupel by Accretion and Deposition

It has already been pointed out that simultaneous growth by condensation and accretion is an important factor in the warm rain process. This is also true when graupel pellets grow in a supercooled cloud, but this is a somewhat more complex problem. The accreted supercooled cloud drops freeze, releasing the latent heat of fusion that raises the temperature of the

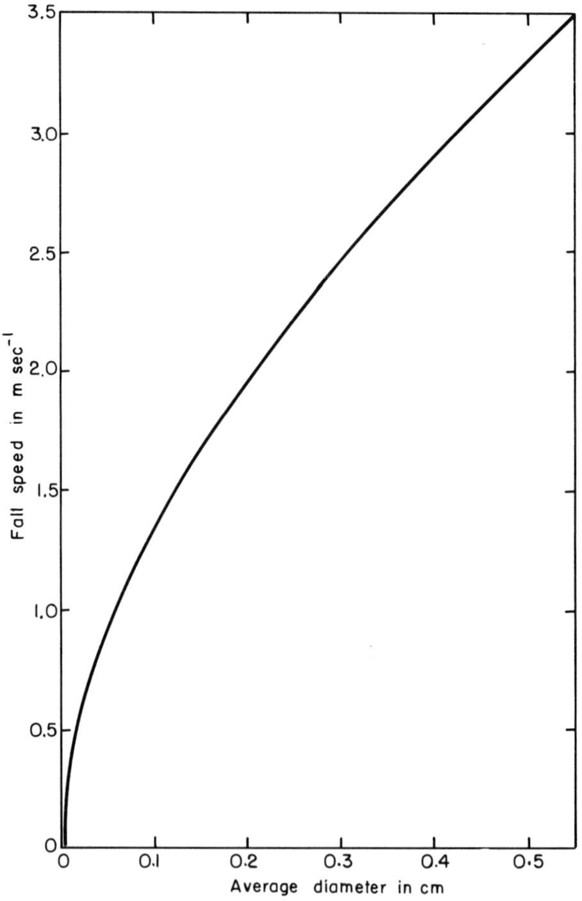


Figure 8.7 Fall speeds of graupel pellets versus average pellet diameters—a rough average curve based on Locatelli and Hobbs (1974).

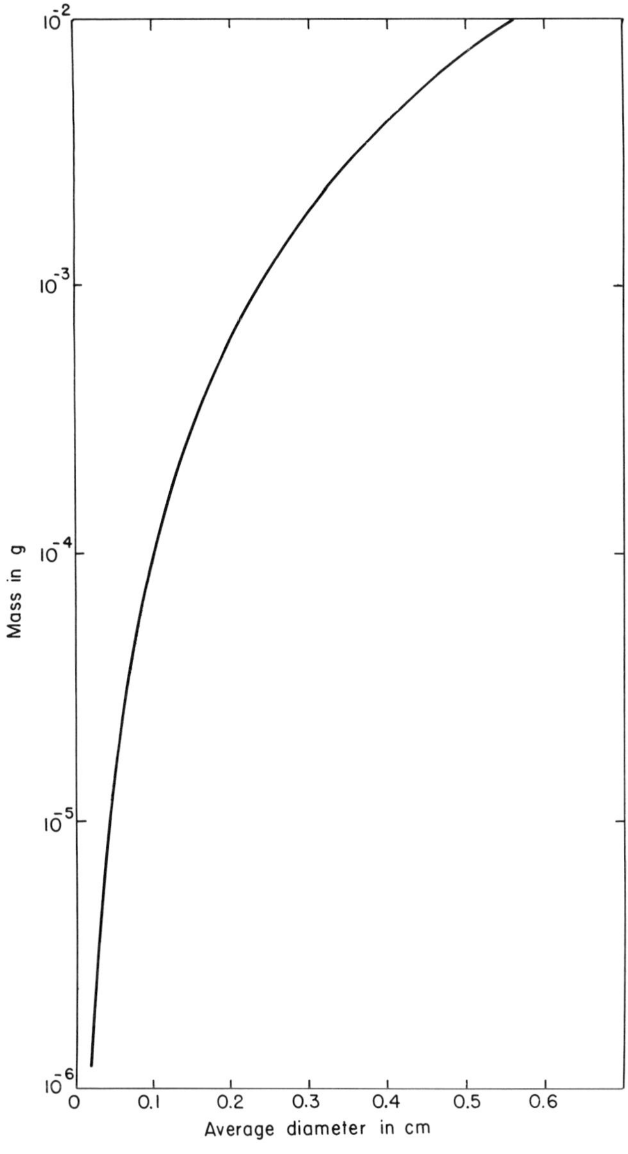


Figure 8.8 Mass of graupel pellets as a function of the average pellet diameter—a rough average curve based on Locatelli and Hobbs (1974).

graupel particles. This reduces the growth rate by deposition and, with a sufficiently large accretion rate, may change deposition to sublimation (evaporation).

The heat balance of a graupel pellet growing simultaneously by deposition and accretion may be written

$$L_s(dm/dt)_d + (L_f - c\,\Delta T)(dm/dt)_a = 4\pi C c_v K \Delta T, \tag{8.1}$$

where $L_s$ and $L_f$ are the latent heats of sublimation and fusion at the particle temperature,[2] $c$ is the specific heat of supercooled water, $\Delta T$ is the difference between the temperature of the pellet and the ambient temperature, the subscripts $d$ and $a$ stand for deposition and accretion, $C$ is the shape factor, $c_v$ is the ventilation coefficient, and $K$ is the heat conductivity of air. Assuming that the ventilation coefficients for heat and mass transfer are equal and that the radii and fall speeds of the cloud drops are negligible compared with those of the graupel pellets gives

$$(dm/dt)_d = 4\pi C c_v D\,(\rho_{am} - \rho_{oi}), \tag{8.2}$$

$$(dm/dt)_a = AEVw = Gw, \tag{8.3}$$

where $\rho_{am}$ is the vapor density of the environment (taken as equilibrium over water at the ambient temperature), $\rho_{oi}$ is the vapor density over ice at the pellet temperature, $A$ is the horizontal projected area of the pellet, $V$ is its terminal velocity, $w$ is the liquid water content of the cloud, and $G$ is the accretion kernel. Combining (8.1)–(8.3) gives

$$w = \frac{4\pi C c_v [K\Delta T - L_s D(\rho_{am} - \rho_{oi})]}{G(L_f - c\Delta T)}. \tag{8.4}$$

The value of $w$ at which the depositional growth just vanishes is found by setting $\rho_{am}$ equal to $\rho_{oi}$:

$$w^* = \frac{4\pi C c_v K \Delta T^*}{G(L_f - c\Delta T^*)}, \tag{8.5}$$

where $\Delta T^*$ is the temperature increment above ambient at which $\rho_{am}$ and $\rho_{oi}$ are equal. It may be derived from the Clausius-Clapeyron equation, but it is easily found from tables of the equilibrium vapor densities over ice and water. It ranges from about 0.6°C at −5°C to 2.2°C at −20°C.

---

2. This assumes that the latent heat of fusion is released at the pellet temperature. This yields essentially the same result as the more usual path, in which the accreted water is first warmed to 0°C, is frozen at 0°C and the ice is then cooled to the pellet temperature.

Table 8.1
The limiting values of liquid water content at which depositional growth of graupel pellets just vanishes

| | Limiting liquid water content in $g\,m^{-3}$ | |
|---|---|---|
| Pellet radius (cm) | $-5$°C and 900 mbar | $-15$°C and 600 mbar |
| 0.025 | 4.6 | 13.0 |
| 0.05 | 2.6 | 7.2 |
| 0.10 | 1.5 | 4.2 |
| 0.15 | 1.1 | 3.1 |
| 0.20 | 0.9 | 2.5 |
| 0.25 | 0.8 | 2.1 |

A few numerical values of $w^*$ are given in table 8.1, where it is assumed that the shape factor $C$ is equal to the mean radius. The numerical results are dependent on the values selected for $V$, $c_v$, and $G$. The pellet fall speeds were taken from figure 8.7, the ventilation coefficients from (7.17), and $G$ from the work of Pflaum and Pruppacher (1979). To the extent permitted by approximations and the choices of parameters, table 8.1 shows that, in view of the relatively large liquid water contents of convective clouds (say 1 to 2 $g\,m^{-3}$), depositional growth of graupel pellets will be suppressed, or reversed, only for the larger pellets and at relatively warm temperatures.

By defining $F$ as the fraction of the total growth rate due to accretion, it is found from (8.2) and (8.3) that

$$\frac{1}{F} = 1 + \frac{4\pi C c_v D(\rho_{am} - \rho_{oi})}{Gw}. \tag{8.6}$$

Since the temperature of the pellet does not exceed ambient by more than 2°C, a simple empirical linear expression for $\rho_{oi}$ may be used instead of the exponential that arises from the Clausius-Clapeyron equation:

$$\rho_{oi} = \rho_{iam} + k\,\Delta T, \tag{8.7}$$

where $\rho_{iam}$ is the vapor density over ice at the ambient temperature and the coefficient $k$ may be found by inspection of a table of the equilibrium vapor density over ice.

By defining the ratio of the liquid water content to $w^*$ as $f$ and inserting (8.2), (8.3), and (8.7) in (8.1), it is found after some algebra that

$$\frac{1-F}{F} = \frac{kD(L_f - c\,\Delta T)}{K + kDL_s}\,\frac{1-f}{f}. \tag{8.8}$$

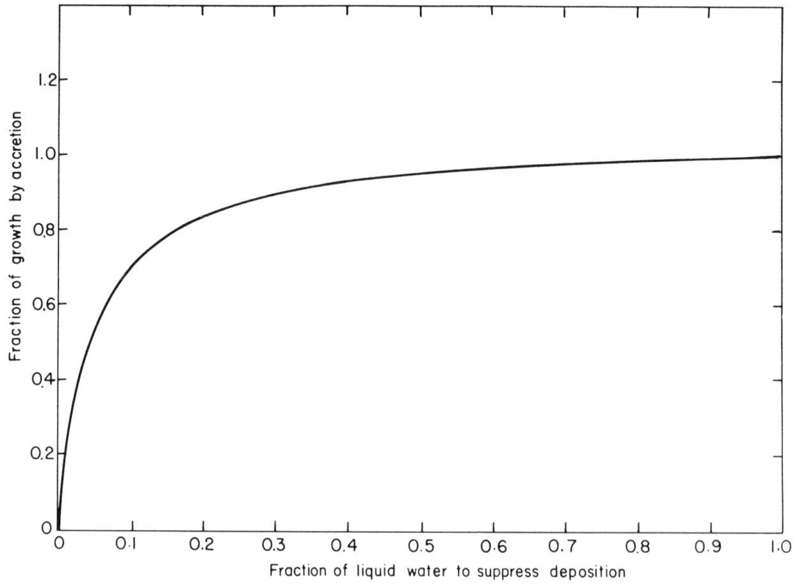


Figure 8.9 Fraction of graupel growth by accretion as a function of the fraction of the liquid water required to just suppress deposition. Pressure is 600 mbar and temperature is −10°C.

Equation (8.8) is plotted in figure 8.9 for a temperature of −10°C and a pressure of 600 mbar. Although most of the quantities in (8.8) are dependent on temperature and/or pressure, compensating effects make the dependence of (8.8) on temperature and pressure small, permitting the use of figure 8.9 for most ambient conditions.

The rapid rise of the curve in figure 8.9 from the origin shows that accretional growth becomes dominant already at a small fraction of the liquid water content required to suppress growth completely by deposition. This, together with the large condensation rate in convective clouds, which maintains a supercooled cloud of relatively high liquid water content, seems to explain why graupel pellets are the most common type of precipitation in cool clouds.

## 8.15 Hail

Hail falls from thunderclouds, but only a small fraction of thunderstorms yields hail at the surface. In the United States this fraction ranges from near zero in Florida to about one-fifth in the Great Plains and the eastern slopes of the Rocky Mountains, where there are 5–9 days with hail per year.

Relatively rare as it is, hail causes severe crop damage, and many growers find it desirable to purchase hail insurance. Largely as a result of crop damage, hail has been extensively studied in the U.S.S.R., the United States, France, Italy, and Switzerland. Various schemes, both fanciful and sound, for reducing damaging hail have been proposed and tried. The scientific literature on hail is extensive, but only a brief account can be given here. The reader is referred to Foote and Knight (1977) for an extended discussion of hail and possible methods for its supression.

The largest hailstones at the surface typically are 1 to 2 cm diameter. Emphasis in the public press is usually placed on hailstones of record size, which may have diameters of 15 cm or even more, but these are very rare. In a given region of a hail streak the hailstones are of rather uniform size, the ratio of the largest to the smallest hailstones seldom exceeding about three. This is probable due to sorting by gravity and wind shear. In partial confirmation, the last hailstones to fall are smaller than the first. It would be anticipated that a broader spectrum of hail size would be found well up in the cloud, before sorting can take effect, and there is some observational evidence for this.

Shapes of hailstones have been classified as spheroidal, ellipsoidal, conical, and irregular. The most common shapes appear to be spheroidal and ellipsoidal, but the very large stones are apt to be irregular. The large stones often have rounded proturberances or lobes. A well-known feature of large hailstones is their layered or onionlike structure. The layers may be composed of nearly clear ice, opaque ice, or bubbly ice. The bulk density of the stones ranges from about 0.8 to 0.92, the latter being that of solid ice. Most hailstones are relatively hard and bounce on impact, but a few are spongy, and some of these contain unfrozen water.

An important property of hailstones is their free-fall velocities. Because of their irregular shapes, hailstones in free fall may rotate around one or more axes, oscillate, and gyrate. Because of these and other complexities, the terminal velocities of hailstones cannot be reliably calculated from drag coefficients for smooth spheres. To give some idea of the fall speeds of hailstones, figure 8.10 has been computed for a drag coefficient of 0.55, considered to be a reasonable value for hail, and an average air density corresponding roughly to the region of hail growth.

## 8.16 Growth of Hailstones

Hail grows by the accretion of supercooled cloud and precipitation water. Examination of a section through the center of a hailstone almost always reveals an embryo that served as the particle on which the accretion leading to hail began. The hail embryo or nucleus is not to be confused with the ice-forming nuclei discussed earlier. Hail embryos are large ice particles of a few millimeters to a centimeter across. In the majority of hailstones the

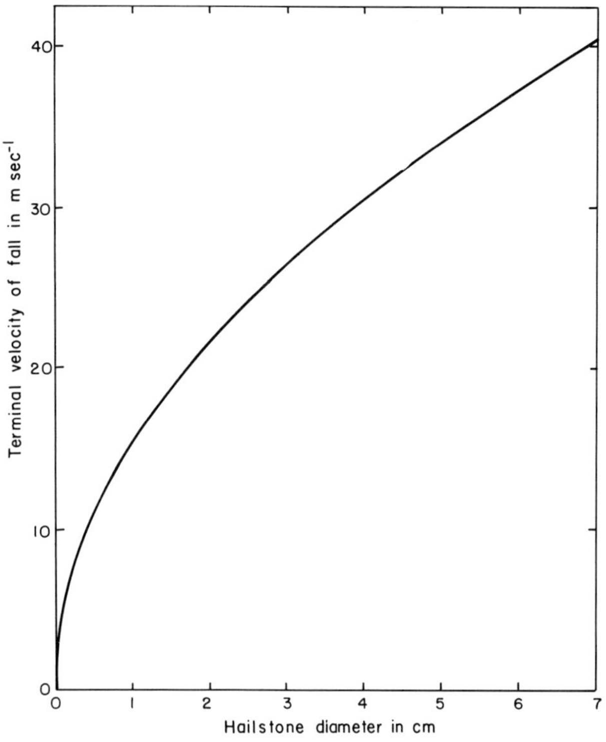


Figure 8.10 Approximate terminal fall speeds of spheroidal hailstones.

embryo appears to be a graupel particle, but large frozen drops and ice crystals are also found. The most obvious feature of hailstones is their size, by far the largest precipitation particles. Their large size must be attained in deep supercooled clouds of relatively large liquid water content and by a rather long sojourn in the region of growth. Because of their large fall speeds, strong updrafts are needed to keep them from falling out of the supercooled region of the cloud prematurely. There is reason to believe that, in many cases, the hailstones make excursions into more than one convective cell.

The basic accretional growth process of a hailstone is the same as that of graupel particles, but the details are much more complex. Heat transfer to a rotating, oscillating, and/or gyrating hailstone is different from that to a smooth sphere. The many uncertainties limit the accuracy of computations of the growth of hail to a demonstration that hailstones of the observed sizes can be formed in realistic clouds.

Inspection of (8.5) and table 8.1 for graupel shows that depositional growth of hailstones, which are larger than graupel pellets, is unimportant. Rather, it is likely that most hailstones sublime, and this must be considered in the heat balance. A hailstone gains heat from the release of the latent heat of fusion of the accreted water and loses heat by conduction to the air and sublimation. If the accretion rate is sufficiently large, the temperature of the hailstone surface may reach 0°C, and the hailstone is unable to freeze all of the accreted water and enters the wet growth regime. It is believed that the wet regime leads to a clear ice layer, while the more opaque, whitish layers are a result of dry growth. It has been observed that the larger supercooled drops may bounce from a hailstone in wet growth and that portions of the water layer may tear off, both processes acting to reduce the net accretion.

Equation (8.4) may be used to determine the onset of wet growth by setting $\rho_{oi}$ equal to the vapor density over water at 0°C and $\Delta T$ equal to the supercooling, $\Delta T = (273 - T_{amb})$, where $T_{amb}$ is the ambient temperature in degrees Kelvin. Because of uncertainties about the values of the ventilation coefficients and the collection kernel, the ratio of (8.4) to (8.5) is more reliable than $w_*$. This ratio ranges from about 15 to 25 for typical ambient conditions; it increases slowly with decreasing pressure and temperature. These ratios seem large, but $w_*$ is also quite small for moderate to large hailstones at relatively warm temperatures. Uncertain as they are, computations of the limiting liquid water content for wet growth make it plausible that wet growth is not uncommon on hailstones of typical size at relatively warm temperatures. Rather similar results were reported by Ludlam (1958) many years ago.

Most computations of hail growth, including those just presented, assume spherical hailstones and that the nascent water layer is uniform over the sphere. It is more reasonable to expect that the accreted water will be non-uniformly distributed and that there will be surface temperature gradients and possibly both dry and wet areas on the same hailstone.

## 8.17 Precipitation from Cold Clouds

By the classification adopted here, cold convective clouds are colder than 0°C throughout. Cold clouds are to be found primarily in the winter in midlatitudes. Because of the low temperatures of such clouds, the convection is less vigorous than in cool clouds, and the condensation rate is also smaller. Examples of cold clouds are those that produce snow showers at the surface and the so-called generating cells often observed by radar at midtropospheric levels. The streamers of ice crystals falling from such convective cells are markedly affected by wind shear because of the relatively slow fall speeds of the crystals, usually about 1 m $sec^{-1}$. On occasion the streamers fall into a lower stratiform supercooled cloud, releasing a substantial shower either of snow or rain.

It is plausible that the precipitation from a cold cloud is both initiated and grown in the ice phase. The precipitation is typically in the form of aggregates of snow crystals and individual crystals. Riming of the crystals is commonly observed particularly in the more vigorous cold clouds.

## 8.18 Precipitation Efficiency

Although other definitions have been proposed, the precipitation efficiency as used here is the ratio of the mass of the precipitation to the mass of the condensate. On the basis of the observations made during the Thunderstorm Project, Braham (1952) estimated the precipitation efficiency of a typical air mass thunderstorm to be 19 %. In arriving at this efficiency, Braham estimated that 45 % of the condensate was evaporated in the downdrafts and 36 % represented evaporation from the sides of the cloud plus that left behind in the residual cloud. From other information in Braham's paper I estimate that one-third to one-half of the 36 % represents the condensate in the residual cloud. This says that nearly as much condensate is left behind and subsequently evaporates as is precipitated.

With the possible exception of cold clouds, most of the mass of the precipitation is gained by accretion both in the warm rain process and in the formation of graupel particles in cool clouds. Even in cold clouds, accretion of ice crystals that forms aggregates is an important part of precipitation growth. Accretion can be thought of as a sweeping process, and the fraction of the condensate that is swept out is an important part of the precipitation efficiency. Assuming that 15 % of the condensate is left in the residual cloud, the estimates of Braham suggest a sweeping efficiency of 19/(19 + 15), or about one-half.

This matter may be pursued further with the aid of observed raindrop size spectra and the fall speeds of the raindrops. The fraction of a horizontal area swept per second is[3]

$$\Lambda = \pi \sum Enr^2V, \tag{8.9}$$

where the summation is to be taken over the raindrop size spectrum and the other quantities have been defined earlier. Houghton (1968) had computed $\Lambda$ for a number of spectra of rain from convective clouds ranging from warm showers to thundershowers. The accretion efficiency $E$ was taken as unity for the present purposes. The quantity $\Lambda$ is plotted against rainfall rate in figure 8.11.

As an example, if the rainfall rate from a convective cell of a thunderstorm is 15 mm hr$^{-1}$ and its duration is 0.5 hr, $\Lambda$ is about 0.0034, the horizontal area is swept six times, and all but 0.2 % of the cloud drops is

3. This quantity is called the washout efficiency in studies of the scavenging of particulates by rain.

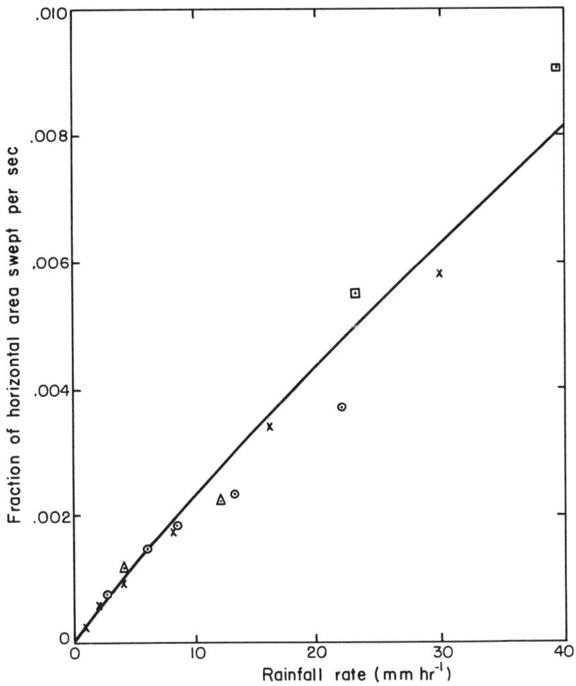


Figure 8.11 Fraction of a horizontal area geometrically swept per second by rain from convective clouds as a function of the rainfall rate. Symbols identify points from different locations and, usually, different methods of measurement.

captured (negative exponential of the product of $\Lambda$ and the duration). For a modest warm shower with a rainfall rate of 4 mm hr$^{-1}$ and a duration of 10 minutes, only about half of the horizontal area is swept and less than half of the cloud drops is collected.

From these and other examples it is clear that the sweeping efficiency is high for moderate to heavy rainshowers. However this applies only at or near cloud base because figure 8.11 is based on data collected beneath the clouds. In a steady state, the product $Vn$ is constant, and (8.9) then shows that $\Lambda$ is proportional to $r^2$. Thus further up in the cloud, where, for example, the radius is only half of its final value, $\Lambda$ will be only a quarter of that at cloud base. The purpose of this simplified analysis is to show that incomplete sweeping in the earlier stages of accretional growth is inherent in this precipitation process.

## 8.19 Numerical Models of Convective Clouds

The interplay between the cloud dynamics and the microphysical processes is very complex, and simple numerical models represent an important means of examining some of the results of this interaction. A considerable number of numerical models of convective clouds have been developed, and no attempt will be made here to review them. An excellent critical survey of these models has been prepared by Cotton (1975).

The characteristics of individual models depend on the use to which they are put and often on the size and speed of the available computers. For some purposes detailed microphysics must be incorporated, while for others the dynamics is considered more important. Relaxation of the details of the microphysics is commonly achieved by parameterization. The dynamical processes may be simplified by parameterization, approximations in the basic equations, and, most drastically, assuming one-dimensionality. As with other numerical models, the aim is to capture the essence of the physics in as simple a form as possible. As the model becomes more complex, it becomes more difficult to understand its workings, and a basic purpose of modeling is lost. Thus the selection of the simplifications and approximations becomes the most difficult task of the modeler. This must be based on careful and perceptive observations and measurements of the prototype, which also serve as the basis for the verification of the model.

## 8.20 One-Dimensional Models

The simplest numerical models of convective clouds are one-dimensional and time dependent. Typical examples are the models of Simpson and Wiggert (1969) and Weinstein (1970). These are based on the vertical component of the equation of motion, the first law of thermodynamics, and parameterizations of entrainment and the conversion of cloud water to precipitation, although some models incorporate more complete cloud microphysics.

The fractional entrainment is parameterized by

$$\frac{1}{\mathrm{M}}\frac{dM}{dz} = \frac{c}{R}, \tag{8.10}$$

where $M$ is the mass of air in the cell, $R$ is the radius of the cell, and $c$ is a constant. This expression is based on laboratory models, observations of real clouds, and Levine's (1959) spherical vortex model. The constant $c$ is often taken as 0.2, but larger values up to 0.6 have been employed.

The conversion of cloud water to precipitation has usually been parameterized through Kessler's (1969) autoconversion expression

$$-\frac{dw}{dt} = k(w - w_0), \tag{8.11}$$

where $w$ is the cloud liquid water content and $k$ and $w_0$ are constants. The loss of cloud water to precipitation, $-dw/dt$, is also the rate of increase of precipitation water content. when $w \leq w_0$, $w - w_0 \equiv 0$. The form of (8.11) was suggested to Kessler (1969) by the observation that rain did not seem to form until the cloud liquid water content exceeded some limit on the order of 1 g per cubic meter of air. A typical value of $w_0$ is 0.5 g m$^{-3}$ and of $k$ is $10^{-3}$ sec$^{-1}$. The conversion rate is more sensitive to the choice of $w_0$ than of $k$.

A more complex autoconversion equation has been proposed by Berry (1968) and used in some models:

$$-\frac{dw}{dt} = \frac{w^2 \times 10^6}{120 + 1.596 n/w \times 10^6 \delta_0} \text{ g cm}^{-3}\text{ sec}^{-1}, \tag{8.12}$$

where $n$ is the cloud drop concentration per cubic centimeter of air and $\delta_0$ is the relative dispersion of the cloud drop size spectrum.

In these one-dimensional vertical models, the time dependence is introduced through the entrainment and the increasing weight of the condensate and the precipitation. The buoyancy is finally overcome by these factors, and downward motion ensues. The principal virtues of these models are their simplicity and concomitant modest computational demands and that the initial condition may be an actual vertical sounding. Their disadvantages are due to their one-dimensionality, which requires an arbitrary choice of the cell radius and eliminates much of the cloud dynamics. In the hands of their creators, these models have had modest success in predicting the cloud top height and the increase in cloud height due to cloud seeding designed to release the latent heat of fusion.

## 8.21 Two-Dimensional Models

Nearly all of the more recent models are two-dimensional. They are essentially models in the $(X, Z)$ plane, but are often referred to as cylindrical models in which the properties are the same in all radial directions. Some examples are the models of Wilhelmson and Ogura (1972), Koenig and Murray (1976), and Takahashi (1978a).

Introduction of the second dimension greatly improves rendition of the cloud dynamics, at the expense of much greater computational demands. Grid sizes of a few hundred meters and time steps on the order of 10 sec are required. It is important to note that no semiempirical expression like (8.10) is required to introduce entrainment.

The dynamical equations are usually in the form of a vorticity equation, and the Boussinesq approximation is often used, particularly for fairly shallow convection. It would be expected that the trade-off for the enhanced cloud dynamics would be a rather cavalier treatment of the cloud microphysics. However, the rapidly increasing power and speed of the computer has made it possible to incorporate many microphysical processes, though usually in parameterized form. Thus there are models that incorporate the ice phase simultaneously with the liquid phase, growth by deposition as well as by accretion, evaporation, sublimation, and elaborate schemes for developing size spectra of the particles. In fact, it seems, in some cases, that the details go beyond our knowledge of the microphysical processes and their interactions.

These models have been used to study the details of the precipitation process and also to explore the effects of cloud seeding on cloud dynamics and precipitation. As noted earlier, the models must be tested against the prototypes before they can be used for prediction with any real confidence. Unfortunately, our quantitative observations are still quite limited. Recent developments in measurement techniques offer promise of a more adequate observational base, but the inherent difficulties and expense of field observations will make progress slow. Continued work with models is very desirable because they can pose the significant questions to be answered by field observations.

## 8.22 Stratiform Precipitation

By this term is meant the synoptic-scale precipitation that results from large-scale lifting, typified by the precipitation shield of an extratropical migratory cyclonic system. Much of the precipitation from such storms results from convective activity, but in what follows, only the steadier precipitation resulting from the large-scale lifting will be considered.

The vertical air velocities in stratiform systems do not usually exceed a few tenths of a meter per second. The horizontal extent of the precipitation is large, the system occupies much of the troposphere in depth, and the lifetime is on the order of days, although some stratiform cloud laminae are ephemeral. Because of the relatively small vertical air velocities, the precipitation particles fall with respect to the surface after they have attained moderate size. It can be anticipated that the precipitation will be initiated as ice in the upper portions of the cloud system. The lifetime of the system is large enough that it is not usually a limiting factor in the growth of precipitation particles; rather, the limit on particle growth is set by the rate of condensation imposed by the field of vertical motion and the time the particles take in falling through the regions favorable for growth.

## 8.23 Model of Stratiform Precipitation

Certain important aspects of the growth of stratiform precipitation may be studied with the aid of a simple numerical model. Following Houghton (1968), the model is based on that used by Wexler and Atlas (1958). Their

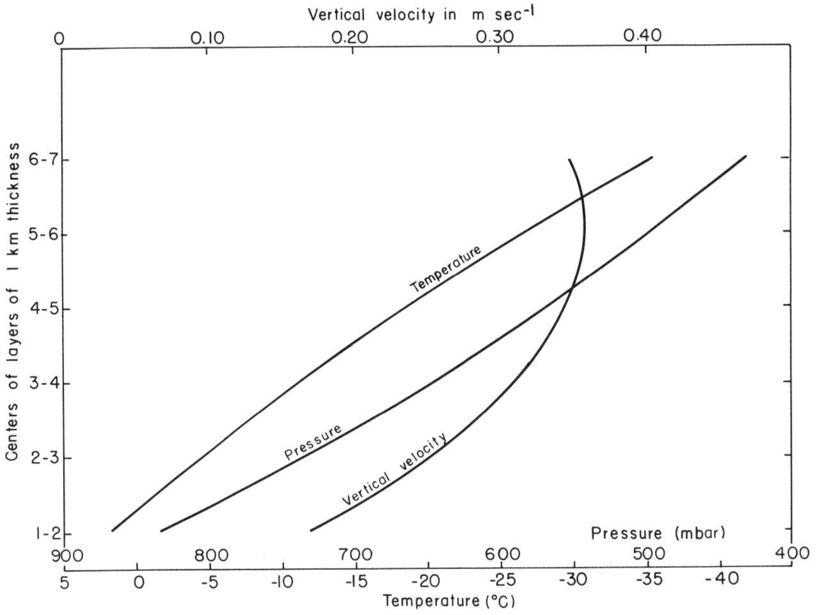


Figure 8.12 Mean properties of the 1-km layers used in the model of stratiform precipitation. Points plotted at center of 1-km layers. Curves drawn through these points are for visual convenience only.

model atmosphere follows the 10°C moist adiabat through the troposphere and is assumed to be at water saturation throughout. On the basis of observations, the product of the vertical velocity and the air density is assumed to vary parabolically with pressure, being a maximum at 600 mbar and zero at 1,000 and at 200 mbar. The magnitude of the vertical velocity was chosen to yield a precipitation rate of 5 mm hr$^{-1}$, assuming that all of the condensate falls out as precipitation. For the present purposes it suffices to consider average values over layers of 1 km depth. Further, the lowest layer was assumed to be cloud-free. The principal properties of this model atmosphere are shown in figure 8.12.

The condensation rate for each layer was computed from a modification of Fulks's (1935) treatment,

$$\text{condensation rate} = \frac{2.167}{T}\frac{p}{p - e_{\text{m}}}|\gamma_{\text{m}}|\frac{de_{\text{m}}}{dT} - \frac{7.40e_{\text{m}}}{T^2}, \tag{8.13}$$

where the condensation rate is in g m$^{-3}$ sec$^{-1}$ for a vertical velocity of 1.0 m sec$^{-1}$, $e_{\text{m}}$ is the equilibrium vapor pressure over water in mbar, $p$ is atmospheric pressure in mbar, and $|\gamma_{\text{m}}|$ is the absolute value of the moist adiabatic lapse rate in °C 100 m. The condensation rate is linearly dependent

Table 8.2
Condensation and deposition rates in model stratiform cloud

| Layer (km) | Mean temperature (°C) | Crystal concentration ($m^{-3}$) | Crystal habit | Condensation rate ($g\,m^{-3}\,sec^{-1}$) | Deposition rate ($g\,m^{-3}\,sec^{-1}$) at indicated time (sec) | | | |
|---|---|---|---|---|---|---|---|---|
| | | | | | 100 | 500 | 1,000 | 2,000 |
| 2–3 | −4.7 | $1.57 \times 10^{-1}$ | Needles | $3.18 \times 10^{-4}$ | $5.2 \times 10^{-11}$ | $1.1 \times 10^{-10}$ | $6.1 \times 10^{-10}$ | $2.7 \times 10^{-9}$ |
| 3–4 | −11.5 | $1.55 \times 10^{1}$ | Thin plates | $2.89 \times 10^{-4}$ | $1.4 \times 10^{8}$ | $7.1 \times 10^{-8}$ | $2.2 \times 10^{-7}$ | $6.5 \times 10^{-7}$ |
| 4–5 | −19 | $2.75 \times 10^{3}$ | Narrow arm stellars | $2.18 \times 10^{-4}$ | $3.0 \times 10^{-6}$ | $4.7 \times 10^{-5}$ | $2.0 \times 10^{-4}$ | $7.2 \times 10^{-4}$ |
| 5–6 | −27 | $3.75 \times 10^{5}$ | Thick plates | $1.37 \times 10^{-4}$ | $3.4 \times 10^{-4}$ | $1.7 \times 10^{-3}$ | $4.9 \times 10^{-3}$ | $1.4 \times 10^{-2}$ |
| 6–7 | −36 | $3.25 \times 10^{6}$ | Hollow columns | $7.1 \times 10^{-5}$ | $1.6 \times 10^{-3}$ | $5.2 \times 10^{-3}$ | $1.53 \times 10^{-2}$ | $4.6 \times 10^{-2}$ |

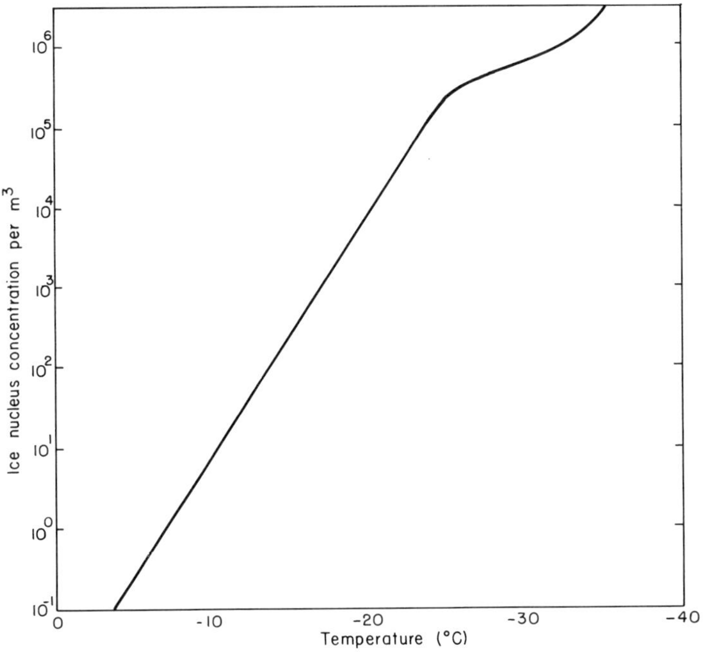


Figure 8.13 Ice nuclei activity spectrum adopted for model of stratiform precipitation.

on the vertical air velocity. The condensation rate for each layer of the model atmosphere is given in the fifth column of table 8.2.

It is assumed that the number of ice crystals nucleated in each layer is equal to the concentration of ice nuclei, since the conditions in the model cloud are generally unsuitable for ice multiplication and graupel is not often found in such clouds. The ice nucleus activity spectrum assumed here is shown in figure 8.13. This is a somewhat arbitrary choice based on an examination of a number of observed activity spectra. The effects of a different choice of the spectrum are considered later.

For each layer the rate of vapor deposition on the crystals in the layer was computed at a number of times after nucleation, with the results shown in table 8.2. These computations were based on procedures and data of the type presented in chapter 7. Ice crystal habits were selected that are appropriate to the layer temperatures. It was assumed that ice crystal growth occurred at water saturation.

## 8.24 Results from the Model

The 1–2-km layer is warm, and the lifting will form a water cloud that will contribute to the precipitation only when particles formed in higher layers arrive and grow by accretion. In the 2–3- and 3–4-km layers, ice crystals will form in the concentrations given in figure 8.13. Even after $2 \times 10^3$ sec the growth rate of these crystals is more than two orders of magnitude smaller than the condensation rate, and a water cloud will form and coexist with rimed ice crystals. This water cloud will be collected, by deposition and accretion, by the larger and more numerous crystals falling from the 4–5-km layer. In this latter layer a similar situation exists for about $10^3$ sec, but thereafter the deposition rate exceeds the condensation rate, leading to the evaporation of the water cloud and the formation of the first precipitation particles that will reach the surface.

In the 5–6- and 6–7-km layers, the deposition rate exceeds the condensation rate shortly after the nucleation of the ice crystals, and only an ice cloud will be formed. The deposition rates given in table 8.2 for these layers are fictitiously large because they were computed for water saturation. Actually, the vapor pressure will adjust so that the deposition rate is equal to the condensation rate.

Examination of table 8.2 shows that the dominant factor is the exponential increase in the crystal concentration with decreasing temperature. This exponential increase seems to be well established by observations, although both more and less steep activity spectra than that in figure 8.13 have been observed. Similarly, the concentration of ice nuclei at a fixed temperature is known to vary over two to three orders of magnitude in space and time. The effects of such changes on the model considered here would be to raise or lower the layer in which the first significant precipitation is initiated by something like 1 km. This is believed to be the principal adjustment to the variability of the natural ice nucleus population.

After the ice particles from the 4–5-km layer fall into the 3–4- and 2–3-km layers, the entire cloud system above the warm 1–2-km layer will consist of ice growing by vapor deposition and aggregation. This leads to the high precipitation efficiency of stratiform systems. There are some losses of condensate in the descending motion to the rear of the system, and some of the ice in the upper layers may evaporate before reaching the surface. As shown in section 7.20, individual snow crystals have masses corresponding to drizzle drops. The much more massive raindrops are formed by the aggregation of snow crystals and accretion in the warm 1–2-km layer. These processes are not included in the model.

Although the model incorporates realistic parameters, it should be considered only as an aid to understanding some of the microphysical precipitation processes. Real stratiform precipitation systems are much more complex and cannot be adequately simulated by a steady state model. In many cases, imbedded convective cells interact synergistically with the stratiform system in the formation of precipitation, so that the two are not separable. Examples of the complexity of real situations of this kind have been given by Matejka et al. (1980) in their studies of rainbands in extratropical cyclonic storms.

## 8.25 Orographic Precipitation

This form of precipitation is of considerable importance in mountainous regions. In its simplest form, an airflow with a component normal to a long mountain ridge is forced up the slope and finally descends on the other side. The forced lifting leads to condensation much like the ascent along a warm front. The height to which the resultant vertical motion extends depends on the height of the mountain range and the vertical stability of the atmosphere. Under some circumstances the vertical motion, and hence condensation, may not extend to levels where the temperature is low enough to activate a sufficient number of ice nuclei. Precipitation will then be formed rather inefficiently, much of the condensate being evaporated in the descending flow on the lee side of the barrier. (See Young (1974) for a useful study of orographic precipitation.) As in the case of warm frontal stratiform precipitation, the precipitation from stratiform orographic clouds is usually rather light. The bulk of orographic precipitation is due to convective clouds. These often result from the release of convective instability by the forced lifting; there is some contribution from the elevated heat source from solar heating of the mountain slope. The most favorable situations for the release of convective activity and hence of substantial precipitation are a deep moist layer and a fairly strong wind component normal to the ridge line. These are most likely to occur within a cyclonic storm. For example, Elliott and Hovind (1964) have shown that the principal orographic precipitation in Southern California results from the passage of occluded fronts extending from low pressure areas centered well to the north. They find that the offshore precipitation is often in the form of convective rainbands that can be traced by radar for 100 miles or more. This convective activity is enhanced by the orographic lifting leading to heavy precipitation on the western slopes. Elliott and Shaffer (1962) showed the striking effect of the vertical stability of the air on the distribution and intensity of the precipitation. When the air is stable, more of the precipitation falls on the coastal plain than in the mountains and is relatively light; the reverse is true when the air is unstable.

## 8.26 Raindrop Size Spectra

One of the earliest cloud physics measurements was that of the size of raindrops. This subject received renewed attention when microwave radar was used to study precipitation. A simplified form of the radar equation for the back-scattered power $P$ received at the radar from rain is

$$P = \frac{C}{D^2} \sum nd^6 = \frac{C}{D^2} Z, \tag{8.14}$$

where $C$ is a constant dependent on the characteristics of the particular radar, $D$ is the range, or distance, to the rain being observed, $n$ is the concentration of drops of diameter $d$, the summation is taken over the drop size spectrum, and $Z$ is the radar reflectivity factor. This equation assumes Rayleigh scattering, since raindrops are usually small compared with the radar wavelength.

If $Z$ is known as a function of the rainfall rate, the latter may be determined from the received power by means of (8.14). Marshall and Palmer (1948) obtained a substantial number of raindrop size spectra of stratiform-type precipitation near Montreal. They showed that the size spectra, for a given precipitation rate, were exponential of the form

$$n = n_0 \exp(-41I^{-0.21}d), \tag{8.15a}$$

where $I$ is the precipitation rate in mm $hr^{-1}$, $d$ is in cm, and $n_0$ is 0.08 $cm^{-4}$, independent of $I$. It was noted that the observed spectra were deficient in small drops compared with the exponential. Best (1950) developed a more complex exponential expression, which seems to fit the observed data somewhat better than the Marshall-Palmer equation, but it has not been extensively used. Some of the Marshall-Palmer raindrop spectra are shown in figure 8.14.

The relation between $Z$ and $I$ is of the form $Z = AI^B$, where $A$ and $B$ are constants for a given type of rainfall. $A$ and $B$ have been evaluated from simultaneous observations of $Z$ and $I$. Typical relations, as summarized by Battan (1973), are given in table 8.3.

It has been observed that individual raindrop spectra taken over a short time interval often deviate markedly from the exponential but that long term averages are close to exponential form. This is illustrated in figure 8.15 from Joss and Gori (1978), where an average 1-min sample is shown to deviate from the exponential, while a 512-min sample is closely exponential.

This and other evidence suggest that raindrop spectra approximate an exponential, particularly when averages are taken over a substantial period. However, $n_0$ in (8.15a) is not constant but ranges over an order of magnitude or more. Similarly, the slope of the exponential in (8.15a) varies by a factor of two or more for a given rainfall rate.

Gunn and Marshall (1958) first showed that the size spectra of snow aggregates was also an exponential of the same form as (8.15a) if $d$ is taken as the diameter of the melted aggregate:

$$n = n_0 \exp(-22.9I^{-0.45}d). \tag{8.15b}$$

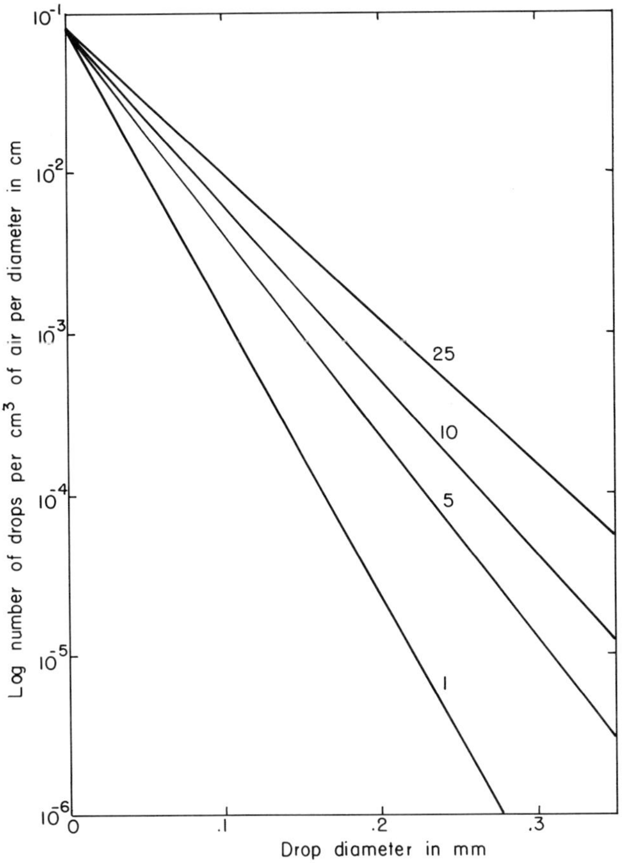


Figure 8.14 Marshall-Palmer raindrop size spectra for rainfall rates of 1, 5, 10, and 25 mm $hr^{-1}$. Based on Marshall and Palmer (1948).

Table 8.3
Empirical relations for $Z$ as a function of $I$[a]

| Type of rain | Relation ($mm^6\,m^{-3}$) |
|---|---|
| Stratiform | $Z = 200I^{1.6}$ |
| Orographic | $Z = 31I^{1.71}$ |
| Thunderstorm | $Z = 486I^{1.37}$ |

a. $I$, the rainfall rate, is in mm $hr^{-1}$.

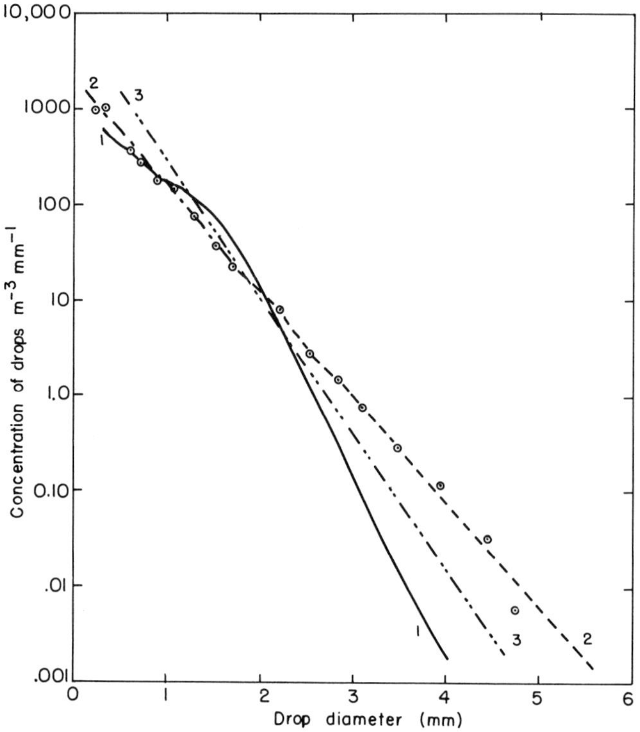


Figure 8.15 Raindrop size spectra showing the effect of a change in the sampling interval. Rainfall rate is 2.9 mm hr$^{-1}$. Key: 1, 1-min sample; 2, 512-min sample; 3, Marshall-Palmer spectrum; ○, detailed data for 512-min sample. After Joss and Gori (1978).

This was confirmed by Sekhon and Srivastava (1970), who gave slightly different values of the numerical constants. This leads to a radar reflectivity factor of about

$$z = 2{,}000 I^{2.5}\ \mathrm{mm}^6\,\mathrm{m}^{-3}. \tag{8.16}$$

## 8.27 Limits to the Size of Raindrops

The largest raindrops observed are about 5–7 mm in diameter and are typically found in thundershowers. The largest drops observed in warm rain showers are 2.5–3.0 mm in diameter. Based in part on the early work of Lenard (1904), it was assumed that aerodynamic forces caused drops larger than 6 mm to break up, and Langmuir (1948) based his chain reaction theory of warm precipitation on this assumption. In an experimental study, Blanchard (1950) found it was relatively easy to suspend

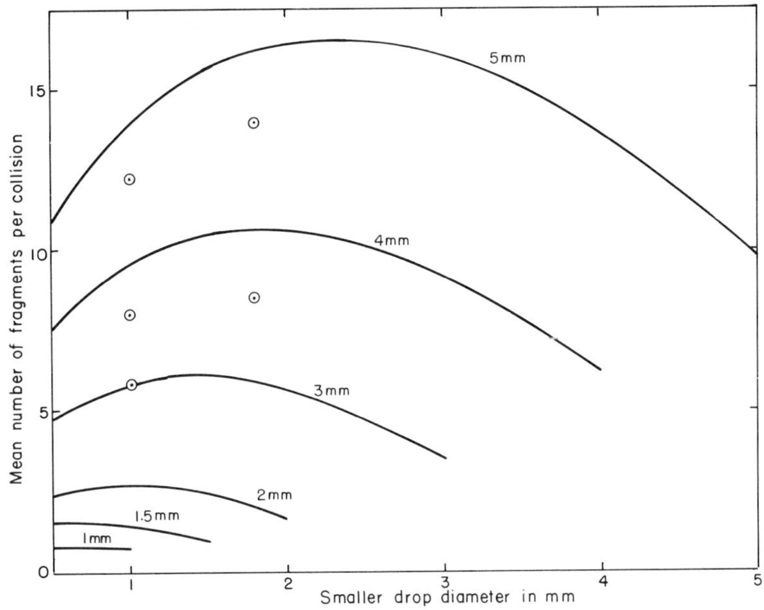


Figure 8.16 The mean number of fragments formed by a collision between raindrops. The numbers on the curves are the diameters of the larger drop. The circles are the experimental data for the five drop pairs. After List (1977).

water drops of 9 mm diameter in a low-turbulence wind tunnel. He conjectured that drop breakup in the free atmosphere was due to turbulence and drop collisions. His photographs show that large drops are deformed to approximate an ellipsoid with a flattened base. The drop diameters mentioned here are the diameters of a sphere of the same volume.

List (1977) and his collaborators studied experimentally the drop breakup resulting from raindrop collisions, duplicating as nearly as possible the conditions in the free atmosphere. Five different pairs of raindrops were used—4.6 and 1.8 mm, 3.6 and 1.8mm, 4.6 and 1.0 mm, 3.6 and 1.0 mm, and 3.0 and 1.0—and 712 collisions were photographed. The mean number of the fragments per collision are shown in figure 8.16. When the smaller drop of the pair was larger than 1 mm and the larger exceeded 3 mm, a collision always caused a breakup. When a drop pair of 0.72 and 1.8 mm collided, coalescence occurred in only 36% of the collisions, and the balance caused breakup.

On the basis of the experimental data, List computed the evolution of the raindrop spectra for a variety of conditions. It was found that the drop spectra all tended toward the exponential or Marshall-Palmer spectrum. The time, expressed as the distance of fall, required for a close approach to the Marshall-Palmer spectrum depends on the shape of the initial spectrum, the rainfall rate, and the size of the largest drop. As an example, this distance of fall was found to be 2 km for a rainfall rate of 29 mm $hr^{-1}$ and would be greater for smaller rainfall rates. When the larger drops are 2.5 mm or less, this distance of fall increases rapidly as the drop size and rainfall rate decrease. As the drop size increases beyond 2.5 mm, the required distance of fall becomes only a few kilometers or less. In view of these results, List feels that collisional drop shattering is primarily responsible for the upper size limit of warm raindrops. He attributes the larger drops in thunderstorms to the melting of hailstones and snow aggregates too close to the surface for the collisional process to destroy them.

There can be little question that rupture due to collision is an important process in the evolution of drop size spectra, although more data are desirable. It should be remembered that warm clouds are limited to the height of the melting level and that it continues to be rather difficult to explain drops as large as 2.5 mm in diameter from warm clouds. It is probably too early to write off aerodynamic breakup in thundershowers; more information on the turbulence spectra in such clouds and its disruptive action is needed.

Raindrop size spectra at the surface have already been modified by a variety of processes. Among these are the sorting effects of wind shear and the fall speeds of raindrops, translation of the cloud with respect to the ground, and evaporation, which may eliminate rain at the surface.

The smallest observed raindrops depend strongly on the relative humidity in the subcloud layer. At one extreme is the stratus cloud, which extends nearly to the surface and releases drizzle drops as small as 0.1 mm in diameter. In the more usual case, an unsaturated subcloud region, the smallest raindrops are perhaps 0.3–0.5 mm in diameter.

## 8.28 Terminal Fall Velocities of Raindrops

Small raindrops fall at the same velocity as solid spheres of the same size and density. Their fall speeds may be calculated from the well-known drag coefficients of spheres. The solid sphere approach is valid up to a Reynolds number of about 200, which corresponds to a raindrop diameter of about 0.9 mm at sea level. As the drop diameter increases, the drops progressively lose the spherical shape, becoming flattened at the bottom. Larger drops have internal circulations, and very large drops pulsate. For these reasons the fall speeds of drops larger than about 1 mm diameter are determined by direct measurements.

The terminal velocities of fall for five different atmospheric conditions are shown in figure 8.17. The lowest curve in figure 8.17 was drawn from the measurements of Gunn and Kinzer (1949). The other curves are based on the data of Gunn and Kinzer by an approximation method given by

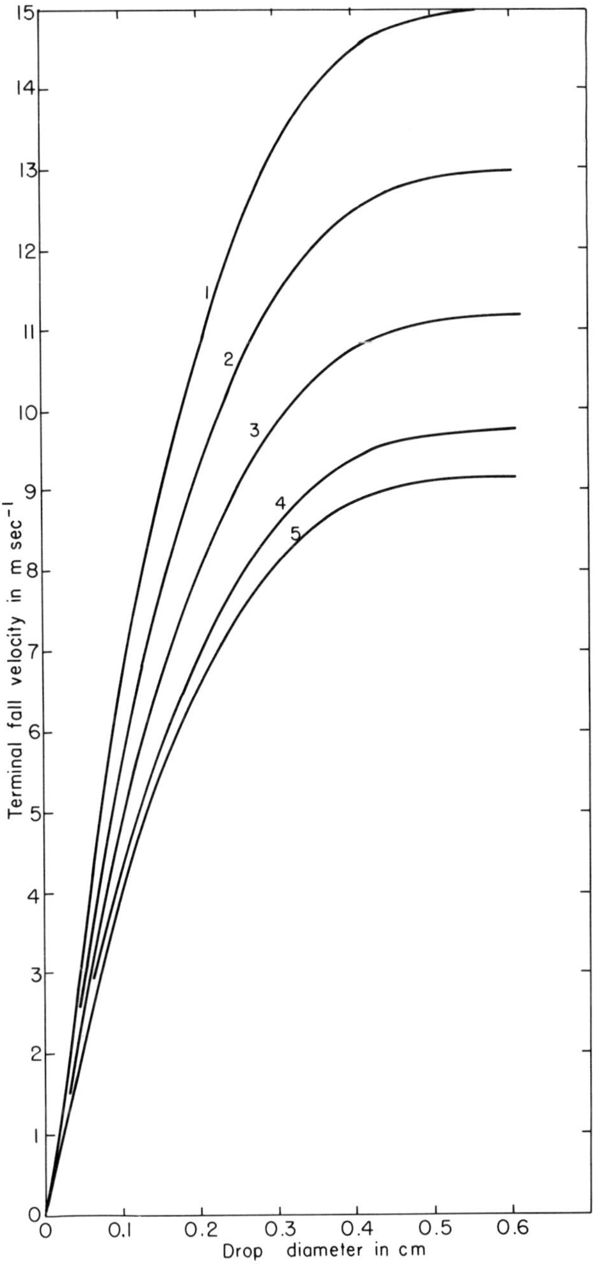


Figure 8.17 Terminal velocities of fall of raindrops in five atmospheres. Key: 1, 250 mbar and −20°C; 2, 380 mbar and −10°C; 3, 575 mbar and 0°C; 4, 850 mbar and 10°C; 5, 1013 mbar and 20°C.

Foote and de Toit (1969) that is probably good within a few percent. It is seen that the fall speeds of raindrops of typical size increase with elevation in the troposphere.

## 8.29 Comments on Cloud Seeding

No attempt will be made to review the extensive literature on precipitation modification by cloud seeding. Rather, the possible effects of seeding will be considered briefly on the basis of the preceding discussion of natural precipitation processes.

Convective clouds are attractive targets for cloud seeding because of their low precipitation efficiency. In part, this low efficiency is due to incomplete sweep-out of the cloud water by the precipitation in the upper portions of the cloud. In a warm cloud the sweep-out might be enhanced by introducing drizzle-size water drops or hygroscopic salt particles, but the total weight of material required makes this uneconomical.

Cool clouds may be seeded with ice nuclei either near cloud base or at the cloud top, depending on the vertical temperature structure, to increase the number of precipitation particles. Since the time required to grow precipitation is about the same as the active lifetime of a convective cell, the seeding should be done early in the lifetime of the cell, a difficult task. If the artificial nuclei are introduced near the top of a mature cell, the resultant ice particles can feed only on the cloud water remaining after the fallout of the natural precipitation. Ice multiplication appears to be very important in clouds with top temperatures in the vicinity of $-5°C$, and such clouds are not attractive seeding candidates.

Heavy seeding, $10^3$–$10^4$ nuclei per liter, in the supercooled region of cool convective clouds will convert them to ice in a few minutes. The concomitant release of the latent heat of fusion and the latent heat of sublimation, as the cloud comes to a new thermodynamic equilibrium, leads to a temperature rise of 1–2°C. This added buoyancy is often sufficient to increase significantly the cloud height and size. Generally larger clouds release more rain, but the "icing-out" of the cloud eliminates the supercooled water on which graupel pellets feed, and the large concentration of ice particles makes it difficult for any to grow to precipitation size. However, the enhanced convection creates more supercooled water and dilutes the ice crystal concentration. If the increase in cloud volume is large, it seems probable that more precipitation results.

The apparent high precipitation efficiency of stratiform systems makes it unlikely that the total precipitation can be increased by seeding. However, it may be possible to affect the timing of the precipitation. The detailed structure of stratiform systems is both complex and poorly known. The observations of Matejka et al. (1980) show the kind of information required to assess the possibility of useful modification of stratiform systems.

It seems probable to me that it will be necessary to have real time observations of the cloud structure prior to and during a seeding operation.

It has long been believed that orographic precipitation can be increased by seeding, and this has been confirmed by a number of successful cloud-seeding projects. A very simple form of an orographic cloud is the cloud cap or lenticular cloud that forms over a mountain and remains stationary while a strong wind blows through it. This cloud is formed by lifting on the upstream slope and evaporates in the descending downstream flow. In the larger and more complex orographic situation over an extensive mountainous barrier, it is believed that a part of the condensate is evaporated in the descending flow beyond the barrier.

Cloud seeding upstream of the barrier may initiate the precipitation process earlier, thus giving the ice particles more time to fall out before the descent and evaporation begin. This is due to the higher activation temperature of the artificial nuclei. In some cases the lowest temperatures in the orographic cloud are such that only a very few natural ice nuclei will be activated.

Studies of the precipitation process over the west coast of the United States and the extensive mountain barriers show that most of the important precipitation occurs in convective bands found in cyclonic storms. The convection is enhanced by the forced lifting. It is not clear how seeding would affect this convective activity, but it is probably still true that some of the precipitation results from the forced lifting and is subject to evaporation in the region of descent.

## 8.30 Acid Rain

In the absence of all other solutes, the pH of water in the atmosphere is expected to be 4.6–5.6 due to dissolved carbon dioxide. Many other substances are found in rainwater that affect its acidity. Of these the more important ones are sulfuric and nitric acids and ammonia and calcium carbonate. Sulfur is released into the atmosphere as gases, typically sulfur dioxide, hydrogen sulfide, and organic sulfur compounds. After some hours to a very few days, these gases are oxidized to sulfur trioxide, which rapidly becomes sulfuric acid. The details of the oxidation process are still being debated. It may occur in the gas phase or after solution in a cloud drop. The process followed will probably affect the time required. The time of oxidation must also be affected by the cloud microphysics and whether there is a cloud. The time is important because, with the wind speed and direction, it determines the region affected by acid rain from a given source.

Similarly, the oxides of nitrogen lead to the formation of nitric acid. Generally, sulfuric acid is dominant, but nitric acid will always add to the acidity. Given time, the acids will be partly or completely neutralized by ammonia or other strong bases. It will be noted that the processes leading to acid rain are the same as those leading to hygroscopic cloud condensation nuclei.

It is often important to locate the source of acid rain. In general, meteorological trajectory analysis is much too crude for this purpose, and it is unlikely that any major improvements will be made in this technique. The only procedure that seems to offer hope is the introduction of tracers at the source. The key to this method is the existence and identification of suitable tracers.

Acidic ponds and lakes are widely considered to be a result of acid rain. This introduces new problems. The rain percolates into the soil; some may find its way to the water table, and some may run off to brooks and streams and into lakes. At every step the rainwater is exposed to the land and its pH may be significantly changed. Some of the lakes are buffered against acidity and may resist acidification for many years.

Acid rain must have occurred for many years in susceptible regions, such as the northeastern United States, but there are few data available. I made a number of pH measurements of cloud water on the summit of Mount Washington, New Hampshire, in 1939. The mean pH of 24 samples was 4.6, while the minimum was 3.0. No quantitative comparison is possible, but the inference is that there was acid rain in New Hampshire some 50 years ago.

## Problems

1.
Which precipitation process—convective, stratiform, or orographic—has the largest precipitation efficiency and why?

2.
Convective clouds are sometimes seeded at cloud base and sometimes at cloud top to increase the rainfall. Which would you choose and why?

3.
What does the presence of graupel in cool or cold convective clouds tell us about the precipitation process in such clouds?

4.
At one time the presence of the ice crystal anvil in a thundercloud shortly before the rain reached the surface was taken as evidence of the operation of the Bergeron-Findeisen ice crystal process. What is your evaluation of this reasoning?

5.
One of the most successful cloud-seeding projects was carried out on orographic clouds in southern California. What are some of the features of these experiments that led to its success?

# 9

# Common Optical Phenomena in the Atmosphere

## 9.1 Introduction

Those aspects of the optical properties of the atmosphere of importance in the transfer of solar and thermal radiation are discussed in preceding chapters. The purpose of this chapter is to examine and explain common optical phenomena, such as mirages, rainbows, and halos, that are of little importance in radiative transfer, but are of considerable general interest. A meteorologist is often expected to be knowledgeable about these phenomena. A brief discussion of visual range, or visibility, is also included.

As will be seen, most of these optical phenomena can be analyzed and explained quantitatively with the aid of geometric optics. Partly for this reason, meteorological optics has been a well-explored area of meteorology for about a century. The books of Mascart (1889–1894), Pernter and Exner (1922), and Humphreys (1929) are still valid as references in spite of their age. Newer books are those of Tricker (1971) and Fraser (1975).

As indicated by the chapter title, only the more common optical displays will be considered here. Descriptions and explanations of rare phenomena, such as those that occasionally accompany a halo, or unusual mirages will be found in most of the texts cited. Even in the case of the common phenomena, the treatment offered here is less complete than that available in the references. The objective is to acquaint readers with the basic causes of the selected phenomena without burdening them with undue detail.

## 9.2 Index of Refraction of Air

The index of refraction of air at optical wavelengths is about 1.0003 and is negligibly different from unity in comparison with those of water and ice. For sufficiently long paths or in the case of extreme atmospheric stratifi-

Table 9.1
Index of refraction of dry air at 1,013.25 mbar and 15°C

| Wavelength ($\mu$m) | Index of refraction |
|---|---|
| 0.35 | 1.0002861 |
| 0.40 | 1.0002828 |
| 0.45 | 1.0002805 |
| 0.50 | 1.0002790 |
| 0.55 | 1.0002778 |
| 0.60 | 1.0002770 |
| 0.65 | 1.0002763 |
| 0.70 | 1.0002758 |
| 0.75 | 1.0002754 |

cation, the refraction of light by air is sufficient to produce optical effects of some significance. For dry air at 1,013.25 mbar and 15°C containing 0.03% $CO_2$ by volume, the index of refraction $n_0$ as a function of wavelength $\lambda$ in micrometers is given empirically by (9.1) (after Edlén, 1953):

$$(n_0 - 1) \times 10^8 = 6{,}432.8 + 2{,}949{,}810/(146 - \lambda^{-2}) + 25{,}540/(41 - \lambda^{-2}). \tag{9.1}$$

Table 9.1 contains a few values of the index of refraction computed from (9.1). The spectral dispersion of air is seen from this table to be small. Consequently color is not an important feature of refractive phenomena within the atmosphere, although it may be over long paths (for example, the color of the eclipsed moon). For most purposes the index of refraction of air for visible light and standard conditions may be taken as 1.000278. For small changes in air density, such as occur near the surface, $n - 1$ varies nearly directly with density. Inserting the value of the density of air at the standard condition $n - 1 = 0.227\rho$, where $\rho$ is the air density in g cm$^{-3}$.

The presence of water vapor results in a slightly smaller index of refraction than for dry air. As an example, at 1,013.25 mbar and 15°C, a relative humidity of 80% (vapor pressure 13.6 mbar) reduces $n - 1$ for dry air by 0.2%. It may be noted that the index of refraction of air at microwave frequencies is strongly dependent on the water vapor content of the air.

## 9.3 Astronomical Refraction

When light from a star traverses the atmosphere along a slant path, the light ray is bent by refraction. Snell's law of refraction shows that the light ray, which is normal to the wave fronts, is concave down, as shown in

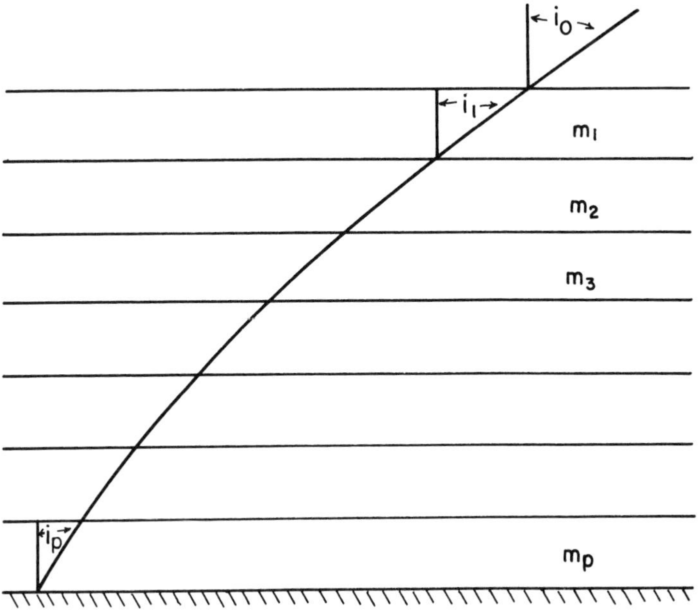


Figure 9.1 Illustration of astronomical refraction.

figure 9.1. Since the atmospheric density, and hence the index of refraction, decreases with height, the curvature of the ray increases downward from the top of the atmosphere, as shown schematically in figure 9.1. The apparent zenith angle of a star is slightly smaller than its true angle, and consequently corrections must be applied to astronomical observations. This effect is called astronomical refraction.

The atmosphere depicted in figure 9.1 is assumed to consist of a series of horizontal layers each thin enough so as to have a nearly uniform index of refraction. A light ray from a star is incident at the angle $i_0$ on the top layer. From Snell's law

$$\sin i_1 = \frac{\sin i_0}{n_1}, \tag{9.2}$$

where $i_0$ and $i_1$ are the angles of incidence on the top layer and on the second layer and $n_1$ is the index of refraction of the first layer. Similarly,

$$\sin i_2 = \frac{n_1}{n_2} \sin i_1 = \frac{\sin i_0}{n_2} \tag{9.3}$$

and the angle of incidence on the $p$th layer is

$$\sin i_{p-1} = \frac{\sin i_0}{n_p}. \tag{9.4}$$

Therefore the apparent angle of incidence at the surface depends only on the initial angle of incidence at the top of the atmosphere and the index of refraction of the surface layer. The actual atmosphere may be replaced by a homogeneous atmosphere of the density of the surface layer.

The true zenith angle of the star is $Z + \Delta Z$, where $Z$ is the observed zenith angle at the surface and $\Delta Z$ is the reduction due to refraction. Applying (9.4) and expanding give

$$\sin \Delta Z = (n - \cos \Delta Z) \tan Z. \tag{9.5}$$

At zenith angles less than about 45°, for which the flat earth approximation is valid, $\Delta Z$ is less than 1′ of arc and (9.5) reduces to

$$\Delta Z = (n - 1) \tan Z. \tag{9.6}$$

At larger zenith angles the curvature of the earth must be taken into account. Extensive tables of atmospheric refraction are available, which are based in part on observations.

## 9.4 Refraction over Quasi-Horizontal Paths

In a typical atmosphere in which the temperature decreases 6.5°C per kilometer of height, the radius of curvature of a horizontal ray is about six times the radius of the earth. Therefore, the visual horizon is just a bit farther away than the geometric horizon. For standard conditions (6.5°C $km^{-1}$, 1,013 mbar, and 15°C), the angle of refraction in the horizontal is about 33.9′ of arc. This acts to make sunrise earlier and sunset later than they would be without the atmosphere. Refraction lengthens daylight by a few minutes.

Because $\Delta Z$ changes rapidly when the zenith angle is near 90°, the refraction of a ray from the upper limb of the sun or moon is less than that from the bottom limb when the latter is close to the horizon. The elliptical shape of the sun or moon under these conditions is readily observable.

Just before the upper limb of the sun disappears into the sea at sunset, the last tiny bright spot is sometimes seen to change rapidly in color from yellow to green to bluish. The green is most striking, and this phenomenon is called the "green flash." It results from the fact that the refracted light disappears in the order of the refrangibility (see table 9.1).

The atmosphere is not a homogeneous medium, and local variations in the index of refraction, carried along by the wind, distort the incoming wave fronts, resulting in the well-known scintillation or twinkling of the

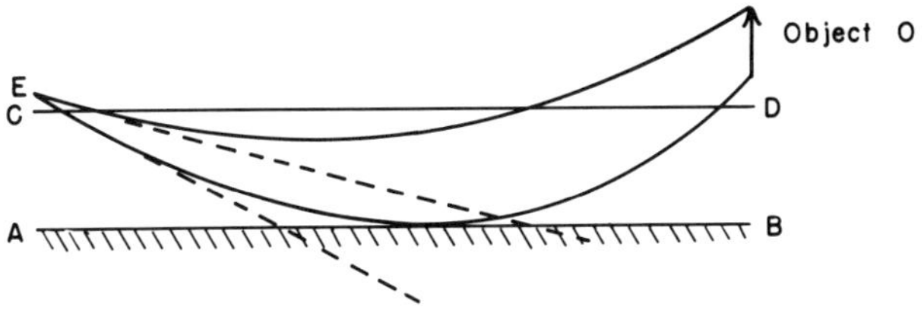


Figure 9.2 Course of the refracted rays during an inferior mirage. The dashed lines are tangent to the rays at the eye *E* and show the apparent direction to the object. The refracting layer extends from level *AB* to *CD*.

stars. The scale of the distortions of the wave fronts is small enough so that the scintillations are averaged out by large telescopes and are seldom apparent when observing planets.

## 9.5 Mirages

Mirages[1] result from very large changes of sign in the variation of atmospheric density with height, usually in rather thin layers near or at the surface. In spite of apparently complex optical effects, mirages can be explained adequately by simple ray optics. The rays are everywhere normal to the wave fronts. When the air density decreases with height, the wave speed does also, and horizontal rays are concave down. Similarly, the rays will be concave up when the density increases with height. The magnitude of these effects obviously depends on the rate of change of the density with height.

These simple principles will be applied first to the inferior mirage, which is seen over strongly heated and fairly level surfaces, such as deserts and paved roads. Evidently the strong heating at the surface results in a marked increase of air density with height. Conventional static stability considerations predict a rapid overturning of the layer. Extending earlier work of Rayleigh, Baum (1951) suggested that the inclusion of eddy heat transfer makes it plausible that such layers may persist. Objections may be made to Baum's theory, but it would be very difficult to explain the common inferior mirage without the persistence of a thin layer in which density increases with height.

A schematic representation of the inferior mirage is shown in figure 9.2. The surface layer, in which density increases with height, extends from *AB* (the surface) to *CD*, above which the density is assumed nearly constant. An observer's eye at *E* looks toward an object at *O* and sees two images. A ray that is above *CD* is not appreciably refracted, and the object is seen in

---

1. The treatment given here closely follows Tricker (1971).

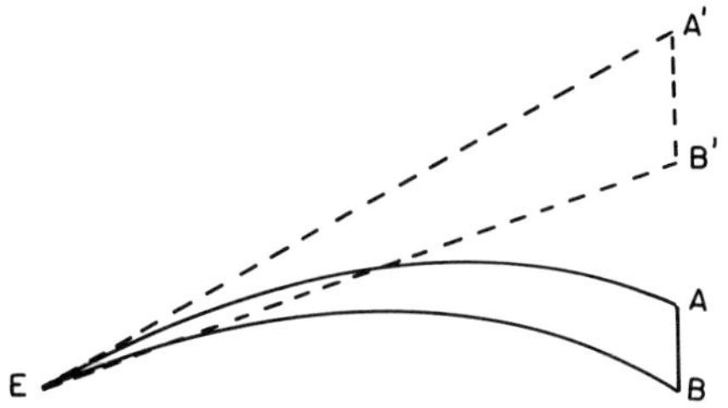


Figure 9.3 Illustrating looming. Solid curves are rays connecting the eye *E* with the real object *AB*. Dashed lines are tangents to rays at *E* and show the direction to the apparent elevated and stretched object *A′B′*. After Tricker (1971).

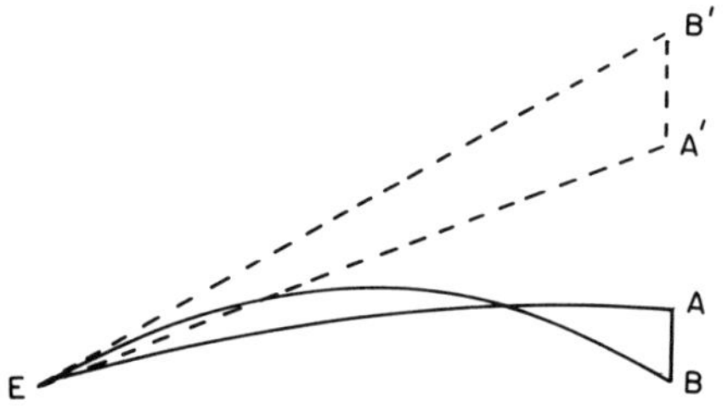


Figure 9.4 Illustrating an inverted superior mirage. Solid curves are rays from eye *E* to real object *AB*. Dashed lines are tangents to rays at *E* and show direction to apparent elevated and inverted object *B′A′*. After Tricker (1971).

its proper position. The lower ray lies in the refracting layer (except at the ends) and is concave up. The apparent location of the object is along the tangent to the ray at the eye. A glancing reflection from the surface intersects the horizon sky and leads to the characteristic appearance of a water surface. If the observer's eye is below *CD*, only the refracted image will be seen.

## 9.6 Superior Mirages

Superior mirages are caused by strong temperature inversions in which the temperature increases rapidly with height. The light rays are concave down, which results in an apparent elevation of the images. The simpler effects are vertical stretching of the image, called looming or its converse, stooping. The more complex displays include inverted as well as upright images and lateral as well as vertical distortions. The variety of such displays is great, and no attempt will be made here to recount them.

An adequate qualitative explanation of superior mirages may be given with the aid of the refracted light rays. Figure 9.3 illustrates looming. The eye at point *E* views the object *AB*, and the solid curves are the rays

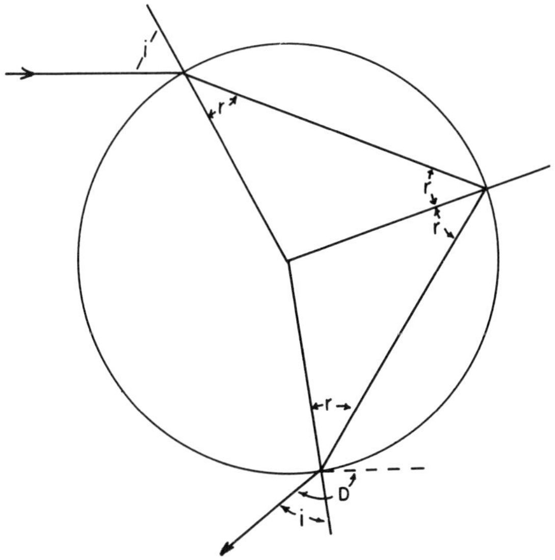


Figure 9.5 Optics of the primary rainbow involving one internal reflection and two refractions. Key: *i*, angle of incidence; *r*, angle of refraction; *D*, angle of deviation.

connecting the eye and the object. The dashed lines are tangents to the rays at the eye and show the direction in which the object appears to lie. The apparent image $A'B'$ is erect, elevated, and vertically stretched. It is evident that the elevation of the image depends on the curvature of the rays and that the stretching results from an increase of the curvature with height.

An inverted image may appear when the curvature of the rays decreases with height, as illustrated in figure 9.4. The inverted image may also be stretched or shrunken. An adjacent layer may simultaneously form an upright image. A variety of images can result from a complex inversion in which the curvature of the rays changes rapidly with height.

## 9.7 Rainbows

The geometry of the primary rainbow is illustrated in figure 9.5. A ray from the sun to the left enters the drop at an angle of incidence $i$. On entering the drop the ray is refracted at an angle $r$ from the normal, is reflected from the far side of the drop, and exits after a second refraction. The ray is rotated clockwise by an angle $D$, the deviation:

$$D = (i - r) + (180 - 2r) + (i - r) = 180 + 2i - 4r. \tag{9.7}$$

The axial ray ($i = 0$) is reflected back on itself from the far side of the drop without refraction, and this ray is deviated by 180°, the maximum possible for one internal reflection. The deviations of rays at increasing distances from the axial ray decrease steadily until a minimum deviation is reached, beyond which the deviation again increases. This minimum deviation, well-known in prism spectroscopy, is responsible for the rainbow because the refracted light is concentrated near this angle. Without such an angular concentration the refracted light would be broadly distributed over the sky and no bow would be seen. The minimum deviation may be found by setting the derivative of (9.7) equal to zero:

$$\frac{dD}{di} = 0 = 2 - \frac{4\,dr}{di}. \tag{9.8}$$

Using Snell's law to evaluate $dr/di$ and to eliminate $r$, one obtains

$$\cos i_{\mathrm{m}} = \left[\frac{n^2 - 1}{3}\right]^{1/2}, \tag{9.9}$$

where $i_{\mathrm{m}}$ is the angle of incidence corresponding to minimum deviation. By taking the index of refraction of water in the visible spectrum as 4/3, $i_{\mathrm{m}}$ is 59.4° and $r_{\mathrm{m}}$ is 42°. By reference to (9.5) and figure 9.5, this makes $D_{\mathrm{m}} = 138°$. The rainbow is centered about the antisolar point, the opposite of the direction to the sun. The angle of the emerging ray of minimum deviation with respect to the antisolar point is then $180° - 138° = 42°$. This is approximately the angular radius of the primary rainbow. The ray drawn in figure 9.5 is intended to represent the ray of minimum deviation. If additional rays are drawn, it will be found that they all have larger deviations.

The colors of the rainbow are due to the change of the index of refraction of water with wavelength. For red light of $\lambda = 0.6678\ \mu$m, the index of refraction is 1.3316 and the angular radius of the rainbow is 42° 17′; similarly, for violet light of $\lambda = 0.4047\ \mu$m, the index of refraction is 1.3435 and the angular radius is 40° 35′. Thus the red appears on the outer part of the rainbow and the violet, or blue, appears near the inner boundary.

In the foregoing it has been assumed that the rainbow appears at the angle of minimum deviation. The concentration of the reflected and refracted light at this angle is the fundamental reason for the appearance of a rainbow. Even an elementary discussion of the rainbow should demonstrate the concentration of the light at or near the angle of minimum deviation.[2] The angle of incidence of the parallel solar beam on a spherical raindrop varies from 0° at the drop's "equator" to 90° at the "poles." In

2. The treatment given here follows Tricker (1971).

consequence, the radiative flux per unit angle of incidence varies with the cosine of the angle of incidence just as it does on the spherical earth. If the radiant flux per unit area of the sunlight is $E_0$, it is easy to see that the total flux incident on the drop surface at a given angle of incidence is

$$F_{\mathrm{i}} = 2\pi R^2 \sin i \cos i\, \delta i = \pi R^2 (\sin 2i)\, \delta i, \tag{9.10}$$

where $R$ is the radius of the drop.

Only a fraction of this flux will emerge from the drop in the rainbow direction because, at each contact with the water-air interface, some of the light escapes in another direction. If $A(i)$ is the coefficient of reflectivity as a function of the angle of incidence, the fraction $A(i)$ is lost by reflection at the drop surface and $1 - A(i)$ proceeds to the internal reflection, at which only a fraction $A(i)$ is reflected. Similarly, a fraction $1 - A(i)$ of this finally emerges from the drop to produce the rainbow. Putting this together leads to the conclusion that the fraction of the incident radiation that escapes after two refractions and one internal reflection is $A(i)[1 - A(i)]^2$. $A(i)$ as a function of $i$ may be computed from the index of refraction by means of the Fresnel coefficients. It is of some interest here to note that the reflection coefficient is not the same for light polarized in the plane of incidence as that for light polarized perpendicularly to the plane of incidence. A few values of these several quantities are given in table 9.2.

In that region of interest around the angle of minimum deviation ($i = 59.4°$), the fraction of the incident radiation escaping near the rainbow direction is 3–5%. It is of interest to note that, in the rainbow direction, the emerging light is strongly polarized in the plane of incidence.

We turn now to the angular distribution of the emerging light. The brightness as sensed by the eye is proportional to the flux per unit solid angle. The light entering the drop in the increment $\delta i$ at the angle of incidence $i$ emerges at an angle of deviation $D$ to $D + \delta D$. The solid angle subtended by this emerging radiation is

$$2\pi \sin(180 - \delta D)(\delta D) = 2(\sin D)\, \delta D. \tag{9.11}$$

By combining (9.11) with (9.10), the relative visual intensity $I$ is seen to be given by

$$I = E_0 \frac{R^2 \sin 2i}{2 \sin D} \frac{\delta i}{\delta D} [A(i)(1 - A(i)^2)], \tag{9.12}$$

where the losses due to reflection and refraction have been incorporated. This shows that the visual intensity from a single raindrop is proportional to the square of the drop radius. More realistically, the visual intensity is proportional to $\sum NR^2$, where $N$ is the number of drops of radius $R$ illuminated by the sun in the direction to the rainbow.

Table 9.2
Coefficients of reflectivity

| Angle of incidence (degrees) | Reflectivity: $A(i)$ | | | Fraction escaping: $A(i)[1 - A(i)]^2$ | | |
|---|---|---|---|---|---|---|
| | Parallel | Perpendicular | Average | Parallel | Perpendicular | Average |
| 0 | 0.0204 | 0.0204 | 0.0204 | 0.0196 | 0.0196 | 0.0196 |
| 30 | 0.0309 | 0.0119 | 0.0214 | 0.0290 | 0.0116 | 0.0203 |
| 40 | 0.0432 | 0.0059 | 0.0246 | 0.0396 | 0.0058 | 0.0227 |
| 50 | 0.0669 | 0.0005 | 0.0337 | 0.0582 | 0.0005 | 0.0294 |
| 59.4 | 0.1111 | 0.0035 | 0.0573 | 0.0878 | 0.0035 | 0.0457 |
| 65 | 0.1572 | 0.0172 | 0.0872 | 0.1117 | 0.0166 | 0.0642 |
| 70 | 0.2199 | 0.0473 | 0.1336 | 0.1338 | 0.0429 | 0.0884 |
| 80 | 0.4573 | 0.2389 | 0.3481 | 0.1347 | 0.1384 | 0.1366 |

Table 9.3
Relative brightness of primary rainbow

| Angle of incidence (degrees) | 180 − D (degrees) | Relative brightness | Angle of incidence (degrees) | 180 − D (degrees) | Relative brightness |
|---|---|---|---|---|---|
| 50 | 40.28 | 0.0825 | 59.39 | 42.04 | ∞ |
| 53 | 41.20 | 0.1299 | 59.7 | 42.04 | 4.381 |
| 55 | 41.64 | 0.2000 | 60 | 42.02 | 1.746 |
| 57 | 41.92 | 0.3935 | 61 | 41.96 | 0.6854 |
| 58 | 41.96 | 0.7055 | 62 | 41.88 | 0.4423 |
| 58.5 | 42.00 | 1.492 | 63 | 41.72 | 0.3351 |
| 59 | 42.03 | 2.593 | 65 | 41.28 | 0.3102 |

The most important factor in (9.12) is $\delta i/\delta D$, the reciprocal of which is given by (9.8). At minimum deviation $dD/di$ is zero and (9.12) becomes infinite. This physically unrealistic result is due to the simple first-order solution used. More insight can be had by evaluating (9.12) over a range of angles of incidence. Results for an index of refraction of 4/3 and omitting the common factor $E_0R^2/2$ are given in table 9.3.

The tabular values of the relative brightness show that the brightness is small until the angle of incidence reaches within about 1° of the rainbow angle, whereupon it rises very rapidly toward infinity. This demonstrates the angular concentration of the brightness that is required to explain the rainbow.

In the region where the deviation slightly exceeds the minimum deviation, there are two coherent rays, one resulting from a deviation smaller than the minimum deviation and the other a deviation larger than the minimum. The resultant interference pattern is sometimes visible as supernumerary bows inside the primary rainbow. The simple first-order theory used here is incapable of showing such interference patterns.

## 9.8 Higher-Order Rainbows

In addition to the primary rainbows, there are, in principle, higher-order bows resulting from two, three, ... internal reflections. They may be analyzed in the same way as the primary bow. Only the second-order rainbow is seen. The higher-order bows are increasingly weak, and the third- and fourth-order bows are centered on the sun, where there is too much general illumination for them to be seen. The angular radius of the second-order rainbow, for an index of refraction of 4/3, is about 51° around the antisolar point, and it is therefore seen outside the primary bow.

The color sequence of the secondary bow is the reverse of that in the primary bow, with violet on the outside. The second internal reflection leads to an approach to minimum deviation from larger toward smaller angles of incidence. Consequently none of the doubly reflected light is found inside the secondary bow, the opposite of the primary bow in which no once reflected light is found outside the bow. As a result, the region of the sky between the primary and secondary bows is noticeably darker than the regions inside the primary and outside the secondary bows.

## 9.9 Finer Details of Rainbows

Although the simple geometric optics approach used here is adequate to explain the principal features of the rainbow, it is incapable of dealing with the finer details that are due predominantly to diffraction. Use of the wave theory of radiation permits a more complete analysis of the rainbow at the expense of greater complexity. Reference is made to the books cited earlier for accounts of the development of the wave theory of the rainbow. It will suffice here simply to mention a few of the results of the wave theory. It is found that the deviation corresponding to maximum intensity is slightly greater than the minimum deviation, as might be expected. It is also found that the relative visual intensity is proportional to $R^{7/3}/\lambda^{1/3}$ instead of to $R^2$ as given in (9.12). The wave theory predicts the supernumerary bows as a series of maxima of decreasing amplitude and decreasing separation.

The width of the rainbow is found to increase as the size of the drops decreases; at the same time the separate colors overlap more and more. For drops of the size of large cloud or drizzle drops, the bow becomes almost white and is usually called a fog bow. Intense red is formed only by raindrops; red and even orange may be lacking in a bow formed by small raindrops. Thus the wave theory provides a reasonable explanation of supernumerary bows and of some of the variations in coloration and width observed in real rainbows.

The wave theory and Huygens's principle are applicable only when the radius of the drop is large compared with the wavelength of the light. In his detailed consideration, van de Hulst (1957) finds that the wave theory is valid only for drops of radius greater than about 0.2 mm. Undoubtedly some of the results are qualitatively useful for smaller drop radii. The Mie theory provides a complete theory for spherical drops of any size, but computational difficulties have restricted its use for drops of radius larger than about 10 $\mu$m until recently.

All of the theories assume that the raindrops are spheres. It is well-known that raindrops of radius larger than about 0.5 mm are deformed by aerodynamic forces, the more so the larger the drop. The larger drops are greatly flattened on the bottom. Such nonspherical drops cannot form a rainbow. It follows that the rainbow must be due to drops of radius less than about 0.5 mm. However, larger drops may have nearly circular cross sections in the horizontal plane, even though they are flattened in the vertical. Fraser (1972) has used this concept to explain his observation that pure reds are seen only in the near-vertical portions of the rainbow and hence near sunset. His argument is that pure red is formed only near the horizontal because of drop flattening. The small near-spherical drops that form the rest of the rainbow often do not produce any red at all. This is an interesting concept, but, as yet, it has not been supported by detailed computations involving the actual shapes of the raindrops. It should also be noted that the flattened drops oscillate in shape (Volz, 1960) and that the larger drops probably are not circular even in the horizontal.

Rainbows can be seen only when a rain shaft is directly illuminated by the sun and only when the elevation of the sun is less than the rainbow angle of 42° (51° for the secondary bow). It is for these reasons that rainbows are most commonly observed during late afternoon thundershowers. It now appears, for the reasons already discussed, that thundershowers are favored because of their relatively large concentration of raindrops and not, as often held, because of their large drops. Rainbows are often observed in the spray from lawn sprinklers or waterfalls.

## 9.10 Glories

A glory consists of one or more colored rings encircling the shadow of the observer's head, cast on a cloud or fog layer by the sun. Individual observers see the glories around their own shadows and not those of their companions. The conditions necessary for observing a glory formerly obtained only on mountain peaks at low solar elevations when a cloud layer lay below the summit. An alternative name, Brocken bow, comes from the name of a mountain at which glories were often observed. Now, any regular air traveler will occasionally see a glory surrounding the shadow of the airplane on a lower cloud layer.

The glory is evidently centered around the antisolar point. The angular radii of the rings are typically only a few degrees, with the red on the outer portion of the ring. The angular radii of the rings differ from case to case, and some variation with time may be apparent. The central field is usually bright and is often colored. It is now generally agreed that the glory is formed by liquid cloud or fog drops and not by ice crystals or drops of precipitation size. Good, quantitative observations of glories are few, and none include the drop size spectrum of the cloud or fog.

Since a glory occurs around the antisolar point and since a fog bow is often observed simultaneously, at a much larger angular radius, it was first thought that the glory could be explained by geometric optics in much the same way as the rainbow. It will be recalled that a light ray in water cannot emerge if the angle it makes with the normal to the surface is equal to or greater than the critical angle; such a ray is totally internally reflected. For water of index of refraction 4/3, the critical angle is 48.6°. Equations (9.7) and (9.9) show that if angle $r$ is 48.6°, the angle of deviation is 165.6°, or 14.4° from the antisolar point. This is the minimum angle from the antisolar point at which a ray can emerge, and there is no concentration of light at this angle. Glories that usually have much smaller angular radii cannot be explained in this way.

More recent and more satisfactory explanations of the glory attribute it to backward diffraction as a result of interference in the cusped or, more accurately, toroidal shape of the emerging wave front. The well-known corona, to be discussed in section 9.13, is also a diffraction phenomenon. Because of this and an impression that the angular radii of the glory minima were much the same as those of the corona maxima, a glory was often called an anticorona.

A fairly complete discussion of the theory of the glory is given by van de Hulst (1957). His results are incomplete in that the necessary ratio of the two coefficients in his expressions for the intensity of the glory was not determined. He deduced a probable value of this ratio from scanty observations. Tricker (1971) has extended van de Hulst's analysis and believes that he has been able to avoid this limitation. An analysis similar to that of van de Hulst was given earlier by Bucerius (1946). His results were based on an asymptotic form of the Mie equations, presumed to be valid for the rather large scattering size parameters involved ($x = 2\pi r/\lambda$ on the order of 100). Tricker and van de Hulst employed wave optics, which should also be valid for large $x$, and the latter also showed that identical expressions could be derived from the Mie equations for large $x$ and small scattering angles from the backward direction. It is somewhat surprising that Bucerius finds a minimum (essentially zero) intensity at 0°, while van de Hulst and Tricker find a maximum there of magnitude greater than the succeeding maxima; van de Hulst explains this difference as a result of the choice of the ratio of the intensity coefficients mentioned earlier. There is therefore still an important uncertainty in the quantitative theory of the glory, even though the observations seem to support the van de Hulst and Tricker results.

## 9.11 The Glory from Mie Scattering

The glory, as well as all other phenomena due to scattering by spheres, is contained in the complete Mie equations. In the past these equations have not been used in the study of the glory because of the formidable computational problems at the large-size parameters involved; but this changed with the availability of large electronic computers, which has led to many tabulations of numerical solutions of the Mie equations. One of these, due to Howell (1969), seemed marginally adequate for a quantitative study of

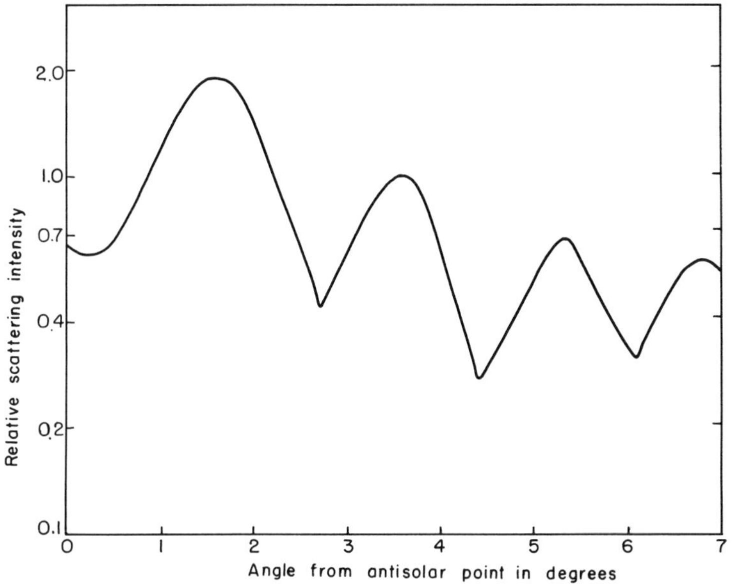


Figure 9.6 Relative Mie scattering intensity as a function of the angle about the antisolar point for scattering size parameter 105 and index of refraction 4/3. After Howell (1969).

the glory. In view of the uncertainties in theory noted earlier, I have decided to use Howell's tabulations, even though this involves more detail than is devoted to the other optical phenomena treated in this chapter.

Howell tabulated the Mie intensity functions by angular increments of 2°, too coarse to resolve the glory. However, he presented graphical data for increments of 0.2°, and the results given here were read from those graphs. Figure 9.6 is a sample curve for $x = 105$ in the angular region where glories are observed. The curve is the sum of the two orthogonally polarized Mie angular scattering functions usually, denoted $i_1$ and $i_2$. Curves for other values of $x$ show the same general features as figure 9.6, although the amplitudes of the maxima and minima do not always decrease in the relatively regular fashion of figure 9.6. Of the 21 available curves for $x \geq 21$, all but one show a maximum at 0°. This tends to support the theoretical curves of van de Hulst and Tricker, but in 17 of these 20, the magnitude of the first maximum exceeds that of the maximum at 0°. Tricker and van de Hulst show that the intensity at 0° is much larger than the first maximum. It may be concluded that the two theories explain most of the observed features of the glory, but some discrepancies remain.

The angular radii of the bright and dark rings that presumably represent the glory are plotted in figure 9.7 against the scattering size parameter as read from the Howell curves. The first minimum was not plotted because of

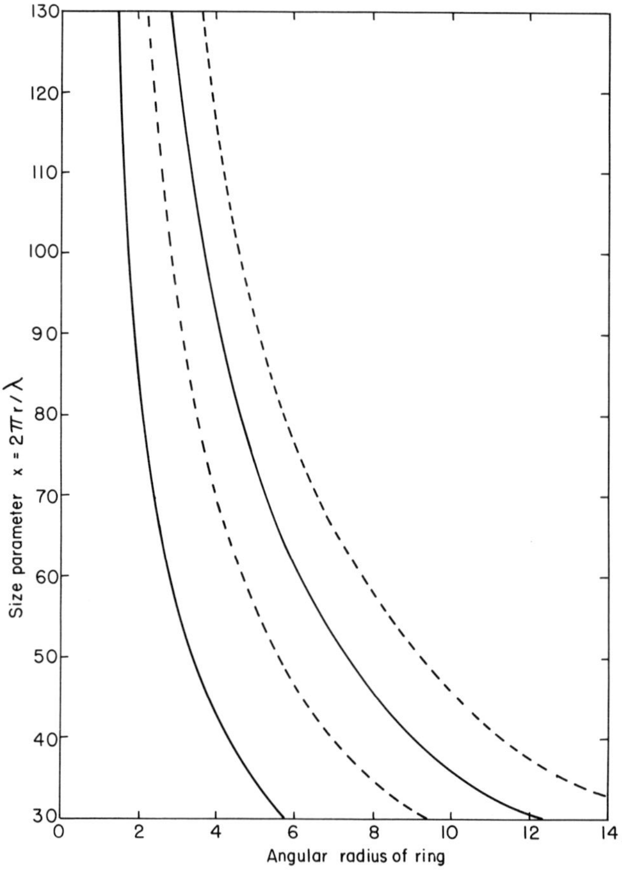


Figure 9.7 Angular radii of bright and dark rings about the antisolar point according to Mie theory as a function of the scattering size parameter. Solid curves are for bright rings, and dashed curves are for dark rings.

the inadequate angular resolution of the data. Figure 9.7 shows that the radii of the rings increase with decreasing $x$, or the outer portion of a bright ring should be red, as is observed (see figure 9.8). It will also be noted that the curves in figure 9.7 become nearly vertical for large $x$. This is important when a cloud drop size spectrum is considered since it suggests that the resultant range of the size parameter will not average out the rings when $x$ is large. In the visible spectrum, the upper portion of figure 9.7 corresponds to droplet radii of 5–10 $\mu$m, typical of natural clouds.

The results of a simple experiment to show the effects of more than one drop size and two different wavelengths are presented in figure 9.8. It was assumed that there are equal numbers of drops of 6 and 9 $\mu$m radius present. Separate curves are given for blue ($\lambda = 0.45$ $\mu$m) and red ($\lambda =$

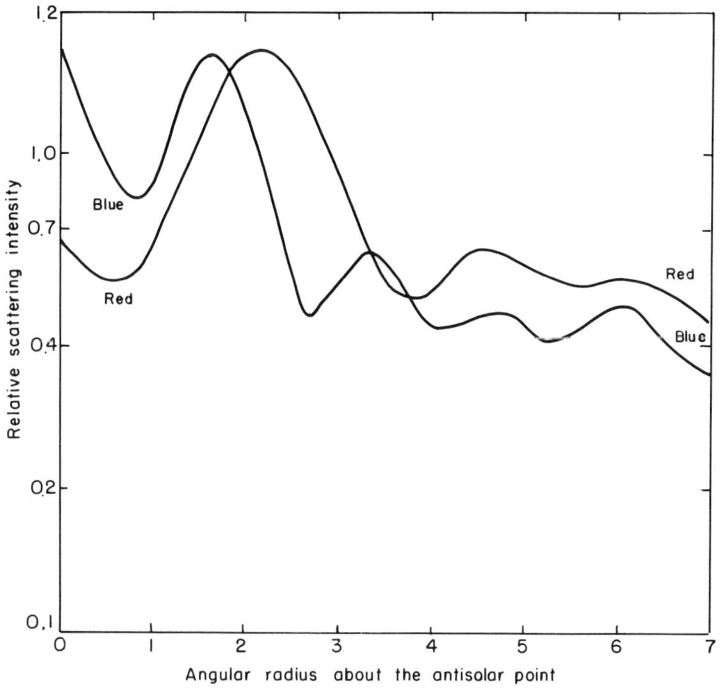


Figure 9.8 The glory in a cloud consisting of equal numbers of drops of radii 6 and 9 $\mu$m. The wavelength of blue light is taken as 0.45 $\mu$m and of red light as 0.63 $\mu$m.

0.63 $\mu$m) light. One minimum and one maximum are seen to survive. This suggests that only one bright ring will be seen in a glory unless the drops are larger than those assumed here or the cloud is nearly monodisperse. Caution should be exercised in drawing firm conclusions from this simple experiment.

## 9.12 Heiligenschein

This phenomenon, for which there is no generally accepted English name, is the glow seen around the shadow of the observer's head when cast on a heavily bedewed lawn. As in the case of the glory, observers see only their own heiligenscheins. At first sight these similarities might suggest that the heiligenschein is formed in the same way as the glory. However, neither rings nor colors are observed, and the dew drops on the grass are found to be much larger than cloud drops.

It is now generally agreed that the heiligenschein is the result of the same process as the reflection from a glass-beaded projection screen. Close examination of dew drops on blades of grass shows that many of the drops are supported in nearly spherical form on the tiny hairs on the grass blade. As shown in standard texts on optics, the focal length of a spherical thick lens in air, measured from the center of the sphere, is $nR/2(n-1)$, where $R$ is the radius of the sphere and $n$ is its index of refraction. If a water drop is located just above the grass leaf on one or more hairs, it will focus the sunlight onto the leaf. In turn, the drop will serve as a lens in the reverse direction and will collimate the light reflected from the leaf back toward the sun. Tricker (1971) has performed some simple experiments with water-filled spherical flasks and a matte screen that illustrate the effect in striking fashion.

This explains why the heiligenschein has such a small angular radius; the reflected light is directed back toward the sun and hence toward the observer's eyes. Only a few drops will be suspended at the exact focal distance from a leaf, but the effect will be present in a less brilliant form from drops that are somewhat out of focus. It would also be expected that the light reflected from the leaf would be greenish, and, in fact, this is confirmed by careful observation. The general appearance is white probably because the background is also green and because bright spots of light look white unless they are strongly colored.

## 9.13 Coronas

The corona is a series of colored rings of relatively small angular radii seen around the sun, moon, or other light source veiled by thin cloud or fog. The color of the coronal rings ranges from red on the outside to blue on the inside; most often only the reddish color is apparent. In spite of some lingering counteropinions, it seems almost certain that coronas are formed only by liquid droplets and not by ice crystals. Bishop's ring is a brownish-reddish corona of relatively large angular radius that is thought to be formed by aerosol particles of volcanic origin.

The corona is a diffraction phenomenon whose theory is quite well-known. Babinet's principle states that the diffraction pattern formed by an opaque circular disk is the same as that produced by a circular hole of the same radius in an opaque screen. (Actually, the two diffraction patterns are of opposite phase, so that their sum is everywhere zero, as is required when the hole and the disk are superimposed.) A sphere of water is assumed to act as an opaque disk, since all of the light incident on the drop is deviated at relatively large angles from the axis. The diffraction is due to the edge of the disk or hole, and it is easily understood that it makes no difference whether the edge surrounds a disk or a hole.

If the drop is large compared with the wavelength of the light (large $x$), the diffraction pattern can be deduced from wave optics and is commonly called Fraunhofer diffraction. As shown by, for example, Humphreys (1929), the intensity $I$ (flux per unit solid angle) of the diffraction pattern

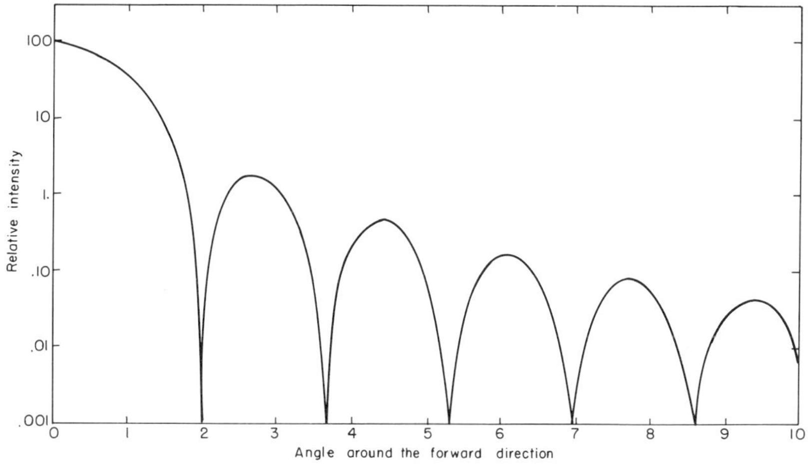


Figure 9.9 Fraunhofer diffraction pattern for a scattering size parameter of 110.

formed by a hole of radius $R$ is

$$I = \pi^2 R^4 \left[ \frac{1}{m} J_1(2m) \right]^2, \tag{9.13}$$

where $m = x \sin \phi$, $\phi$ is the angle measured from the forward direction, and $J_1$ stands for the first-order Bessel function.

The relative intensity of the diffraction pattern is shown in figure 9.9 for a scattering size parameter of $x = 110$. The first maximum is centered on $\phi = 0°$ and represents the bright aureole observed around the light source in a scattering atmosphere. The succeeding maxima, separated by sharp minima of zero intensity, are the coronal rings. The pattern shown in figure 9.9 is formed only in a monodisperse scatterer. When a spectrum of drop sizes is present, the higher-order rings are averaged out. Often no rings at all are seen, and two or three rings are rather rare phenomena.

It is found that the angular radii of the minima are given closely by an empirical formula,

$$\sin \phi = \left( \frac{\pi}{x} \right) (p + 0.22), \tag{9.14}$$

where $p$ is the order of the minimum (1,2,3, . . .).

The coronal rings are also evident in the Mie calculations of Howell (1969). Although, as noted earlier, the angular resolution of the Howell results is not fully adequate to define scattering maxima and minima, it is of interest to see how close they are to Fraunhofer diffraction. Such a comparison is presented in figure 9.10, in which the angular radii of the first

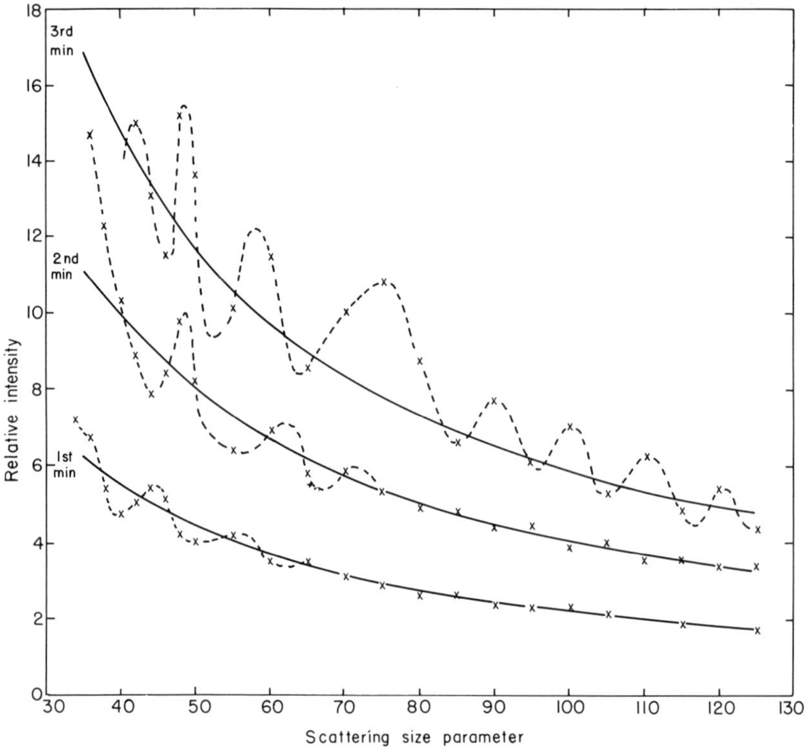


Figure 9.10 The first three minima of the forward scattering pattern as a function of the scattering size parameter. Solid curves depict Fraunhofer diffraction. Crosses are from the Mie calculations of Howell (1969). Dashed curves connecting crosses are suggestive only, since the number of points is not sufficient to define the oscillations.

three minima are plotted against the scattering size parameter for Fraunhofer diffraction (solid curves) and the crosses are from Howell's Mie calculations. The dashed curves connecting the crosses are intended only to help visualize the deviations of the Mie results from Fraunhofer diffraction. Clearly, the number of crosses is inadequate to resolve the oscillations, particularly at the smaller size parameters. However, the Mie results approach the Fraunhofer curves as the scattering size parameter becomes large, more rapidly for the first minimum than for the second and third minima.

Aside from its interest as an optical phenomenon, the corona has been used, notably by Koehler (1925), to gauge an "average" size of cloud droplets. In this application the angular radius of the outer portion of the red ring is measured, and the corresponding drop radius is computed from

(9.14). Koehler took the wavelength as 0.571 $\mu$m, the so-called average wavelength of white light. This is a somewhat questionable choice, and it might be better to select a wavelength of red light, say, 0.63 $\mu$m. He frequently was able to measure the angular radii of the first two minima, thus obtaining two independent estimates of the droplet radius; these were found to agree quite well. As can be deduced from figure 9.10, a rather narrow droplet spectrum is required for a clearcut corona; a broad spectrum, particularly of rather small drops, will result in a blending of the coronal rings and colors, giving only a broad white aureole around the light source.

There is no definitive evidence that coronas can be formed by ice crystal clouds. Small spherical ice crystals are, at best, a transitory result of the freezing of supercooled cloud drops. Almost always ice crystals have platelike or columnlike forms with hexagonal symmetry. Although such crystals will diffract light, it is doubtful that they could form a circular corona. A telling point is that there are no reports of the simultaneous observation of a corona and a halo (due to ice crystals) in the same cloud.

The colors of mother-of-pearl (or nacreous) clouds and iridescent clouds have been attributed to segments of coronas. Mother-of-pearl clouds are at 20–30 km and are presumed to be ice clouds. Their coloration is most marked when the sun is several degrees below the horizon. Evidently the angle at which they are seen is more nearly 90° than 0° to the direction of the illuminating solar beam. Thus the colors are not due to coronas. Examination of Howell's Mie scattering curves reveals maxima and minima at many angles from the axial ray, and a highly monodisperse cloud can be expected to show colored bands at 90° or many other angles. In spite of the low temperatures of these clouds, it is likely that they are composed of water drops and not ice crystals if the colors arise from scattering.

Iridescent clouds show brilliant spots or borders of color, usually red or green, up to 30° from the sun. These clouds are usually at cirrus levels. The colors have been attributed to segments of coronas of large angular radius. If correct, this implies clouds of supercooled water drops and not ice crystals, as usually assumed. Coronal rings of 20°–30° radius correspond to drop radii of 1.0–0.7 $\mu$m (first ring). There is no direct evidence that nearly monodisperse drops of this size are found in these clouds. It seems most likely that iridescent clouds owe their colors to refraction in ice crystals. This question could be answered by directly determining the nature and size of the particles in iridescent clouds, a difficult task.

Bishop's ring is almost certainly a corona due to volcanic particles in the stratosphere that have become fairly monodisperse due to (probably) sedimentation and coagulation. This ring was first observed and measured after the eruption of Krakatoa in 1833 and has been seen after other major eruptions. The outer angular radius of this brownish or reddish ring was found to be 22°–23°. Note that the halo is the same size. The corresponding particle radius is about 0.9 $\mu$m, which does not seem unreasonable for residual volcanic dust.

## 9.14 Optical Effects in Ice Crystal Clouds

The most common optical phenomenon due to ice crystal clouds is the 22° halo, a ring of 22° angular radius centered on the sun or moon when veiled by a high ice cloud layer, commonly cirrostratus. Because the halo is by far the most frequent optical manifestation of ice clouds, many of the optical effects of such clouds are called halo phenomena. The 22° halo may be distinguished from a corona by its larger angular radius and, if colored, the occurrence of red on the inner part of the ring. The halo is evidence for the presence of an ice crystal cloud and is the "ring around the moon" (or sun) which is said, with considerable justification, to be the precursor of bad weather.

The 22° halo is the result of refraction by the faces of ice crystals and is therefore more akin to the rainbow than to the corona. The eight faces on simple hexagonal crystals, the additional faces on pyramidal forms, and the variable orientation of the crystals lead to a considerable number of refraction phenomena. Among these are parhelia or sun dogs, the circumzenithal arc, the 46° halo, and the arcs of Lowitz. These are rather rare phenomena, and the occurrence of a number of them in a single display is usually recorded in the scientific literature. Explanations of the more common displays are well established and adequately confirmed by observations. Some extremely rare manifestations are not fully explained, and a few predicted effects have not been observed.

Internal and external reflections from the prism faces are responsible for a few optical phenomena, notably sun pillars and the parhelic circle. The former are vertical columns of white light above and below the sun and are due to the reflection from the faces of tabular ice crystals (hexagonal plates) falling in their stable horizontal orientation. The parhelic circle is a white partial or full circle passing through the sun and the parhelia, parallel to the horizon. It is due to reflection from the vertical faces of the crystals and is usually quite faint.

## 9.15 Refraction by Ice Crystals

The prismatic refraction that forms parhelia and the 22° halo can be examined by ray optics in much the same way as the rainbow. As in the case of the rainbow, the parhelia and halo occur at (or very near) the ray of minimum deviation. In its simplest form the process is illustrated in figure 9.11, which shows the refraction at alternate faces of a hexagonal prism. A ray is shown entering from the left, suffering refraction at the two inter-

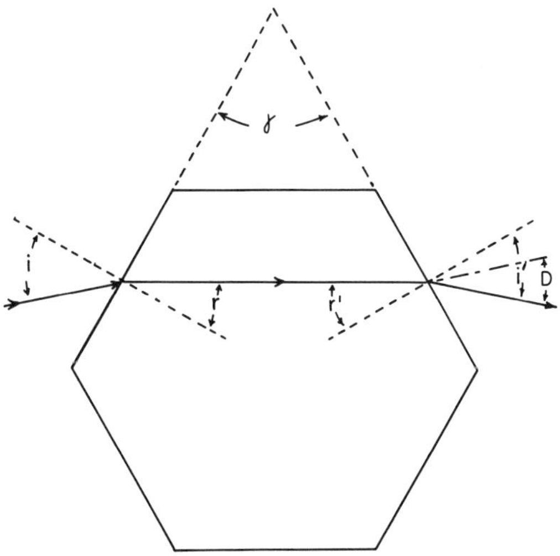


Figure 9.11 Refraction by alternate faces of a hexagonal prism. The dashed lines are normals to the prism faces. Note the symmetric course of the ray through the prism.

faces, and leaving at the right after a deviation $D$. From the geometry of the figure it is evident that

$$D = (i - r) + (i' - r'), \tag{9.15}$$

$$A = (r + r'), \tag{9.16}$$

and hence

$$D = (i + i') - \gamma. \tag{9.17}$$

Setting the first derivative of (9.15) to zero and using (9.16), give $di = di'$ and $dr = -dr'$. From Snell's law (9.17) becomes

$$D = \sin^{-1}(n \sin r) + \sin^{-1}(n \sin r') - \gamma. \tag{9.18}$$

Setting the first derivative of (9.18) with respect to $r$ equal to zero and noting that $dr'/dr = -1$ give

$$\frac{\cos r}{(1 - n^2 \sin^2 r)^{1/2}} = \frac{\cos r'}{(1 - n^2 \sin^2 r')^{1/2}}. \tag{9.19}$$

Evidently the condition for minimum deviation is $r = r'$, and hence $i = i'$. Thus the ray of minimum deviation passes through the prism symmetri-

Table 9.4
Conditions for minimum deviation

| | | |
|---|---|---|
| $\gamma$ | 60° | 90° |
| $i_m$ | 40.92° | 67.87° |
| $r_m$ | 30.00° | 45.00° |
| $D_m$ | 21.84° | 45.73° |

cally. From (9.16) and (9.17) the angle of minimum deviation can be written as

$$D_m = 2\sin^{-1}\left(n \sin\frac{\gamma}{2}\right) - \gamma. \tag{9.20}$$

The maximum possible angle of incidence is grazing incidence. For ice of index of refraction 1.31, this leads to $r = 49.76°$, which is also the critical angle from the normal beyond which no light may escape from within the prism. From (9.16) with $r = r' = 49.76°$, $\gamma$ is 99.52°, which is then the maximum value of $\gamma$ for which refraction as depicted in figure 9.11 can occur, and it also corresponds to maximum deviation. The angle between adjacent prism faces of a hexagonal column exceeds the maximum. However, the angle between alternate prism surfaces is 60°, and the angle between the prism faces and the basal plane is 90°. These are the possible values of $\gamma$ for simple hexagonal columns or plates. In some cases ice crystals have pyramidal shapes or extensions, thus providing additional values of $\gamma$. The angles for minimum deviation for an index of refraction of 1.31 are given in table 9.4 for the two principal values of $\gamma$.

One is immediately tempted to conclude that the two values of the minimum deviation explain the halos of 22° and 46°, but this is premature. In the first place, it has been tacitly assumed that the incident rays lie in a plane normal to the face of the prism, that is, the plane of the paper in figure 9.11. Depending on the orientation of the crystals in the atmosphere and on the solar zenith angle, the incident rays are typically skewed. Also, it is apparent that stationary crystals would form spots of light at minimum deviation and not circular halos. Thus the analysis given earlier applies more to a prism spectrometer than to ice crystal clouds.

## 9.16 Refraction of Skewed Rays

The usual treatment of the refraction of skewed rays is due to Bravais and may be found, for example, in the book by Tricker (1971). A derivation given by Humphreys (1929) seems easier to visualize and will be followed here. The procedure is based on figure 9.12 which shows a simple triangular

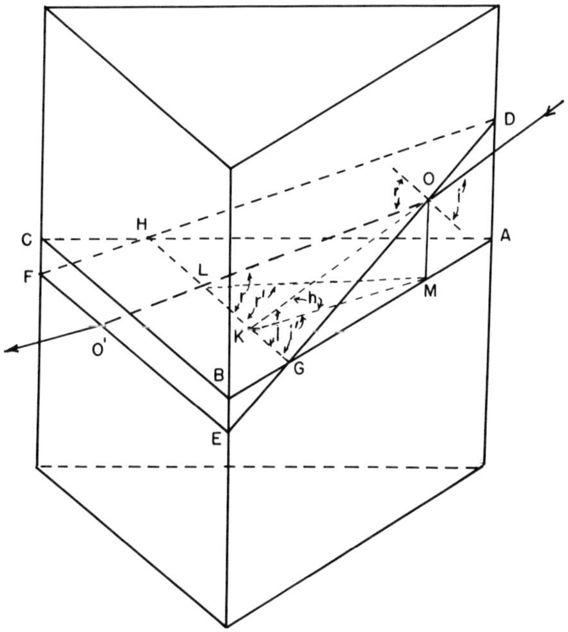


Figure 9.12 Refraction of a skewed ray in a simple triangular prism. *ABC* is a principal plane of the prism; *DEF* is perpendicular to the face on which the ray is incident; *OM* is normal to the principal plane *ABC*; *ML* is the intersection of the two planes. Lines in the sloping plane are dashed for clarity.

prism. The light ray is incident on the prism at $O$ and leaves at $O'$. In the interior of the prism the ray lies within the sloping plane $DEF$. The plane $ABC$ is normal to the vertical axis of the prism and is called a principal plane. It intersects the sloping plane along $GH$. For clarity, all the lines drawn in the sloping plane are dashed; all those in the principal plane are solid. The line $OM$ is vertical and lies in the face of the prism on which the ray is incident. The angles $i$ and $r$ are the angles of incidence and refraction. The line $OK$ is drawn as an extension of the incident ray outside the prism, and the angle $h$ is therefore the elevation angle of the ray with respect to the principal plane.

From the trigonometry,

$$\frac{LO}{\sin i} = \frac{KO}{\sin r'} \qquad \text{or} \qquad \frac{LO}{KO} = \frac{\sin i}{\sin r}. \tag{9.21}$$

From Snell's law,

$$\frac{LO}{KO} = \frac{n \sin r}{\sin r} = n. \tag{9.22}$$

From the two vertical triangles, $OLM$ and $OKM$,

$$\sin h = \frac{MO}{KO} \quad \text{and} \quad \sin k = \frac{MO}{LO}, \tag{9.23}$$

where

$$\frac{\sin h}{\sin k} = \frac{LO}{KO} = n, \tag{9.24}$$

or

$$\sin h = n \sin k. \tag{9.25}$$

Similarly, it can be shown that $\sin h' = n \sin k'$, where $h'$ and $k'$ (not drawn) are the angles between the principal plane and the plane defined by the exterior and interior rays. Since $k' = k$ and $h' = h$, the incident and exit rays are equally inclined to the principal plane.

The triangle $LMK$ is evidently the projection of triangle $LOK$, and the angles $i'$ and $r'$ are the projections of $i$ and $r$. From trigonometry and (9.24),

$$n \sin r' \cos k = \sin i' \cos h,$$

or

$$\frac{\sin i'}{\sin r'} \doteq n \frac{\cos k}{\cos h} = n', \tag{9.26}$$

and $n'$ can be thought of as the index of refraction that yields the refraction of the ray projected on the principal plane. An alternative expression for $n'$ in terms of $h$ can be readily derived from (9.26) and (9.25):

$$n' = (n^2 - \sin^2 h)^{1/2}/(1 - \sin^2 h)^{1/2}. \tag{9.27}$$

This analysis permits one to convert the skew ray into its projection on a principal plane by the use of an adjusted index of refraction. Equation (9.25) for a ray in the principal plane may be used for the skew ray by introducing $n'$ in place of $n$:

$$\frac{D_0' + \gamma}{2} = \sin^{-1}\left(n' \sin\frac{\gamma}{2}\right). \tag{9.28}$$

The maximum possible deviation corresponds to grazing incidence and emergence, or to $(D + \gamma)/2 = 90°$.

From (9.27) and (9.28),

Table 9.5
$D_0'$ as a function of $h$ for $\gamma = 60°$ and $n = 1.31$

| $h$ | $D_0'$ | $\Delta_0$ | $h$ | $D_0'$ | $\Delta_0$ |
|---|---|---|---|---|---|
| 0° | 21.84° | 21.84° | 40° | 36.33° | 27.63° |
| 10° | 22.48° | 22.17° | 50° | 51.51° | 32.43° |
| 20° | 24.58° | 23.07° | 60.75° | 120° | 50.07° |
| 30° | 28.70° | 24.82° | | | |

$$1/\sin^2\frac{\gamma}{2} = \frac{n^2 - 1 + \cos^2 h}{\cos^2 h},$$

$$\cos h = (n^2 - 1)^{1/2} \tan \gamma/2. \qquad (9.29)$$

## 9.17 22° Parhelia

For a prism angle $\gamma$ of 60°, (9.29) gives $h = 60.75°$, and for $\gamma = 90°$, $h = 32.20°$. These represent the maximum permissible elevation angles of the sun with respect to the principal prism plane.

The minimum deviation is given by

$$D_0' = 2\sin^{-1}\left[\frac{(n^2 - \sin^2 h)^{1/2}}{(1 - \sin^2 h)^{1/2}}\right]\sin\frac{\gamma}{2} - \gamma. \qquad (9.30)$$

$D_0'$ as a function of $h$ for $\gamma = 60°$ and $n = 1.31$ is given in table 9.5. The third column, $\Delta_0$, is the angular radius measured on the arc of a great circle between the sun and either parhelion. It is easy to show from spherical trigonometry that $\sin(\Delta_0/2) = \cos h \sin(D_0'/2)$. The angle $\Delta_0$ is the one that is observed. Table 9.5 may be considered as giving the angular distance from the sun to either parhelion at minimum deviation as a function of the sun's elevation angle $h$ when the prism faces of the crystal are vertical. One visualizes thick hexagonal plates settling in their stable horizontal orientation. Note that the angular distance of these so-called 22° parhelia increase with solar elevation. At solar elevations of less than 10° or 15° the parhelia appear to be on the 22° halo, if one is present. At higher solar elevations the parhelia are readily observed to lie outside the halo and to be at the same elevation as the sun.

The index of refraction of ice has been taken as 1.31, which is appropriate for the center of the visible spectrum. Color, particularly red, is sometimes observed in parhelia (and halos) and results from the variation of the index of refraction with wavelength. Ice is slightly birefringent, but it will suffice to give the index of refraction for the ordinary ray. The index increases slightly with decreasing temperature and the values in table 9.6

Table 9.6
Index of refraction of ice at $-30°C$

| $\lambda$ ($\mu$m) | $n$ | $\lambda$ ($\mu$m) | $n$ |
|---|---|---|---|
| 0.40 | 1.3196 | 0.60 | 1.3096 |
| 0.45 | 1.3159 | 0.65 | 1.3082 |
| 0.50 | 1.3132 | 0.70 | 1.3071 |
| 0.55 | 1.3112 | | |

are for $-30°C$. With the sun near the horizon the minimum deviation is 22.28° at $\lambda = 0.45$ $\mu$m (blue) and 21.70° at $\lambda = 0.65$ $\mu$m (red). Thus red is seen on the edge of the parhelion nearest the sun. Since many of the crystal faces are not oriented at the angle of minimum deviation, they will contribute light of different colors at somewhat larger angles; consequently only the red, which has the smallest minimum deviation, is relatively pure. For the same reason, the parhelion may extend to greater angles from the sun than indicated earlier and, indeed, as far as maximum deviation, in which case the parhelion is drawn out to a width of about 20°.

## 9.18 22° Halos

The parhelia discussed are formed by crystal faces oriented vertically. Although this is the stable orientation of relatively large tabular crystals, such as thick hexagonal plates, there are many situations in which the crystal faces are randomly oriented about the vertical. When this obtains, it is easy to see that the parhelia are converted into a circle around the sun, the 22° halo.

Simple crystals such as plates and columns fall stably in any orientation when they are small and viscous flow obtains. Larger crystals fall stably, with their long axes horizontal. Even larger crystals begin to shed vortices and oscillate, and finally they tumble. Crystals caught by Weickmann (1957) in cirrus clouds were very often radial assemblages of prisms. Such crystals would show various orientations of the crystal faces, the more so since they probably tumble as they fall. Turbulence has also been invoked as a cause of random orientation. This does not seem necessary, and it is not known whether the requisite fine grain turbulence exists at the cirrus level. It would appear that a more or less random orientation of the prism faces is more frequent than a vertical orientation and thus that halos would be more common than parhelia, as is observed to be the case. These phenomena are not mutually exclusive. There may exist both oriented and disoriented crystals in the same cloud. This can lead to the simultaneous occurrence of a halo and parhelia.

## 9.19 Visual Range

The visual range, more commonly called the horizontal visibility, is an important quantity in aviation meteorology and is dependent on scattering by the aerosol. Ideally, the daytime visibility is the maximum distance at which a dark object subtending an angle of at least 0.5° can be seen against the horizon sky. In practice it is often necessary to use less than ideal visibility markers. At night, the visibility is the maximum distance at which an unfocused light of about 25 cp (candlepower) can be seen by a dark-adapted observer. At principal airports the visibility is derived from a transmissometer, an instrument that measures the transmissivity over a horizontal path of known length. The transmissivity is related to the observed visual range through a series of direct comparisons. By its nature the visual range involves both the optical properties of the atmosphere and the physiology of human vision.

## 9.20 Air Light

On viewing dark objects against the horizon sky at increasing distances, it is observed that the apparent brightness of a dark object increases with its distance from the observer. Finally, at a sufficient distance, the brightness of the object becomes so nearly equal to that of the adjacent sky that it can no longer be distinguished. This veiling light is called the air light, and it is due to the scattering of sun and skylight by particulates in the air path into the eye of the observer.

The first physically sound expression for the visual range was derived by Koschmieder (1924). With the wisdom of hindsight, his result can be obtained by a simpler approach. It is helpful to define some photometric quantities. These differ from their physical counterparts in that they refer to visual radiation only and incorporate the spectral response of the human eye. It is not necessary in what follows to take specific cognizance of this difference between radiant energy and luminous energy. Luminous flux, $F$, is the rate at which luminous energy flows, and hence it has the dimensions of power. The illuminance, $E$, sometimes called flux density, is luminous flux per unit area. Luminous intensity, $I$, is luminous flux per unit solid angle. The luminance, $B$, formerly called brightness, is the luminous flux emitted (or reflected) from an extended source per unit solid angle and per unit area projected normal to the viewing direction.The luminance corresponds closely to the visual impression of brightness and is therefore the quantity relevant to the problem to be discussed here. It may be noted that a perfectly diffuse extended source or a perfectly diffuse reflector is one for which the luminance is independent of the angle at which the source is viewed. From the definition of luminance, the luminous intensity of a

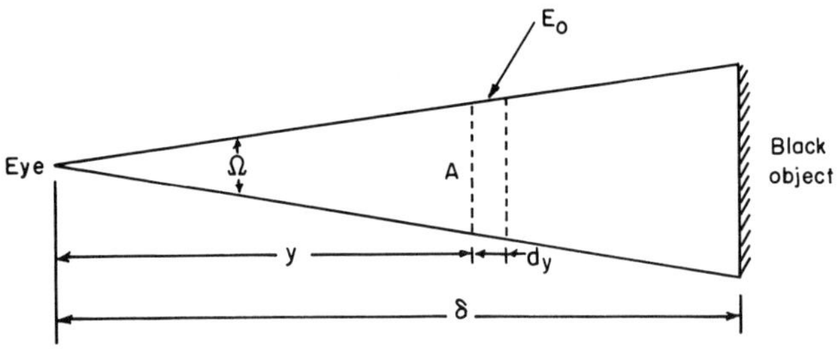


Figure 9.13 Illustration of the evaluation of the air light.

perfectly diffuse source varies with the cosine of the angle of incidence, and this is known as Lambert's law.

Consider a black object at a distance $\delta$ from the observer, seen against the horizon sky, as illustrated in figure 9.13. The object subtends a solid angle $\Omega$ at the observer. Consider an element of the volume of air, defined by the object and $y$, of area $A$ and length $dy$. Assume that the concentration and size spectrum of the aerosol is independent of $y$ (homogeneous aerosol). Assume further that the sun- and skylight illuminating the air are also independent of $y$. It will be shown in section 9.22 that these assumptions are not as restrictive as they seem. If the incident illuminance from the sun and the sky is $E_0$, the luminous intensity of the light scattered in the direction of the observer from the elementary volume is

$$dI_0 = ck_s E_0 A\, dy, \tag{9.31}$$

where $k_s$ is the scattering coefficient and $c$ is the fraction of the scattered light per unit solid angle in the direction toward the observer. The value of $c$ will depend on the angular distribution of $E_0$ and the phase functions of the scatterers. In view of the assumptions, $c$, $k_s$ and $E_0$ are independent of $y$. The luminous intensity of the elementary volume will be diminished by the attenuation over the distance $y$:

$$dI = ck_s E_0 A \exp[-(k_e y)]\, dy, \tag{9.32}$$

where $k_e$ is the extinction coefficient (absorption plus scattering). The luminance (brightness) is, by definition, the luminous intensity per unit area normal to the direction of the view. Hence the luminance at the eye due to the elementary volume is

$$dB_0 = cE_0 k_s \exp[-(k_e y)]\, dy. \tag{9.33}$$

The total luminance of the black object at the distance $\delta$ from the observer is the integral of this expression. From the assumption of a homogeneous atmosphere, uniformly illuminated, $c$, $k_s$, $E_0$ and $k_e$ are independent of $y$ and

$$B_0 = cE_0 \frac{k_s}{k_e} [1 - \exp(k_e \delta)]. \tag{9.34}$$

Koschmieder pointed out that the integral of (9.33) from zero to infinity is the luminance of the horizon sky, $B_h$,

$$B_h = \frac{cE_0 k_s}{k_e}, \tag{9.35}$$

whence

$$B_0 = B_h\{1 - \exp[-(k_e \delta)]\}. \tag{9.36}$$

This shows how the apparent luminance of the black object increases with $\delta$ and that it approaches the luminance of the horizon sky as $\delta$ becomes large. The contrast between an object and its background is

$$C = \frac{B_h - B_0}{B_h} = \exp[-(k_e \delta)]. \tag{9.37}$$

The minimum contrast detectable by the human eye under stated conditions is denoted $\varepsilon$. Thus the visual range $Y$ of a black object viewed against the horizon sky is

$$\varepsilon = \exp[-(k_e Y)],$$

or

$$Y = \frac{(\ln 1/\varepsilon)}{k_e} \tag{9.38}$$

## 9.21 Minimum Contrast

The disarmingly simple equation (9.38) obscures the difficulties that arise in the specification of $\varepsilon$. The luminance contrast threshold is dependent on the angle subtended by the object, the adaption luminance of the observer, the sharpness of the boundary between object and background, and the presence of disturbing luminance in the field of view. Even with all these held constant there is some variance between observers, and it is also necessary to specify the probability of detection of the object. In practice, the probability of detection is essentially unity, but is typically one-half in experiments designed to evaluate $\varepsilon$.

Koschmieder assumed $\varepsilon = 0.02$ under all conditions, and this was accepted for some time. A very extensive series of experimental determinations was reported by Blackwell (1946). Even a casual examination of these results shows that $\varepsilon$ cannot be taken as a constant. For values of the luminance in the typical daylight range, values of $\varepsilon$ for a probability of detection of one-half range from 0.007 to 0.04, depending largely on the angular dimensions of the object. Both larger and smaller values of $\varepsilon$ have been reported. The standard conversions of transmissometer readings, which determine $k_s$, to visual range correspond to an $\varepsilon$ of about 0.055.

## 9.22 Practical Considerations

In the derivation of (9.36) and (9.38) it was assumed that the extinction coefficient and the sky illumination were invariant with horizontal distance. When the black object is at the visual range, the luminances of the object and the sky background are nearly the same. Each of these luminances is an integral of (9.33), differing only in the upper limit of integration. It is evident that any variation of $k_e$ or $E_0$ with $y$ will affect the two adjacent optical paths in the same way. It is therefore to be expected that the visual range of a black object against the horizon sky is not sensitive to variations of $k_e$ and $E_0$ along the path in most cases.

In practice it is often impractical to provide visibility marks that are completely black and/or marks seen against the horizon sky. A case of special importance is the visual range of objects on the ground as seen from an airplane. Examination of these more complex cases in detail is beyond the scope of this introductory treatment. The interested reader is referred to Middleton's (1952) excellent book. If the mark is not black, it will have an illuminance when viewed close up that depends on its reflective properties and its orientation with respect to the sun. The apparent luminance of such an object seen at a distance will be the sum of the air light and of its own luminance, attenuated by the intervening atmosphere. It is apparent that the visual range of the object against the horizon sky will be less than that of a black object. Middleton shows that the visual range of a nonblack or grey object viewed against the horizon sky is

$$D = \frac{\ln(C_0/\varepsilon)}{k_e}, \tag{9.39}$$

where $C_0$ is the contrast of the object against the horizon sky when viewed at the object. For a black object, $C_0$ is unity and (9.39) reduces to (9.38). Also, if $C_0 = \varepsilon$, the object is invisible at all distances. In most cases $C_0$ will be 0.5 or more, and the visual range will be smaller than that of a black object. Middleton (1952) presents a tabular comparison showing that objects with diffuse reflectivities less than about 0.1, viewed against the horizon sky, may be used as visibility marks without causing significant errors.

When both object and background are at the surface and are viewed from above, the visual range depends on the difference in reflectivities of the mark and its background, the illumination, and the variation of $k_e$ with height in the atmosphere. If the contrast between background and object is very large, as when a nearly black object is seen against fresh snow, the visual range will generally exceed that of the standard black object against the horizon sky in the same atmosphere. In the more usual case, the contrast between the object and background is relatively small, and the visual range is correspondingly smaller.

Particularly in rolling or mountainous terrain, it is often necessary to select marks that are viewed against the terrain. Middleton notes that a dark object seen against a terrestrial background at least 1.5 times as far away as the object is almost equivalent to a horizon background. Qualitatively, it is easy to see that an object of almost any kind at a distance beyond the visual range will have a luminance very close to that of the horizon sky.

## 9.23 Nighttime Visual Range

At night the visual range is the maximum distance at which an unfocused light of about 25 cp can be seen by a dark-adapted observer. The illuminance $E$ (luminous flux per unit normal area) of a point source of luminous intensity $I$ (candlepower) at a distance $\delta$ in a turbid atmosphere is

$$E = \frac{I}{d^2}\exp(-k_e\delta). \tag{9.40}$$

This is simply the product of the inverse square law and the exponential extinction law. If $E_m$ is the threshold illuminance of the eye and $Y$ is the distance at which $E$ becomes $E_m$,

$$Y = \frac{1}{k_e}(\ln I/E_m - 2\ln Y). \tag{9.41}$$

This is usually solved graphically. The value of $E_m$ depends on the background luminance, the degree of dark adaption of the observer, and the presence of other lights in the field of view. For observers whose eyes are adapted to the background luminance, $E_m$ may range from about $3 \times 10^{-8}$ to $5 \times 10^{-7}$ lm/m$^2$ (lumens per square meter), the larger value corresponding to a background luminance typical of a clear night with a full moon.

It is important to define the objectives of nighttime estimates of the visual range. It may be desired to have the nighttime estimates closely the same as those that would be made in the same atmosphere in the daytime. If suitable values of $\varepsilon$ and $E_m$ are selected, (9.38) and (9.41) may be solved for $I$, the candlepower of the light. If this is done for various values of the visual range, it will be found that the candlepower of the lamps must increase with distance from the observing point. If, on the other hand, it is desired to make estimates of visual range that will let pilots know how far away they will be able to see runway approach lights, it is necessary to consider the candlepower of these lights in the pilots' viewing directions. In general, candlepower of the lights will be substantially larger than that of the unfocused lights of about 25 cp specified in the observer's instructions. With the aid of (9.41) and a knowledge of the candlepower of the light used to estimate the visual range of the approach lights, it is possible to interpret the observations in terms of the visual range of the approach lights. Fortunately, this problem becomes of critical importance only at quite short visual ranges. Under these conditions the value of $I$ is of lesser importance than $k_e$ in determining the visual range. For example, at a visual range of about half a mile, a light of 1,000 cp can be seen only about 30% farther than one of 100 cp.

Increasingly, the transmissometer is used as the basic visibility instrument at airports. Basically, this instrument measures the extinction coefficient over the path of the transmissometer beam. The readings are transformed into visual range by means of standard tables based on comparisons with visual determinations and the basic equations. Evidently the transmissometer provides the same information during both the day and the night. The readings depend only on the extinction coefficient, a purely physical property. The necessary interpretations in the form of visual range introduce physiological factors that are subject to wide variations.

## 9.24 Colored Objects

In all of this brief discussion it has been tacitly assumed that nonblack objects are grey or white and that the lights are white. With reasonably clear air both the horizon sky and black objects appear bluish due to the preferential scattering of blue light by the small aerosol particles. At the visual range the colors of a dark object and of the adjacent horizon sky are much the same; hence the effects of color are minimal. The apparent color of a colored object changes with the distance at which it is viewed. A treatment of the colors of distant objects and lights and the visual range of colored objects and lights is presented by Middleton (1952). No effort will be made to summarize this topic because the concepts involved in chromatic vision are so unfamiliar to most of us that extensive background material would be a prerequisite.

## Problems

1.
Draw the ray of minimum deviation of the secondary rainbow.

2.
Why is it that the majority of rainbows occur in the afternoon?

3.
Two images of a distant ship are seen, a lower inverted image and an elevated upright image. Sketch the vertical profile of density or temperature that could cause this display.

4.
You observe simultaneously two sun dogs and a 22° halo. What inferences can you draw about the ice crystals that form the display?

5.
What would limit the visual range if there were no atmospheric aerosol?

# 10

# Atmospheric Electricity

## 10.1 Introduction

The title of this chapter immediately brings to mind the spectacular electrical manifestations of thunderstorms. There is little doubt that these are the most important aspects of atmospheric electricity. However, it is first necessary to gain some understanding of the unseen but pervasive electrical properties of the atmosphere that provide the milieu within which thunderstorms develop. This ubiquitous electrification of the atmosphere is referred to as "fair weather" electrification to distinguish it from thunderstorm electrification. The fair weather electric fields and currents are orders of magnitude smaller than those of the thunderstorm. However, the area of the earth covered by thunderstorms at any given time is also orders of magnitude smaller than the total area; the globally integrated properties are of the same order. It is generally believed that thunderstorm electrical generators provide the source of the fair weather electrification and, in turn, the fair weather electrification is often used as the initial condition in proposed thunderstorm charge generation mechanisms.

The fair weather vertical electric field is about 130 $V\,m^{-1}$ (volts per meter) the surface and decreases exponentially with elevation. The total potential difference between the ground and the ionosphere is about $3 \times 10^5$ V, the earth being the negative pole. The atmosphere is a conductor of electricity, although a very poor one relative to metals, as a result of the presence of ions that are formed by cosmic radiation and radioactive substances in the earth. The ions drift in the vertical electric field and form the conduction current that flows in a direction to discharge the quasi-spherical condenser comprised of the ground surface and the upper conducting layer. The electric charge on this condenser that will give a field of 130 $V\,m^{-1}$ at the surface is about $6 \times 10^5$ C (coulombs). The "leakage" current averages about $3 \times 10^{-12}$ $A\,m^{-2}$ (amperes per square meter), or some 1,500 A for the entire earth. It is easy to show that this condenser would be discharged by the leakage current in less than half an hour. Since the field remains approximately constant, there must exist a supply current equal to the discharge current, which, as stated, is apparently provided by the thunderstorms acting as generators. Because of the relatively low resistance of the upper conducting layer and the planetary surface, the charge is distributed over the globe in a time that is short compared with the decay time. This combination of thunderstorm generators and the fair weather discharge, or conduction, current is often referred to as the global electrical circuit.

Over the years many theories of thunderstorm electricity have been proposed, but there is still no fully accepted mechanism of thunderstorm charge generation. It may well be that there is more than one charge separation mechanism at work and that different processes are dominant in different thunderstorms. Many of the theories depend on the large vertical air velocities and the fall of precipitation particles; some depend on the presence of solid precipitation, and many use the fair weather field as a starting point for thunderstorm electrification. The principal reason for the absence of an accepted charge separation process is the lack of crucial observations in real thunderstorms. This is due to the extreme difficulty and danger of measurements within an active thunderstorm. Many observations have been made of the electric fields from outside a thundercloud, but these are of limited value because of shielding charges and the inability of external measurements to define the geometry of the charged regions. A few observations within or near thunderstorms by aircraft or balloons have raised nearly as many questions as they have answered. Purely electrical observations are of limited value without simultaneous information of the dynamics and cloud physics of thunderclouds.

## 10.2 Sources of Atmospheric Ions

The sources of the small atmospheric ions that make the atmosphere electrically conducting are radioactive matter in the earth and cosmic radiation. In the ionosphere, above about 80 km, solar ultraviolet and x radiation are the important ionizing agents. Radioactive substances in the surface of the ground emit $\alpha$, $\beta$, and $\gamma$ radiations, which produce ions. The $\alpha$ and $\beta$ radiations penetrate only a few centimeters into the atmosphere, but $\gamma$ radiation may reach 100 m or so. Of more importance are the radioactive emanations that are gaseous decay products of radioactive substances. The emanations of principal importance are radon and thoron. Being in gaseous form, these emanations diffuse through the soil and escape into the atmosphere. The half-lives of the several elements in the radon and thoron series range from a fraction of a second to years; for example, radon 222 has a half-life of 3.823 days. The longer-lived emanations are distributed in the lower atmosphere by turbulent diffusion and are found in significant concentration within the planetary boundary layer.

The rate of release of radioactive emanations depends on the radioactivity of the surface soil layers, the porosity of the soil, and meteorological conditons. Rain and freezing tend to reduce the soil porosity and thus to reduce the rate of release. Falling atmospheric pressure tends to increase the rate by "sucking" the emanations from the soil. Radon has been used as a natural tracer of motions in the surface layers of the atmosphere.

Cosmic radiation at the surface increases slowly with geomagnetic latitude and morc rapidly with height in the troposphere and lower stratosphere. It also varies inversely with the 11-year solar cycle, being a maximum at solar minimum. It is usual to express the rate of ionization in ion pairs formed per cubic meter of air per second. The ionization rate due to cosmic radiation at the surface is $1.5 \times 10^6$ in these units. This rate increases rapidly with height to a broad maximum of about $40 \times 10^6$ at around 12 km and then declines slowly at higher levels. At around 80 km and above, ionization by solar ultraviolet and x radiation becomes dominant. In this region, the ionosphere, ionization yields free electrons, which become the principal charge carriers.

When the ionization rate due to cosmic radiation is combined with that due to radioactivity, there is a small decrease from the surface to a minimum at the top of the planetary boundary layer. This results from the sharp decrease in radioactivity with height. Thereafter the ionization rate increases monotonically to the maximum near 12 km noted earlier.

The near-surface radon concentration exhibits diurnal and seasonal variations similar to those of other quantities that are transported by turbulent diffusion. For this reason and the nonuniformity of radioactive matter in the earth's surface, only approximate near-surface ionization rates can be given. A typical value over the continents is $10^7$ ion pairs $m^{-3} sec^{-1}$; variations by a factor of two or three are to be expected. Over the ocean, far from land, the rate is $2 \times 10^6$ and is relatively invariant.

Ionization may also result from fires, the bursting of bubbles at the sea surface, corona discharges from surface objects in thunderstorm electric fields (St. Elmo's fire), waterfalls, and a variety of man-made sources. With the exception of bursting bubbles over the oceans, these sources are only of local significance. The bursting of bubbles at the sea surface is discussed in section 1.12 as a source of sea salt aerosols. Blanchard (1963) showed that the jet drops carried a positive charge and made some calculations that indicated that this mechanism was an important charge transfer process on a global scale. This is not important in the present context because the charged jet drops are so massive relative to small ions that they make a negligible contribution to the conductivity of the air.

## 10.3 Atmospheric Ions

In principle, the ionization processes discussed earlier remove an electron and leave the molecule with a positive charge. In the lower atmosphere the lifetime of a free electron is less than a microsecond before it becomes attached to a molecule. It has been suggested that the positive ions are composed of hydronium $(H_3O)^+(H_2O)_n$ and that the negative ions are $(NO_2)^-(H_2O)_n$, but the chemistry of small ions is incompletely known. These are the small or fast atmospheric ions, which are the dominant current carriers in the troposphere and stratosphere. This is expressed quantitatively by the mobility, which is defined as the terminal velocity of the ion in a unit electric field.[1] At standard temperature and pressure (0°C and 1,013 mbar), the mobilities of small ions formed in the laboratory in clean, dry air are (in $m^2 V^{-1} sec^{-1}$) $1.4 \times 10^{-4}$ for positive ions and $2.0 \times 10^{-4}$ for negative ions. In the atmosphere Hoppel and Kraakevik (1965) found mobilities ranging from 0.4 to $5 \times 10^{-4}$ with averages of 1.15 and $1.24 \times 10^{-4}$ for the positive and negative ions, respectively (all in the same units). Misaki and Kanazawa (1969) found average values of about 1.16 and $1.45 \times 10^{-4}$ for the positive and negative ions, respectively. The ratios of the mobilities of negative to positive ion mobilities are 1.4, 1.08, and 1.25 for these three sets of data. The greater mobility of negative ions is probably due to differences in the molecules comprising the ions. The mobility depends on a balance between the electric force and the resistance due to collisions with the air molecules. For large ions, which are charged aerosol particles, the resistance is given approximately by the Stokes-Cunningham law. The mobility of small ions is inversely proportional to the air density and also depends on the humidity. The mobility at the tropopause is some four times that at the surface.

Once the small ions are formed they collide with each other and with aerosol particles. The latter collisions result in ions of much greater size and hence of much smaller mobility. These ions are called intermediate or large ions depending on their mobilities table 10.1 after Israel (1970) is useful for orientation. Note that the several ranges of mobility and size are contiguous and therefore that the tabulated values represent a continuous spectrum. However, there is observational evidence that there is a gap in the spectrum between the small and the small-intermediate ions. This seems reasonable, since the small ions are of molecular size, while the smallest aerosol particles have radii on the order of $10^{-3}$ $\mu$m.

The mobility of intermediate ions is two orders of magnitude smaller than that of small ions, and the difference is much greater for large ions. It

---

1. The mksA (meter, kilogram, second, ampere) system of units is used in this chapter.

Table 10.1
Mobility and size of atmospheric ions[a]

| Designation | Range of mobility ($m^2 V^{-1} sec^{-1}$) | Approximate radius range ($\mu m$) |
|---|---|---|
| Small | $\geq 10^{-4}$ | $6.6 \times 10^{-4}$ |
| Small-intermediate | $10^{-6}$–$10^{-4}$ | $6.6 \times 10^{-4}$–$7.8 \times 10^{-3}$ |
| Large-intermediate | $10^{-7}$–$10^{-6}$ | $7.8 \times 10^{-3}$–$2.5 \times 10^{-2}$ |
| Large | $2.5 \times 10^{-8}$–$10^{-7}$ | $2.5 \times 10^{-2}$–$5.7 \times 10^{-2}$ |
| Ultralarge | $<2.5 \times 10^{-8}$ | $>5.7 \times 10^{-2}$ |

a. After Israel (1970).

is for this reason that the small ions carry almost all of the current. In heavily polluted air the concentration of intermediate and large ions may be two orders of magnitude greater than that of small ions, and the former may then be significant charge carriers. Large and intermediate ions may also contribute to space charges and thereby modify the potential gradient.

## 10.4 Equilibrium Ion Concentrations

The equilibrium concentration of ions of any species represents a balance between the production and destruction rates. Ions are destroyed by recombination of positive and negative ions or by combining with other ions or neutral particles. In particular, small ions are removed by recombination and combining with aerosol particles. The latter process is the source of intermediate and large ions. Small ions carry single electron charges, as do the great majority of intermediate and large ions.

For simplicity consider an atmosphere containing only small positive and negative ions. Then we may write

$$\frac{dn_1}{dt} = \frac{dn_2}{dt} = \phi - \alpha n_1 n_2, \tag{10.1}$$

where $n_1$ and $n_2$ are the concentrations of small positive and small negative ions; $\phi$ is the rate of production of ions, and $\alpha$ is the recombination coefficient. If $n_1 = n_2 = n$,

$$\frac{dn}{dt} = \phi - \alpha n^2. \tag{10.2}$$

The equilibrium concentration is then $n_0 = (\phi/\alpha)^{1/2}$. The recombination coefficient $\alpha$ depends on pressure and temperature:

Table 10.2
Ion recombination coefficients[a]

| Colliding species | Symbol | Recombination coefficient (standard conditions) ($m^3\ sec^{-1}$) |
|---|---|---|
| Small-small | $\alpha$ | $1.6 \times 10^{-12}$ |
| Small-large | $\eta$ | $2.2–14.9 \times 10^{-12}$ |
| Large-large | $\nu$ | $0.6–17. \times 10^{-15}$ |

a. After Israel (1970).

$$\alpha = \alpha_0(p/p_0)(T_0/T)^{7/2}. \tag{10.3}$$

Simple theory shows that the recombination coefficients are proportional to the sum of the mobilities of the colliding pair. Experiment shows that only about one-quarter of the collisions result in an electron exchange, and the recombination coefficients are therefore determined by experiment. Table 10.2, after Israel (1970), shows typical values of the recombination coefficients.

As shown in table 10.2, the ranges of the coefficients involving large ions is considerable; some of this range reflects the size range of large ions, and some is due to experimental uncertainties. Because of these uncertainties, it is not yet realistic to distinguish between large and intermediate ions; the term "large" in the table embraces all ions larger than small ions. As would be expected, the recombination coefficient for collisions between small and large ions is of the same order as that for small-small collisions, while that for collisions between large ions is three orders of magnitude smaller. For this reason, collisions between large ions may usually be neglected, and large ion equilibrium is seldom attained. Typically, only one-third to one-half of the aerosol particles carry a charge.

Equation (10.1) can be extended to cover the more realistic case of an atmosphere containing both small and large ions:

$$\begin{aligned} dn_1/dt &= \phi - \alpha n_1 n_2 - \eta_1 n_1 N_2 - \eta_3 n_1 N_0, \\ dn_2/dt &= \phi - \alpha n_1 n_2 - \eta_2 n_2 N_1 - \eta_4 n_2 N_0, \end{aligned} \tag{10.4}$$

where $N_1$ and $N_2$ are the concentrations of large positive and negative ions and $N_0$ is the concentration of uncharged aerosol particles. Four different recombination coefficients for collisions between small ions and aerosol particles are used, the meanings of which are clear from the equations. It is to be expected that $\eta_1 \simeq \eta_2$, $\eta_3 \simeq \eta_4$, $\eta_1 > \eta_3$, and $\eta_2 > \eta_4$.

Under the reasonable assumption that the only source of large ions is collisions between small ions and aerosol particles, one can write

$$dN_1/dt = \eta_3 n_1 N_0 - \eta_2 n_2 N_1, \tag{10.5}$$

$$dN_2/dt = \eta_4 n_2 N_0 - \eta_1 n_1 N_2.$$

If the additional approximations are made that $n_1 = n_2 = n$, $N_1 = N_2 = N$, $\eta_1 = \eta_2 = \eta^1$, $\eta_3 = \eta_4 = \eta^{11}$ and that the small and large ion concentrations are at equilibrium, (10.4) and (10.5) become

$$0 = \phi - \alpha n^2 - \eta^1 nN - \eta^{11} nN_0, \tag{10.6}$$

$$0 = \eta^{11} N_0 - \eta^1 N. \tag{10.7}$$

Assuming further that the total concentration of aerosol particles, charged and neutral, $N^*$, is constant gives

$$2N + N_0 = N^* = \text{constant}. \tag{10.8}$$

The solutions of these equations are

$$N = \frac{N^*}{2 + \eta^1/\eta^{11}}, \tag{10.9}$$

$$n = \left(\frac{1}{\alpha^2}\left(\frac{\eta^1 N^*}{2 + \eta^1/\eta^{11}}\right)^2 + \frac{\phi}{\alpha}\right)^{1/2} - \frac{1}{\alpha}\frac{\eta^1 N^*}{2 + \eta^1/\eta^{11}}. \tag{10.10}$$

Equation (10.9) somewhat surprisingly indicates that the fraction of the aerosol that is charged is independent of $q$ and $n$. Mathematically this arose from setting $n_1 = n_2 = n$, which decouples (10.6) and (10.7). Physically, the assumption is approximately satisfied, although the concentration of positive ions usually exceeds that of the negative ions by a small amount. More questionable is the assumption that the large-ion concentration is at its equilibrium value.

Equation (10.10) shows the dependence of the equilibrium small-ion concentration on the aerosol concentration, but it is difficult to interpret the relation without introducing numerical values. Computations were carried out for the open ocean, countryside with minimal local pollution, and an urban region. The total aerosol concentrations were taken as $4 \times 10^8$, $10^{10}$, and $10^{11}$ m$^{-3}$, respectively. The near-surface ion production rates were taken as $2 \times 10^6$ m$^{-3}$ sec$^{-1}$ over the ocean and $10^7$ m$^{-3}$ sec$^{-1}$ for both continental areas. The results were roughly averaged over the range of small ion-aerosol recombination coefficients. The small-ion equilibrium concentrations were found to be $9 \times 10^8$ m$^{-3}$ over the ocean, $7 \times 10^8$ m$^{-3}$ for the countryside, and $8 \times 10^7$ m$^{-3}$ in the urban area. Note that there is little difference between the ocean and countryside regions. The larger aerosol concentration and greater ion source strength over the continent

Table 10.3
Small ion concentrations[a]

| Place | Concentration of positive ions ($m^{-3}$) | Concentration of negative ions ($m^{-3}$) | Ratio of concentrations (+/−) |
|---|---|---|---|
| Seewalchen | 944 × $10^6$ | 797 × $10^6$ | 1.18 |
| Davos | 1,240 | 1,010 | 1.23 |
| Fribourg | 1,013 | 908 | 1.12 |
| Seeham | 650 | 626 | 1.04 |
| Argentina | 566 | 545 | 1.04 |
| Amazon | 375 | 354 | 1.06 |
| Linear mean | 798 | 707 | 1.13 |

a. After Israel (1970).

nearly balance out, leaving the ion concentrations much the same, a result confirmed by measurements. The tenfold increase in the aerosol population of the urban air is not balanced by an increased source strength, and the result is a near-tenfold decrease in the small-ion concentration. The numerical values of small-ion concentration just given for the ocean and the countryside are somewhat larger than the observed mean values, but are within the range of the measurements. In view of the uncertain numerical values of the recombination coefficients, the doubtful assumption of large-ion equilibrium, and the other approximations, the close correspondence between the computed and observed small-ion concentrations should be considered somewhat coincidental. The important result is that the aerosol concentration is as important as the ion source strength in determining the small-ion concentration. This not only explains the near equality of the ion concentrations over the oceans and over the continents but also much of the characteristic diurnal variation of the small-ion population near the surface over the continents.

## 10.5 Observed Ion Concentrations

Many measurements of the small-ion concentrations near the surface have been made. A few values are given in table 10.3, taken from Israel (1970). Israel believes these concentrations are too large because the theoretical capacity of the condenser was used instead of the measured capacity. (See the discussion of methods of measurement in section 10.8.) Israel suggests an average concentration of 480 × $10^6$ $m^{-3}$ for small positive ions and 425 × $10^6$ $m^{-3}$ for small negative ions. These averages apply to the near-surface layer in relatively clean continental areas, and similar values are found over the oceans. In almost all cases the near-surface concentration of small positive ions exceeds that of the negative ions. As shown, the equilibrium concentration of small ions depends on source strength and on aerosol concentration. The source strength decreases with elevation up to the top of the planetary boundary layer and then increases up to the lower stratosphere. The aerosol concentration decreases with height in the troposphere, with a marked decrease at the top of the planetary boundary layer. Consequently, one would expect the small-ion concentration to increase with elevation in the troposphere, except for an inversion in the planetary boundary layer. Observations generally support this conclusion. The large variations in the aerosol concentration in the lower troposphere will result in inverse variations of the small-ion concentration in this region.

## 10.6 Large-Ion Concentration

Large ions are aerosol particles that have captured small ions. Their concentration depends primarily on the aerosol concentration. It has been estimated that about one in two to four carries a charge. This fraction must depend on the aerosol concentration and size distribution. As shown in table 1.3, the near-surface aerosol concentration ranges over at least three orders of magnitude, and a similar range of the large-ion concentration is found. The variation of the large-ion concentration with altitude was measured on 40 flights in relatively unpolluted regions by Sagalyn and Faucher (1954). Figure 10.1 represents data taken on a flight between Bedford, Massachusetts, and Sebago, Maine, at midday on a clear day in February. The concentrations of positive and negative ions were found to be equal within the accuracy of the observations, and hence only the positive concentration is plotted. The most striking feature is the sharp decline of $N_+$ at an altitude $Z$ of about 4,200 ft (1,280 m) and the relatively very low concentration above 6,000 ft (1,830 m). The electrical conductivity, which depends primarily on the small-ion concentration, is seen to follow a nearly inverse course.[2] As indicated by the temperature and absolute humidity profiles, the layer extending from near the surface to about 1,280 m is the turbulent planetary boundary layer, sometimes called the exchange layer. The aerosol, which is formed mostly at low levels, has a roughly uniform distribution in the boundary layer, but is prevented from penetrating farther by the temperature inversion. Basically similar results were found in all flights. In level flight, they observed some sharp maxima of $N_+$ that were positively correlated with the humidity. These were interpreted as convective plumes that carry water vapor and the aerosol up from lower levels.

---

2. To convert the conductivity in electrostatic units as given in figure 10.1 to mksA units, multiply by $1.1133 \times 10^{-10}$.

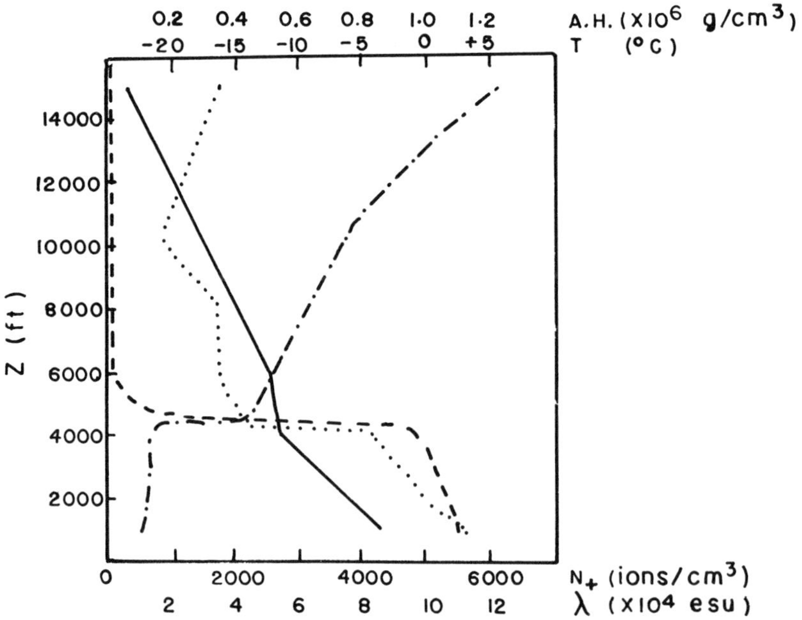


Figure 10.1 The large ion concentration, $N_+$, as a function of altitude in relatively unpolluted air. Also shown are the electrical conductivity ($\lambda$—in electrostatic units, esu), the temperature ($T$), and the absolute humidity (A.H.) The data were taken on a flight from Bedford, MA to Sebago, ME, 18 February 1953, from 1100 to 1500 local time. Key: · · ·, altitude versus absolute humidity; ———, altitude versus temperature; – – –, altitude versus $N_+$; – · –, altitude versus electrical conductivity. From Sagalyn and Faucher (1954), by permission of Pergamon Press.

## 10.7 Electrical Conductivity

The atmospheric ions migrate under the influence of the vertical electric field and constitute an electric current. On occasion, in the presence of convection, there may be variable horizontal components, but this will be ignored here. From the definition of mobility, the current due to a single species of ions in a potential gradient $E$ is

$$J = n_i k_i e E, \tag{10.11}$$

where $n_i$ *and* $k_i$ are the concentration and mobility of ion species $i$, $e$ is the electron charge, and $J$ is the current in A m$^{-2}$. When more than one ion species is present, as is usual, (10.11) becomes

$$J = Ee \sum_i n_i k_i. \tag{10.12}$$

From Ohm's law, $J = \lambda E$, where $\lambda$ is the electrical conductivity.[3] Then from (10.12)

$$\lambda = e \sum_i n_i k_i \text{ mho m}^{-1}. \tag{10.13}$$

Since the mobility of large ions is typically three orders of magnitude less than that of small ions, the contribution of large ions to the conductivity can usually be neglected. Then (10.13) reduces to

$$\lambda = e(m_1 k_1 + n_2 k_2), \tag{10.14}$$

where subscripts 1 and 2 denote positive and negative ions, respectively. If typical near-surface values of $n$ and $k$ are inserted in (10.14), it is found that $\lambda$ is about $2\text{–}3 \times 10^{-4}$ mho m$^{-1}$. Taking the surface potential gradient as 130 V m$^{-1}$, the current is $2.6\text{–}3.9 \times 10^{-12}$ A m$^{-2}$. The conductivity due to small ions is directly proportional to the ion concentration. The conductivities due to the positive and to the negative ions are often given separately and are called polar conductivities.

The column resistance of the atmosphere is given by $R = \int_0^{Z_0} dZ/\lambda\ \Omega$ (ohms) for a vertical column of one square meter cross section, where $Z_0$ is the height of the upper conducting layer. Because of the rapid increase of $\lambda$ with height, $R$ is relatively insensitive to the value chosen for $Z_0$. A typical or average value of $R$ is $10^{17}\ \Omega\,\text{m}^{-2}$, or a resistance of about 200 $\Omega$ for the global atmosphere. In heavily polluted atmospheres $R$ may be several times the average. Even in relatively clean atmospheres a substantial fraction of $R$ is contributed by the planetary boundary layer; in a polluted atmosphere, up to 85–90% of $R$ is contributed by this layer.

## 10.8 Methods of Measurement

The ion concentration or the conductivity is usually measured with a Gerdien (1905) condenser, which is shown schematically in figure 10.2. In practice the inner cylinder is shorter than the outer, and an inlet funnel may be attached. The ions of polarity opposite that of the inner cylinder migrate toward the center under the influence of the applied electric field while moving axially in the air flow produced by the fan. The current, measured by a suitable instrument,[4] gives the number of ions captured at the inner electrode per unit time.

It is shown in standard texts on electricity that the electric field in a

3. The unit of electrical conductivity is the mho and is numerically equal to the reciprocal of the resistance in ohms.
4. The current is of the order of a picoampere ($10^{-12}$ A).

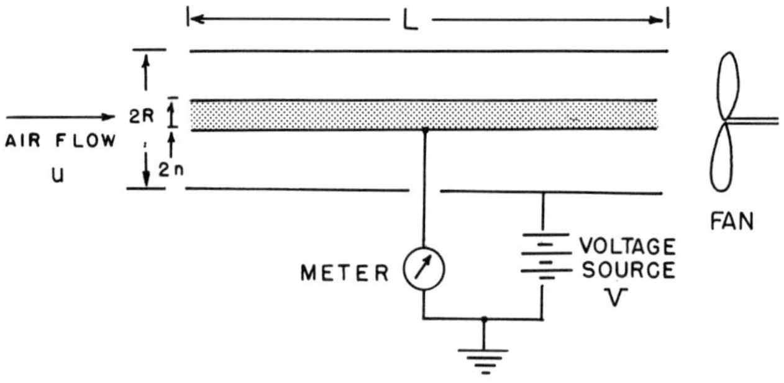


Figure 10.2 Schematic of cylindrical ion collector.

concentric cylindrical condenser of infinite length is $V/y(\ln R/n)$, where $V$ is the applied voltage, $y$ is the radial coordinate, and $R$ and $r$ are the radii of the outer and inner cylinders, respectively. The time required for an ion of mobility $k$ at the outer radius to reach the inner cylinder is

$$\Delta t = \frac{\ln(R/r)}{kV}\int_r^R y\,dy = \frac{\ln(R/r)}{2kV}(R^2 - r^2). \tag{10.15}$$

To ensure the capture of an ion initially at a radial distance $R$ and hence of all ions that enter the instrument, $\Delta t$ must be equal to or less than the time required for the air to transit the instrument, or $L/u$, where $L$ is the length of the cylinder and $u$ is the air velocity, assumed to be independent of $y$. Inserting this limit and solving for the corresponding mimimum mobility, $k^*$, gives

$$k^* = \frac{u(R^2 - r^2)}{2VL}\ln(R/r). \tag{10.16}$$

This shows that the minimum mobility depends on the air velocity, the inverse of the applied voltage, and the dimensions of the instrument. It is customary to introduce the volumetric flow rate, $W = \pi(R^2 - r^2)u$, and the capacitance per unit length of a cylindrical condenser of infinite length, which is $\frac{1}{2}\ln(R/r)$, or $(L/2)\ln(R/r)$ for the finite length $L$:

$$k^* = W/(4\pi VC), \tag{10.17}$$

where $C$ is the capacitance per unit length.

If all of the ions have mobilities greater than $k^*$, the current to the central electrode will be $Wne$, where $n$ is the ion concentration and $e$ is the electron charge. The current is thus a direct measure of the ion concentration and, if the mobility of the ions is known, permits a direct computation of the polar conductivity.

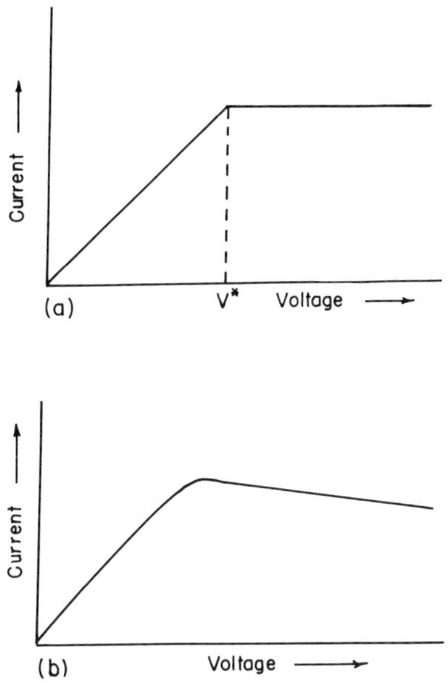


Figure 10.3 Idealized current-voltage curves of a cylindrical condenser instrument. The upper curve is for a single ion mobility without end effects. The lower curve is for a single ion mobility, but end effects are included.

In use, the applied voltage is varied so as to yield a current-voltage curve. An idealized curve is shown in figure 10.3a for a situation in which all of the ions are of the same mobility. The sloping portion of the curve corresponds to $k < k^*$, and only a fraction of the ions is captured. If the axial air velocity is independent of $y$, the radial increment of $W$ is directly proportional to $y$, whereas the field strength is inversely proportional to $y$. This leads to a current proportional to the ratio $k/k^*$ and thus explains the linear slope of the curve. It is easy to show from (10.14) and (10.17) that the slope is directly proportional to the polar conductivity. It is also evident that $k = k^*$ at $V^*$ and that thereafter the current is constant, since all of the ions are captured.

A more realistic current-coltage curve is shown in figure 10.3b in which the slope is not constant, there is no sharp break, and the current declines slightly with increasing voltage. There are several causes for these differences. It may first be noted that even in the ideal case, there would not be a discontinuous slope because the ion trajectories are parabolic. The most important difference is due to the end effects of a capacitor of finite length. Briefly, the field is not confined to the annular space between the cylinders, but extends in distorted form outside the cylinders. An entering ion is acted upon by this external field in such a way as to reduce the number of ions entering the instrument. For the same reason the capacity is not the same as that per unit length of a condenser of infinite length and the internal field is not that of an infinite condenser. Direct measurements on actual instruments yield capacities 20–40% larger than the idealized computed capacities. It is generally believed that better results can be achieved by using the measured capacity in (10.17), although this cannot eliminate all of the discrepancies.

In addition to motion in the electric field the ions also move by molecular diffusion. This affects the time of migration of the ions to the inner cylinder. Corrections for diffusion can be made, and they are usually relatively small. Another factor is the assumption of a uniform air velocity across the annular space between the cylinders. Laminar air flow is sought to minimize turbulent ion transport; the velocity profile is then essentially parabolic. This will cause a decrease in $k^*$ and introduce curvature in the current-voltage curve to the left of the maximum. It appears that the velocity profile has little effect on the measurement of the ion concentration. In the idealized case of figure 10.3a the slope is directly proportional to the conductivity; in the more realistic case of figure 10.3b, the slope is not constant, but it is customary to take the straight line from the origin to the first data point as a measure of the conductivity. Figure 10.3b suggests that measurements of the ion concentration should be made just to the right of the maximum. The decline of the curve in this region is due primarily to the capacitor end effects.

The mobility of the ions can be determined by inspection in figure 10.3a but this becomes more difficult in figure 10.3b. Theoretically the second derivative of the curve should yield the mobility, and this is probably adequate when ions of only a single mobility are present. In the usual case of a mobility spectrum, it becomes more difficult to derive the ion spectrum from the current-voltage curve. This speaks against trying to represent fine details of the spectrum.

Many theoretical, experimental, and empirical methods have been devised to reduce the effects of the several limitations discussed. None of these is completely satisfying, and some uncertainty therefore remains in the interpretation of the data obtained with a Gerdien condenser. It is beyond the scope of the current treatment to explore this question in more detail; the interested reader is referred to Israel (1970) and Chalmers (1957).

In principle, and generally in practice, the Gerdien condenser may be used to measure atmospheric ions of any mobility by selecting appropriate dimensions, air flow, and voltage. Since ion mobilities range over some four orders of magnitude, it is usually not feasible to use a single instrument. The small ions are usually measured by one instrument and the intermediate and large ions by another. The latter may incorporate a condenser designed to remove all of the small ions before the air enters the main portion of the device. It is usually not practical to collect ions of

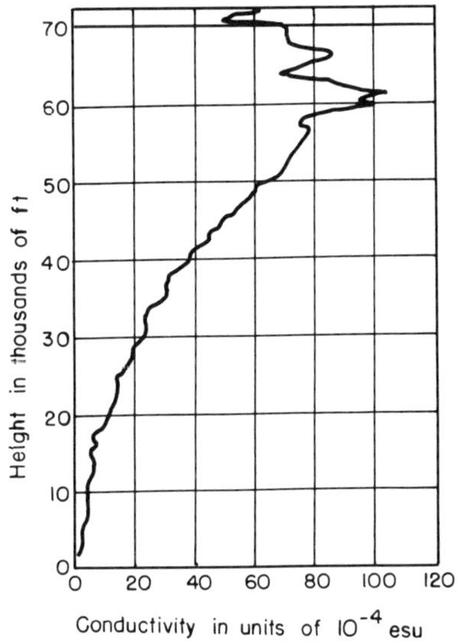


Figure 10.4 Positive conductivity versus height from the manned stratospheric balloon *Explorer II*. From Gish and Sherman (1936), by permission of the National Geographic Society.

mobilities less than about $10^{-8}$ $m^2 V^{-1} sec^{-1}$. Only the relatively small number of ultralarge ions will then be missed.

## 10.9 Observed Conductivity

In the boundary layer, and particularly in the surface layer, the conductivity is inversely proportional to the aerosol or large-ion concentration. The diurnal and annual variations of the near-surface conductivity are therefore expected to be the inverse of the aerosol variations. Maximum conductivity usually occurs in the early hours of the day prior to sunrise. At urban stations the conductivity is higher in summer than in winter. Diurnal and seasonal variations of conductivity are very small over the oceans. Over land areas the conductivity varies in a complex manner, reflecting changes in the surface aerosol due to advection, vertical mixing, local aerosol sources, and the like.

When there is a well-established planetary boundary layer the conductivity within this layer tends to be uniform, as shown in figure 10.1. Above the boundary layer the conductivity increases nearly exponentially with height. This is due to the increase in cosmic radiation, the decrease in the aerosol, and the increasing mobility with height. The beginning of this increase with height is also shown in figure 10.1. A more adequate representation is given in figure 10.4, based on data collected on the flight of the manned stratospheric balloon *Explorer II* (Gish and Sherman, 1936). It shows an increase in conductivity to $6 \times 10^4$ ft (18 km), followed by an irregular decrease, which is not now thought to be typical, but could have resulted from a marked Junge aerosol layer. The conductivity at 10 km is about 10 times the near-surface value, and that at 18 km is some 4 times that at 10 km. Many other vertical profiles of conductivity are available, but the *Explorer II* data were selected for illustration because of their historical interest. It has been shown that conductivities computed from cosmic ray ionization, the small-ion mobilities, and the assumption of negligible aerosol fit figure 10.4 quite closely between about 3 and 18 km. It seems to be generally true that the conductivity profiles above the planetary boundary layer are much the same everywhere, reflecting mainly only the latitudinal and solar cycle variations in cosmic radiation. As already stated, the conductivity within the planetary boundary layer is quite variable in space and time, depending primarily on the aerosol concentration.

## 10.10 Measurement of Potential Gradients

The measurement of the potential gradient at the surface is simple in principle, but the problem is to ensure that the probe is at the same potential as the air at the same level. Any isolated conductor will finally reach the potential of the air if there is no leakage through the insulating support. Several means have been devised to bring the probe to air potential in a short time. The oldest is the water-dropper, due to Kelvin. In this device water or some other liquid is released from the probe, usually by spraying because the charge transfer is proportional to the surface area of the drops. The drops carry charge away from the probe until its potential is the same as that of the air. Another approach is to incorporate a small amount of radioactive $\alpha$-ray emitter, such as polonium, on the probe. The ions formed by the $\alpha$ rays will move in response to the difference of potential between the probe and the air in such a way as to decrease the potential difference. If the $\alpha$-ray emitter is too strong, the ions will seriously distort the field that one is trying to measure; a compromise must be reached between field distortion and time of equalization of the potential. Flames or burning fuses are also used; these both produce ions and products of combustion that act like the water-dropper.

A moderately long horizontal wire with a polonium source or a fuse at the center is often used to calibrate other types of instruments. It does not respond to rapid changes in potential gradient and is therefore not used as the principal instrument.

In all of these devices, the potential between the probe and ground must be read on an electrometer that draws essentially no current; this is now usually done electronically. Electrical leakage from the probe to ground

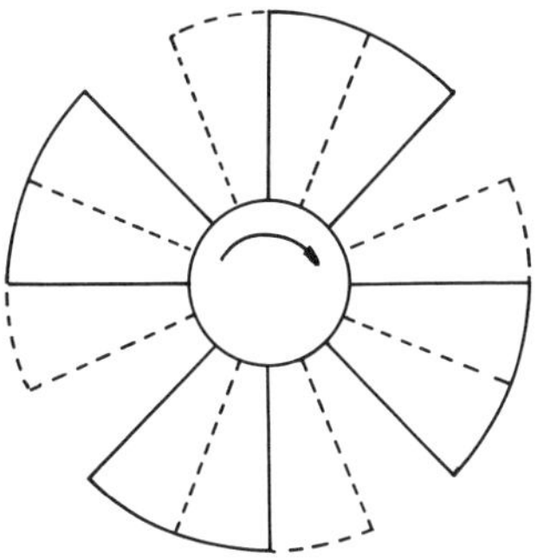

Figure 10.5 Top view of a field mill. Solid lines and curves represent the upper rotating vanes. Dashed lines and curves outline the lower fixed vanes.

limits the accuracy of these instruments. With good design, an error of 5% or less is achieved, and this can be reduced further by corrections. A leak-free device has been developed, but will not be described here. Of even greater importance is the distortion of the natural electric field by buildings and other irregularities. A "reduction factor" is derived by placing a horizontal wire instrument at a distance from the building and the regular device. There is evidence that the reduction factor is not a constant. A very useful device for the measurement of the potential gradient is the field mill. This device typically consists of fixed vanes surmounted by similar vanes that are rotated by a motor so as alternately to shield and uncover the fixed vanes, as indicated in figure 10.5. In measuring the surface potential gradient the upper rotating vanes are flush with the surface and are grounded. When the fixed vanes are uncovered, they aquire a bound charge, which is then released when they are covered. This results in an alternating current between the fixed plate and ground that can be amplified and rectified for recording. It will be remembered that the surface charge on a conductor is related to the potential gradient by

$$E = -\sigma/\varepsilon_0, \tag{10.18}$$

where $\sigma$ is the surface charge in C $m^{-2}$ and $\varepsilon_0$ is the permittivity of free space, $8.854 \times 10^{-12}$ C $m^{-1}$ V (F $m^{-1}$, per meter, in the mksA system).

Field mills may also be used on aircraft. To eliminate the effect of the charge on the aircraft two field mills are used, one on the top surface and the other on the bottom surface.

## 10.11 Observations of the Surface Potential Gradient

The near-surface fair weather potential gradient over the oceans, far from land, averages about 130 V $m^{-1}$. Over the land areas it ranges from about

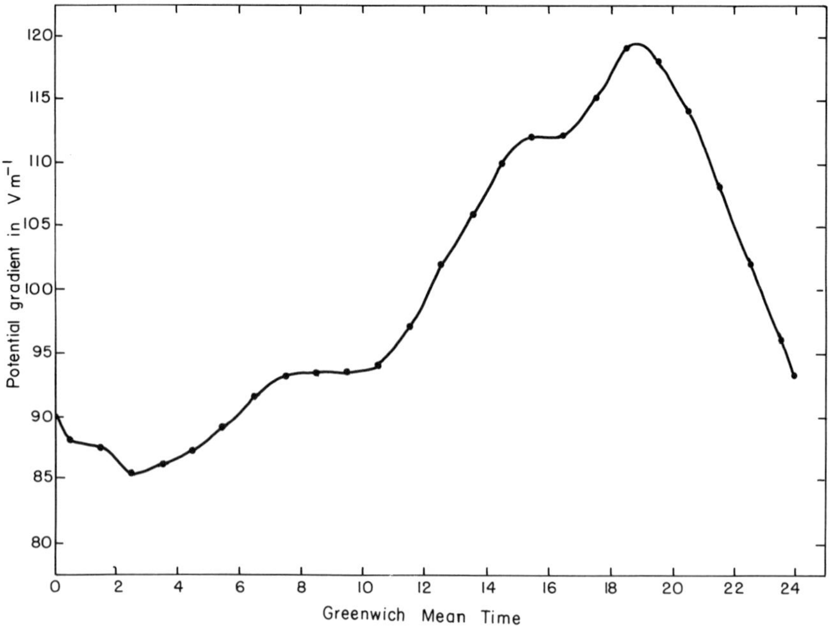


Figure 10.6 The vertical potential gradient over the ocean as a function of Greenwich Mean Time. After Chalmers (1957).

100 $\mathrm{V\,m^{-1}}$ in clean air, including the Arctic and Antarctic, to several hundreds of volts per meter in polluted regions. At Kew Observatory in London, the average value exceeds 300 $\mathrm{V\,m^{-1}}$.

The diurnal variation of the potential gradient is of particular interest. In more or less polluted continental air, the potential gradient varies markedly during the day. At some stations there are single maxima and minima; at other locations two maxima and minima are observed. Analysis shows that these variations reflect the well-known diurnal changes in the concentration of the aerosol, which are tied to local time.

From an analysis of data taken in the polar regions, Hoffman (1923) conjectured that the diurnal variation of the potential gradient was keyed to Universal or Greenwich Mean Time (GMT), with a single minimum around 0400 GMT and a maximum near 1900 GMT. This was soon confirmed by Mauchly (1923) for the oceans on the basis of the measurements made on the several cruises of the R/V *Carneigie*. This universal diurnal variation is shown in figure 10.6. It presumably occurs over land as well, but is obscured by the effects of the aerosol.

Wilson (1922) suggested that thunderstorms were the generators that maintained the fair weather field in the face of the conduction current that acts to discharge the field. He pointed out that maximum potential

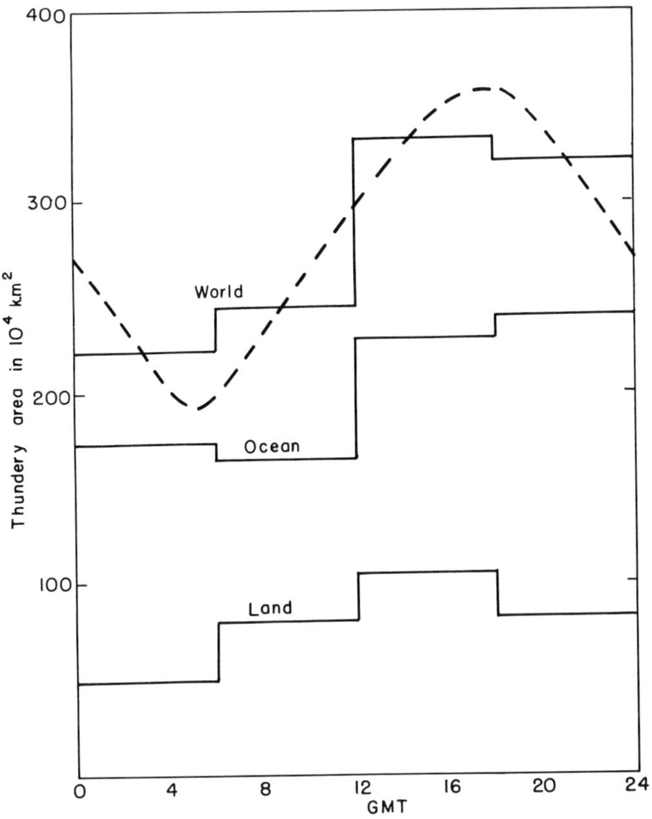


Figure 10.7 Thundery areas over the oceans, the continents, and the world as a function of Greenwich Mean Time. The dashed curve represents a hypothetical continuous variation. After Trent and Gathman (1972).

gradient should occur when the most thundery regions of the earth attain their maximum activity. Brooks (1925) estimated annual, seasonal, and latitudinal frequencies of thunderstorms over the land and the oceans from the rather sketchy observations by weather stations and ships. Whipple (1929) refined Brooks's estimates and judged that the diurnal variation of the potential gradient was similar to that of the number of thunderstorms in progress over the globe.

More recently Trent and Gathman (1972) counted oceanic thunderstorms based on more than 7 million observations made by ships at sea during 1949–1963. Figure 10.7 includes their results for the ocean areas, Whipple's for the land areas, and their combination for the globe. New and, it is hoped, more complete data on thunderstorm frequency are in prospect from satellites that sense lightning flashes.

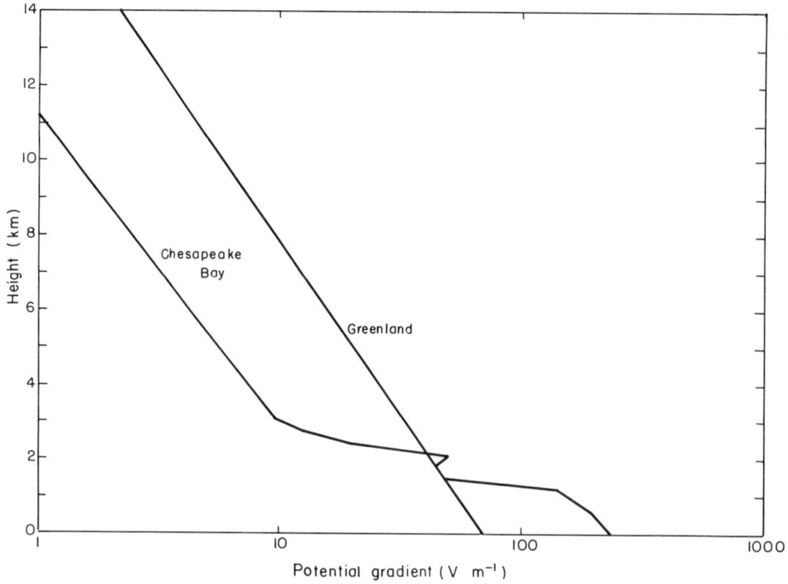


Figure 10.8 Vertical profiles of the potential gradient over Greenland and Chesapeake Bay. After Clark (1958).

It will be seen that there is general correspondence between the global curves in figures 10.6 and 10.7, particularly in the time of the maxima. The maximum is due primarily to thunderstorms in Africa, Europe, and America. The general agreement has led many, if not most, persons to conclude that thunderstorms maintain the earth atmosphere potential difference against discharge by the worldwide conduction current. However, there is a growing feeling that the case has not been proved.

## 10.12 Potential Gradient versus Height

Generally, in relatively clean air the potential gradient decreases exponentially with height, as shown in the Greenland profile in figure 10.8; the other profile is for a rather polluted atmosphere over Chesapeake Bay. The profile over Greenland follows the exponential down to about 100 m from the surface, while the Chesapeake Bay profile is exponential only above 3 km. The planetary boundary layer extended to about 2.3 km, and the profile of the potential gradient in this layer is strongly modified by the aerosol and convection.

The exponential decay of the potential gradient with height shows that much of the positive charge resides in the troposphere and not in the ionospheric upper conductor of the "leaky spherical condenser" analog of fair weather electrification. This is shown in another way in figure 10.9, in

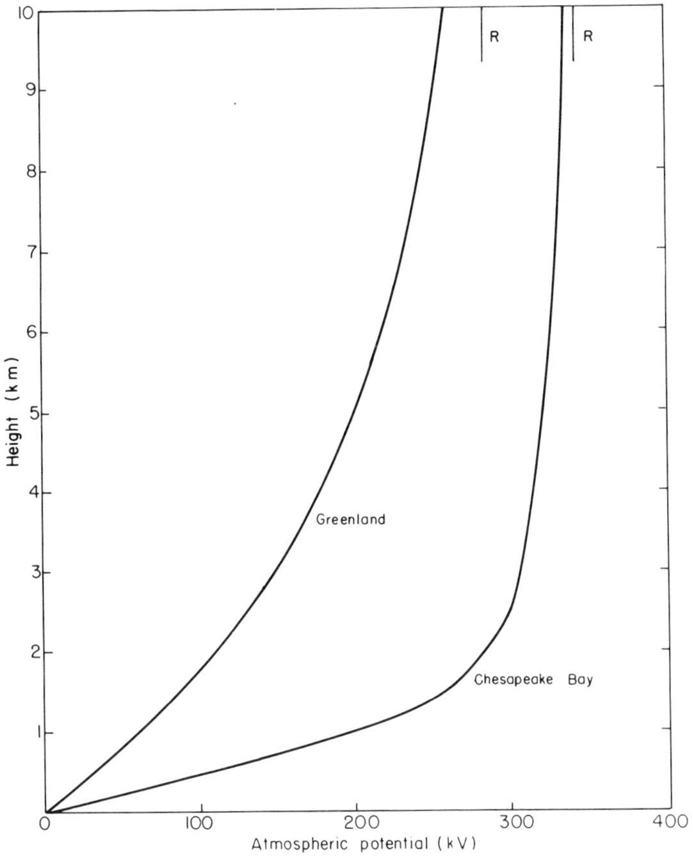


Figure 10.9 Atmospheric potential as a function of height over Greenland and Chesapeake Bay. After Clark (1958).

which the profiles of figure 10.8 have been integrated over height from the surface to yield the atmospheric potential. The limiting values or the so-called ionospheric potential are given at the top of the figure in kilovolts. Evidently nearly 60% of the total atmospheric potential is in the first kilometer and more than 80% in the first 2 km in the Chesapeake Bay profile. Even in the profile over Greenland half of the total potential is found in the first 3 km.

It will be noted in figure 10.9 that the ionospheric potential over Chesapeake Bay is some 20% larger than that over Greenland. The Chesapeake Bay sounding was taken at about 1800 GMT, while that over Greenland was taken at about 1100 GMT. Reference to figure 10.6 shows that the sea surface potential gradient at 1800 GMT is about 25% larger than that at 1100 GMT. This and similar results show that the ionospheric

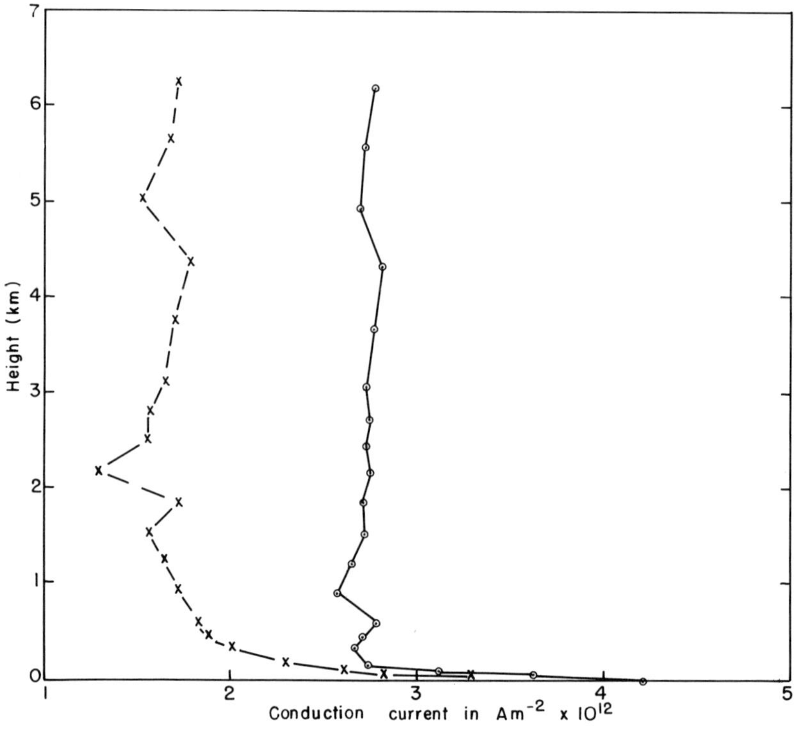


Figure 10.10 Measured profiles of the conduction current. The dashed curve is drawn from data taken near Bermuda, while the solid curve is the average of 18 profiles over the ocean at various locations. After Kraakevik (1961).

potential varies with universal time with a phase and amplitude similar to the surface potential gradient over the oceans.

## 10.13 Conduction Currents

The nearly exponential decrease of potential gradient with height and the exponential increase of conductivity suggest that the conduction current is nearly invariant with height. Indeed, in a steady state, the current must be constant with height.

The current at the surface may be measured directly by means of an insulated plate set flush with the ground surface. Because of boundary layer effects, the surface current is not very representative. The current at elevations above the surface is ordinarily obtained from simultaneous measurements of conductivity and potential gradient. Examples of the measured conduction current are presented in figure 10.10. The oceanic profile is smoother due to the averaging of 18 profiles, but it is notable that the current is nearly constant down to about 200 m height. A planetary boundary layer extending to about 2 km was present during the Bermuda sounding. The rapid decrease in current through this layer is very evident.

It is reasonable to assume that the total current is independent of height. On this basis there must be another current present in the boundary layer flowing in a direction opposite to that of the conduction current. This additional current results from the transfer of space charge by turbulence and convection. This mechanical transfer of charge cannot be measured by the product of conductivity and potential gradient. Its magnitude may be estimated by assuming that the average conduction current above the exchange layer is also the total current within the exchange layer. For example, the average current above the exchange layer in the Bermuda profile is about $1.61 \times 10^{-12}$ A m$^{-2}$. By subtraction, the convection current is about $1 \times 10^{-12}$ A m$^{-2}$ at 100 m and $2.5 \times 10^{-13}$ A m$^{-2}$ at 500 m. This cannot be conventional turbulent diffusion, transporting space charge down the space charge density gradient. The one-dimensional form of Gauss's law is

$$\frac{dE}{dZ} = \frac{\rho}{\varepsilon_0}, \tag{10.19}$$

where $\rho$ is the space charge density and $\varepsilon_0$ is the permittivity of free space. When (10.18) is applied in finite difference form to the vertical profile of $E$ over Chesapeake Bay in figure 10.8, the space charge density in the boundary layer is as shown in figure 10.11. Some of the irregularities are doubtless due to errors in the data, magnified by differentiation, but it is clear that more is involved than simple down-gradient flow. Willett (1975) devised a numerical model incorporating a second-order closure model of convective transfer and the electrical equations. He concluded that with strong convection, such as occurs over intensely heated ground, the convective charge transfer acts as a generator with a voltage of as much as 130 kV (some 40% of the total atmospheric potential) acting in opposition to the normal fair weather potential. Some uncertainty exists in these numerical results because of the lack of pertinent experimental data. It should also be pointed out that the assumed strong convection occurs only over relatively small continental areas, and hence the effect on the global electrical circuit is small.

The conduction current above the exchange layer varies from about 1 to $5 \times 10^{-12}$ A m$^{-2}$. This variation is due to the similar variation of atmospheric columnar resistance, which is relatively high in polluted regions and low in clean air. The current has been measured most often at the surface, but figure 10.10 shows that this is not a reliable measure of the current above the boundary layer. Measurements such as those given in figure 10.10 are relatively much less numerous. Since nearly three-quarters

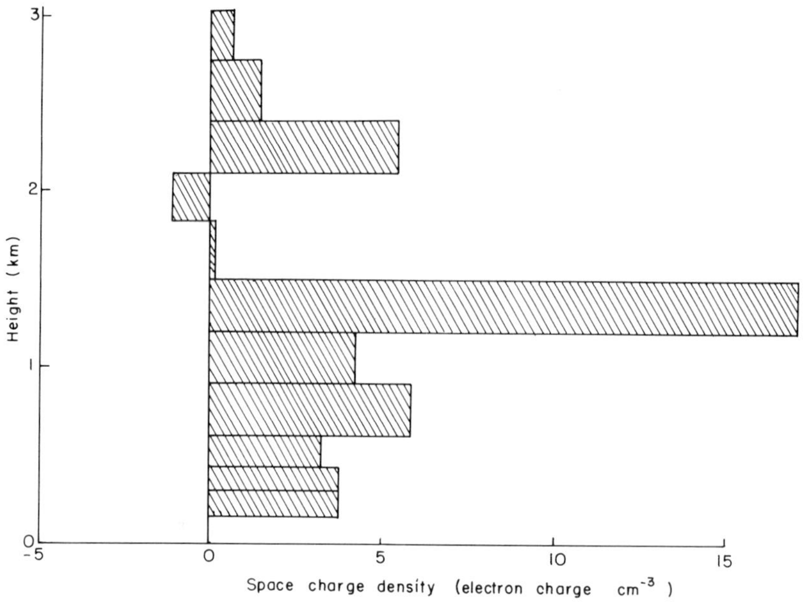


Figure 10.11 Profile of space charge density over Chesapeake Bay by the numerical differentiation of figure 10.8. After Kraakevik and Clark (1958).

of the earth is ocean, it has been suggested that the global mean conduction current is close to that of the mean oceanic value of $2.73 \times 10^{-12}\ \mathrm{A\ m^{-2}}$ in figure 10.10, but some uncertainty still exists. It is also found that the conduction current over the oceans and the polar regions exhibits the same GMT diurnal variation as the atmospheric potential and the potential gradient.

## 10.14 The Electrode Effect

In an ionized gas between two electrodes across which there is a voltage, the positive ions drift toward the negative electrode (which may be the surface of the earth), and the negative ions drift away from the negative electrode. There is a steady flow of positive ions toward the negative electrode, those reaching the electrode being replaced by more from above. However, there is no source of negative ions at the electrode, and those moving up are replaced only by the ion-forming process, which, of course, also forms positive ions. The net result is a relative depletion of negative ions near the negative electrode or the creation of a positive space charge there. This phenomenon is called the electrode effect. It can be produced in the laboratory, but is difficult to detect in the atmosphere. This is due in part to the presence of space charge from other processes and the rapid increase in ionization as the surface is approached due to radioactive matter in the surface soil. It appears that the electrode effect does occur in the atmosphere, but only within the first few tenths of a meter above the surface. Nevertheless, it may be a source of positive space charge of some importance. On occasion the electrode effect has been invoked to explain phenomena extending tens of meters or more above the surface, but it is now clear that this is erroneous.

## 10.15 Cumulonimbus Clouds

By definition, the cumulonimbus is the thundercloud. It is the engine that drives the strong electrical generator manifested in lightning flashes and other electrical manifestations. Detailed discussion of the thermodynamics and dynamics of the cumulonimbus may be found in standard works on meteorology. The associated precipitation processes are discussed in chapter 8. Here it will suffice to review briefly those features of the cumulonimbus that are thought to be pertinent to the electric charge-generating mechanisms.

The cumulonimbus is a cloud of large vertical dimension, often extending from a cloud base that may be from one to several kilometers above the surface to near or slightly above the tropopause. Roughly speaking, the lateral size of an individual cloud is about the same as its height. The vertical air velocities within the cloud characteristically reach some tens of meters per second. Thunderstorms more often occur in groups than singly; the groups are often arranged linearly along a cold front or a squall line.

As shown by Byers and Braham (1949), a typical thunderstorm consists of several more or less vertical columnar cells of strong ascending velocity that release heavy rain. The life of a single cell averages about 20 min, but the sequence of several cells leads to a storm lifetime of an hour or more. Once precipitation forms, the individual cells become clearly evident on radars. Byers and Braham identified three stages in the development of a cell. In the first stage there are updrafts throughout the cell. In the second or mature stage both updrafts and downdrafts are found at least in the lower portion. In the third or dissipating stage weak downdrafts extend throughout the cell. The precipitation at the surface begins with the mature stage.

The thunderstorms sampled by Byers and Braham were predominantly of the so-called air mass type, although some were associated with fronts or squall lines. The giant squall line thunderstorms or supercells described by, for example, Atlas (1963) differ from these in intensity and their attainment of a near steady state for some hours.

The great vertical extent of thunderstorms and the high vertical velocities lead to large liquid water contents and deep supercooled regions in which ice is present in the form of snow, graupel, and hail. As stated in section 8.11, the dominant precipitation process is almost certainly accretion, and mostly the accretion of supercooled cloud drops by ice particles, forming graupel. Most of the ice melts in the lower warm portion of the cloud and reaches the surface as rain; because of its larger size, the hail may reach the surface. The usual presence of ice has led to conjectures that the charging process involves contact between ice particles, between ice and supercooled drops, melting, or freezing. Although the ice phase may be responsible for the principal charging mechanism, observations of lightning from warm clouds by Foster (1950), Moore et al. (1960), and Pietrowski (1960) show that there must also be an all-water charging mechanism. It should be noted that attempts have been made to discount warm lightning as due to errors of observation, reminiscent of similar contentions many years ago about warm rain. Careful reading of the cited references leaves little doubt of the reality of warm lightning. It has also been noted by several observers that small convective clouds exhibit weak electric fields even in the absence of visible precipitation. Lightning is also observed in volcanic eruptions and in blowing sand and snow.

## 10.16 Problems in Measuring Thunderstorm Charge

Thunderstorms present serious barriers to the measurement of their electric charge structure. They are inhospitable to penetration with measuring equipment; moreover, they are not in a steady state and are highly complex in all aspects. Gauss's law calls for the three-dimensional measurement of the Laplacian of the potential to define the charge density; this is clearly a practical impossibility in such a complex and transient system. The most comfortable measurement site is the ground surface, but this places the instruments beneath the intense space charge produced by surface point discharges (see section 10.19). Some brave souls have flown in instrumented aircraft below, around, and in thunderstorms; others have sent instruments aloft in balloons and rockets. These efforts have yielded some useful, even invaluable, data, but they are in the nature of random spot observations.

## 10.17 Deductions from Lightning

The surface potential gradient under a thunderstorm seldom exceeds $10^4$ $V\,m^{-1}$, as compared with the laboratory breakdown gradient of $1\text{–}3 \times 10^6$ $V\,m^{-1}$. Measurements of the surface potential gradient do not define the thundercloud charge structure. However, it was soon found that lightning strokes produced a sudden change in the potential gradient, an example of which is shown in figure 10.12, based on observations by Wormell (1939). The lightning stroke was 26.5 km away. Note the approximately exponen-

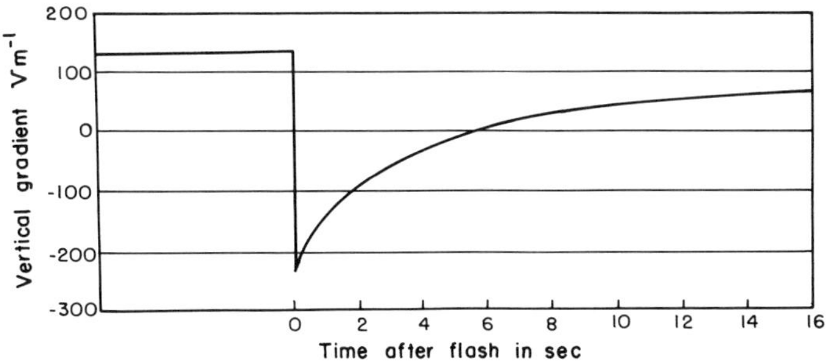


Figure 10.12 Time variation of the vertical potential gradient at the surface after a lightning stroke at time zero. Based on observations by Wormell (1939).

tial recovery after the stroke. More recent instrumentation with a time resolution of 0.05 sec or less shows the complex time history of the lightning discharge, which appears as a nearly vertical line in figure 10.12 and also that the recovery does not usually follow a simple exponential.

If the thunderstorm charge centers are considered to be point charges or spherically symmetric regions of charge, Coulomb's law leads to a simple expression for the vertical potential gradient at the surface. This is illustrated in figure 10.13, where it is assumed that the charges constitute a simple vertical dipole. The vertical component of the field at 0 due to the charge $Q_1$ and its image charge is

$$E = \frac{2Q_1}{4\pi\varepsilon_0(L^2 + Z_1^2)} \sin\omega_1, \tag{10.20}$$

where $\omega_1$ is the elevation angle of $Q_1$ as seen from 0, $L$ is the horizontal distance, and $Z$ is the height. By writing $\sin\omega_1$ in terms of $L$ and $Z_1$, this becomes

$$E_1 = \frac{2Q_1 Z_1}{4\pi\varepsilon_0(L^2 + Z_1^2)^{3/2}}. \tag{10.21}$$

The total vertical field due to the dipole is then

$$E_1 + E_2 = \frac{2Q_1 Z_1}{4\pi\varepsilon_0(L^2 + Z^2)^{3/2}} - \frac{2Q_2 Z_2}{4\pi\varepsilon_0(L^2 + Z_2^2)^{3/2}}. \tag{10.22}$$

The sign convention is that the fair weather field is positive (positive charge on the atmosphere). For a dipole, $Q_1 = Q_2$. Near the storm the field is negative, but becomes positive at a reversal distance $L_\mathrm{R}$, where $E_1 + E_2 = 0$:

$$L_\mathrm{R} = (Z_1 Z_2)^{1/3}(Z_1^{2/3} + Z_2^{2/3})^{1/2}. \tag{10.23}$$

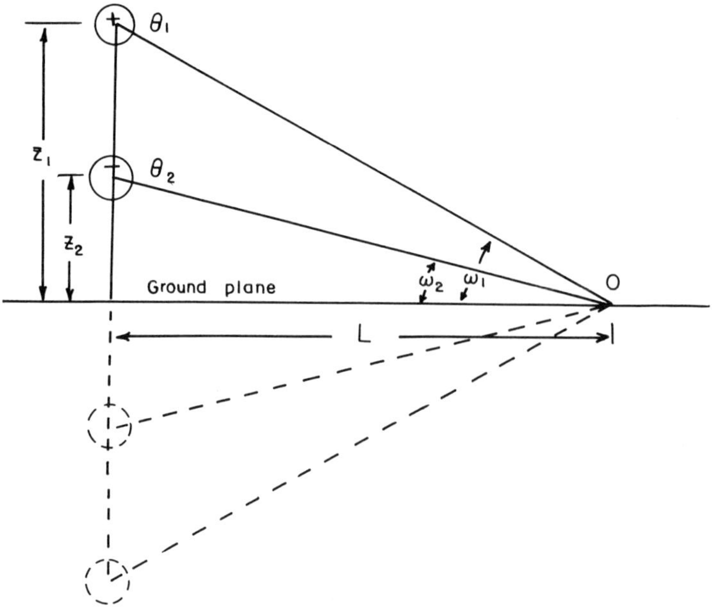


Figure 10.13 Simplified representation of the vertical potential gradient at point 0 due to a vertical dipole charge distribution at a horizontal distance $L$ from 0. The dashed portions are due to the image charges.

When $L$ is large compared with $Z_1$ and $Z_2$, $(L^2 + Z_1^2) \simeq (L^2 + Z_2^2) \simeq D^2$. By setting $Q_1 = Q_2 = Q$ and $E_1 = E_2 = E$, (10.22) becomes

$$E = \frac{2Q(Z_1 - Z_2)}{4\pi\varepsilon_0 D^3} = \frac{M}{4\pi\varepsilon_0 D^2}, \tag{10.24}$$

where $M = 2Q(Z_1 - Z_2)$ is the dipole moment. The sign of the change in $E$ due to a flash may be used to distinguish between intracloud and cloud-to-cloud flashes if the usual dipole orientation is assumed. Observations have confirmed the inverse dependence of $E$ on $D^3$, where $D$ is the mean distance to the dipole.

With an array of field mills at the surface, (10.22) may be used to determine $Q_1$, $Q_2$, $L$, $Z_1$, and $Z_2$. A storm near the network favors a good signal-to-noise ratio, but enhances errors due to nonspherical charge centers. In spite of these and other limitations, most of our knowledge of thunderstorm charge structure has been obtained in this way. More recently, Krehbiel et al. (1974) and Jacobson and Krider (1976) have developed statistical and instrumental means for reducing the errors of the method.

Table 10.4, taken from Jacobson and Krider, is a convenient summary

Table 10.4
Summary of observations of lightning charges, altitudes, and moments in various geographical locations[a,b]

| Location | Charge (C) | Height above terrain (km) | Range of air temperature | Moment (C km) | Authority[c] |
|---|---|---|---|---|---|
| Florida | −10 to −40 | 6–9.5 | −10 to −34 | 100–600 | Jacobson and Krider (1976) |
| | $-\overline{25}$ | | | | |
| England | −11.5 to −46 | 7 | −34* | 33–430 | Wilson (1916) |
| England | −20 | 2 | −0.5* | $\overline{100}$ | Wilson (1920) |
| England | −10 to −40 | 4.5–5 | −16 to −19* | $\overline{220}$ | Wormell (1934) |
| England | — | — | — | $\overline{150}$ | Pierce (1955) |
| South Africa | −15 | 3 | −13* | $\overline{93}$ | Schonland (1928) |
| South Africa | −4 to −40 | 2.5–8.7 | −10 to −48* | 41–495 | Bernard (1951) |
| South Africa | — | 4–8.5 | −7 to −38 | — | Malan et al. (1951) |
| New Mexico | $-\overline{24}$ | 4–7 | −5 to −25 | — | Workman et al. (1942) |
| New Mexico | −5 to −20 | 4.3–7.2 | −7 to −33 | — | Reynolds et al. (1955) |
| New Mexico | −30 to −48 | 4.5–6 | −12 to −23* | — | Krehbiel et al. (1974) |
| New Mexico | −5 to −60 ($-\overline{28}$) | 3–8 | −3 to −36 | $\overline{249}$ | Brook et al. (1962) |
| Japan | −50 to −150 | 4–8 | +4 to −25 | — | Hatakeyama (1958) |
| Japan | −6 to −55 | 6–8 | −11 to −24* | — | Tamura (1954) |
| Japan | $-\overline{20}$ | 3.5–5.5 | +3 to −8* | — | Takeuti (1966) |
| Hong Kong | $-\overline{25}$ | 4 | −1* | $\overline{210}$ | Wang (1963) |
| Australia | $-\widetilde{17}$ | $\tilde{3}$ | +5* | $\overline{150}$ | Mackerras (1968) |

a. After Jacobson and Krider (1976).
b. The Overbar indicates an average, the tilde a median, and the asterisk a temperature estimated from climatological means.
c. References are to be found in Jacobson and Krider (1976).

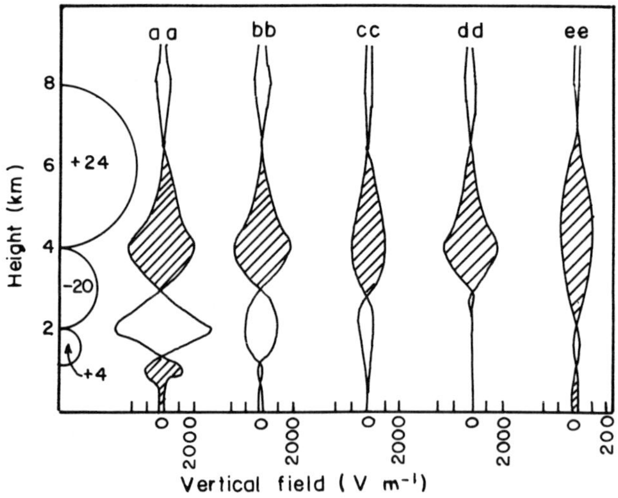


Figure 10.14 Profiles of the electric fields in thunderstorms obtained by Simpson and Robinson (1941) from instrumented free balloons. Regions of positive fields are shaded, while the unshaded areas are regions of negative fields. Semicircles at the left show a charge structure that would lead to the observed fields. The numerals are the charges in coulombs on the spherical regions of radii 0.5, 1.0, and 2.0 km. After Simpson and Robinson (1941).

of most of the results that have been obtained from a network of field strength meters through (10.22). It should be remembered that the magnitudes and positions of the charge centers refer to the charges neutralized by the lightning and that the total charges are probably larger.

## 10.18 Direct Observations of the Charge Structure

The first successful probing of a thundercloud was done by Simpson and Scrase (1937), who attached an alti-electrograph to a free balloon. This instrument consisted of two vertical wires, one above and one below the recorder, which was pole-finding paper driven by clockwork. Originally only the sign of the potential gradient was anticipated, but it was found that the width of the trace on the paper afforded some measure of the magnitude of the field. Some of the results obtained are shown in figure 10.14. The shaded portions of this figure represent positive fields, the unshaded portions are negative fields, and the widths represent field magnitudes.

Simpson and Scrase (1937) found that a thunderstorm has a positive charge on the upper half of the associated cloud, a negative charge on the lower half of the cloud, and usually, if not always, a concentrated positive charge below the negative charge. The temperature of the upper positive charge center was always below $-10°C$ and averaged about $-25°C$. The average temperature of the center of the lower negative charge was $-7°C$ and was above 0°C in 2 of 15 cases. The low positive charge was usually at a temperature above melting. The hemispheres at the left in figure 10.14 are a simplified picture of a charge distribution that agrees qualitatively with the observations.

Chapman (1950) obtained similar results from a balloon-borne probe that telemetered the field strength to the ground. He found that the temperature of the lower negative charge center was near 0°C. The maximum field observed was about $2 \times 10^5$ V m$^{-1}$. Note that both Simpson and Chapman found the negative charge to be somewhat warmer than those in table 10.4.

Winn et al. (1974) made 61 soundings in New Mexico with small rockets instrumented to measure the field strength. The most frequent maximum field found was about $3 \times 10^4$ V m$^{-1}$, but two soundings gave maxima of $4 \times 10^5$ and $1 \times 10^6$ V m$^{-1}$. They concluded that intense electric fields of the order of $4 \times 10^5$ V m$^{-1}$ do exist in thunderstorms but that they are evidently concentrated in relatively small volumes of the cloud. Although the rockets can be better directed into the active regions of thunderstorms than free balloons, these probes have a low probability of entering the regions of maximum field strength.

From observations made by the above methods in England, South Africa, and New Mexico, it may be tentatively concluded that the great majority of thunderstorms have a quasi-vertical dipole charge distribution with the positive charge above the negative. There often is a smaller, concentrated positive charge beneath the main negative charge within the rain shaft.

## 10.19 Point Discharges

On the approach of a thunderstorm the vertical potential gradient at the surface may increase sufficiently to produce a corona discharge from natural and artificial points. This is the St. Elmo's fire of sailing ships. It is now often observed on airplanes and along mountain ridges. It is visible at night and may make a "buzzing" noise when it cannot be seen. A typical point at treetop height goes into corona at a surface potential gradient of 600–1,000 V m$^{-1}$; at 30 m height the necessary surface gradient is about 300 V m$^{-1}$. The point strongly distorts the field and results in breakdown in the immediate vicinity of the point. The electrons ionize the air in the vicinity, and this tends to reduce the potential gradient until the ions migrate or are blown away by the wind, whereupon the corona reappears; it is this intermittency that leads to the buzzing sound. Natural points are tree leaves and needles, grass blades, and the like. In a famous experiment Schonland (1928) cut and insulated a small tree so that he could measure the corona current near thunderstorms. Assuming an average tree spacing of 5 m for his South African location, he estimated that the point discharges brought down 20 times as much charge as lightning. Wormell (1927) measured the corona from an artificial point for 3 years and found that twice as much negative as positive charge reached the earth. His average point discharge current was $4 \times 10^{-9}$ A. He assumed that there were 800 similar points in a square kilometer in his experimental area. This leads to a flow of about 100 C of charge per square kilometer per year, or a current of $3.2 \times 10^{-12}$ A. This is to be compared with the estimate of the conduction current of $3.0 \times 10^{-12}$ made in section 10.13. It is clear that the point discharge current, at least over land, is an important part of the supply current. It is not known what the magnitude of the point discharge current is over the oceans. It may be that discharges take place from filaments in breaking waves at sufficiently high fields, but thunderstorms are much less common over the oceans than over land. Blanchard (1963) has estimated that the positive charge transferred to the atmosphere by the bubbling process yields a current of 160 A over the world ocean, or about $0.4 \times 10^{-12}$ A m$^{-2}$. This is about one order of magnitude smaller than the estimated point discharge current over land.

The point discharge under or near a thunderstorm leads to a substantial space charge, and this is responsible for the relatively small surface potential gradients observed in these circumstances.

## 10.20 Precipitation Currents

A majority of raindrops and snow crystals carry electric charges. Many measurements of the charge on precipitation have been made with quite diverse results. These differences, often of orders of magnitude, may be due in part to measurement difficulties, but are probably real. In view of this, the number of observations is too small to provide a climatology or a global mean.

In the case of both continuous and showery rain the ratio of the number of positive to negative drops is about 1.5. There is some indication that the charge on negative drops is somewhat larger than that on positive drops, but, in the mean, positive charge is brought to the ground. The precipitation current has been estimated to be $10^{-12}$–$10^{-10}$ A m$^{-2}$ and hence is typically larger than the fair weather conduction current and in the same direction. Note that precipitation falls on only a fraction of the earth at any one time. The relatively small number of observations of the charge on thunderstorm rain suggests that it brings negative charge to the surface.

The charges on individual drops near the surface have been reported to range from $3 \times 10^{-15}$ to $10^{-10}$ C, with an average of perhaps $10^{-12}$ C. A few measurements within clouds by Gunn (1950) gave drop charges averaging $1.2 \times 10^{-11}$ C in a convective cloud without lightning and up to $9 \times 10^{-11}$ C in a thundercloud. Limited measurements suggest that snow particles, on the average, bring down negative charge, but there are also contradictory observations.

It was hoped that the measurement of charge on precipitation particles would shed some light on the charge separation process in thunderstorms, but it is now believed that most of the precipitation charge observed at the surface is acquired from the space charges below cloud base.

## 10.21 Lightning

Lightning and the resultant thunder are the most characteristic features of the thunderstorm. Lightning is a spark between cloud charge centers or between a cloud charge and ground that is typically a few kilometers long, but may be, exceptionally, some tens of kilometers long. It has already been noted that changes in the surface potential gradient during a lightning flash destroy an electric dipole of several hundred coulomb-kilometers and that the typical charge is 20–40 C. In most regions there are several times as many intracloud flashes as cloud-to-ground flashes. In general, lower cloud bases lead to more frequent cloud-to-ground discharges.

Lightning and its component strokes occur in times ranging from microseconds to milliseconds, although the entire process may occupy a few tenths of a second or more. The complete details can be revealed only by instruments capable of a time resolution of the order of a microsecond. Measurements are based on high-speed photography or rapid recording of the electric or magnetic fields. The most extensive observations of lightning have been made by high-speed photography. Early studies were able only to reveal that a typical lightning flash is composed of several separate strokes. Sir Charles Boys designed a special camera to photograph lightning in 1902, but he was unable to obtain good photographs of lightning in England, although he tried for 30 years. However, Schonland (1964) was able to use the Boys camera to good effect in South Africa. In 1933 he obtained there a detailed depiction of the cloud-to-ground flash. The original Boys camera consisted of two diametrically opposed lenses that were rotated around a common center and recorded on a fixed film. This arrangement made interpreting the photographs somewhat difficult, however, and so Boys devised another version using a single lens recording on a rapidly rotating cylindrical film; it was a version of this camera that was used by Schonland and his collaborates.

The general results of the studies of Schonland and collaborators, since confirmed by others, are shown schematically in figure 10.15. Note that the time scale shown at the top of the figure is variable to permit depiction of the several features. The first visible phase of a discharge is the stepped leader. This proceeds in short steps of about 50 m length, as suggested by the dashes. There is a brief pause at the end of each step of about 50 $\mu$sec, and the average velocity of the stepped leader is about $1.5 \times 10^5$ m sec$^{-1}$.

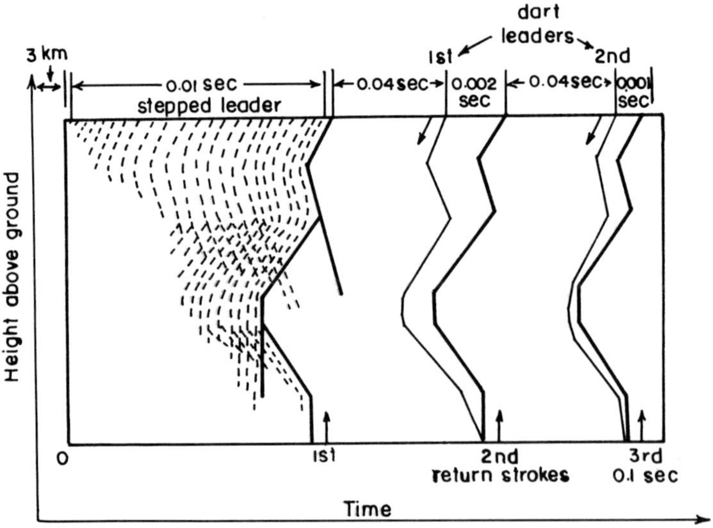


Figure 10.15 Diagram of the time sequences of the components of a lightning flash. Note that the time scale is not the same for all sequences. After Schonland (1956).

Branching of the stepped leader is a common feature, as is the meandering path. When the leader is within 50 m or so of the surface, a streamer arises from a prominent object at the surface to meet the downcoming stepped leader. Immediately thereafter the main or return stroke occurs and proceeds upward at a speed of about $5 \times 10^7$ m sec$^{-1}$, or about one-sixth the velocity of light. The return stroke illuminates and drains the branches formed by the step leader. Typically, after an interval of 3–100 msec, a dart leader traverses the path of the first return stroke, without branching, at a speed of about $2 \times 10^6$ m sec$^{-1}$; this is followed by a second return stroke. The dart leader-return stroke sequence is repeated 3 to 4 times in the mean and, on occasion, 20 to 30 times. If the interval between the first return stroke and the next is more than a few hundred milliseconds, the conductivity of the channel becomes too small to support a dart leader; the discharge process then restarts, with a stepped leader usually along a new path. Successive return strokes appear to tap charge located progressively higher or more distant horizontally. It is the succession of return strokes that gives the visual impression of flickering.

Table 10.5, abstracted from a similar table by Uman (1969), gives representative values and ranges of the velocities and time intervals involved in cloud-to-ground lightning flashes.

The data in table 10.5 and the preceding discussion are based predominantly on observations of cloud-to-ground strokes over land. In mid-

Table 10.5
Representative values and ranges of lightning quantities[a]

| Quantity | Representative value | Range |
|---|---|---|
| Length of leader step | 50 m | 3–200 m |
| Time between steps | 50 $\mu$sec | 30–125 $\mu$sec |
| Leader propagation velocity | $1.5 \times 10^5$ m $sec^{-1}$ | $10^5$–$2.6 \times 10^6$ m $sec^{-1}$ |
| Velocity of dart leader | $2 \times 10^6$ m $sec^{-1}$ | $10^6$–$2 \times 10^7$ m $sec^{-1}$ |
| Velocity of return stroke | $5 \times 10^7$ m $sec^{-1}$ | $2 \times 10^7$–$1.4 \times 10^8$ m $sec^{-1}$ |
| Channel length | 5 km | 2–14 km |
| Return strokes per flash | 3–4 | 1–26 |
| Time between return strokes | 40 m sec | 3–100 m sec |
| Time duration of entire flash | 0.2 sec | 0.01– >2 sec |
| Charge transfer by flash | 25 C | 1–200 C |

a. After Uman (1969).

latitude regions where the cloud base is low, 2 out of 5 strokes may reach the ground. In low latitudes only about 1 in 10 strokes is from cloud to ground. A few, perhaps 1 in 20, cloud-to-ground strokes bring down positive charge.

Thunderstorms are much less frequent over the oceans than over land. On the basis of satellite observations of lightning, Orville and Spencer (1979) estimated that the annual land/ocean ratio of global lightning at dusk is 8/20 and about 4/8 near midnight. This study found that the global rate of flashing is about 100 per sec, confirming earlier estimates based on very sparse data.

Flashes to tall buildings or to mountaintop towers have stepped leaders starting at the tower and branching upward and hence look like upside-down flashes.

## 10.22 Lightning Currents

Currents in a return stroke rise to a maximum of typically 20,000 A in a few microseconds and decline more slowly, reaching half-peak current in around 50 $\mu$sec. Peak currents range from 3,000 to 260,000 A. In about one-fifth of all strokes there is a continuing current that flows for up to 200 msec. These are thought to be responsible for most of the fires set by lightning and are often called "hot" lightning.

## 10.23 Channel Temperatures

Orville (1968) has determined channel temperatures from slitless spectra. They range from about 10,000 to 40,000°K and decrease rapidly with time.

Most of the intense emissions of lightning are attributed to singly ionized nitrogen, but also come from the hydrogen spectral line H$\alpha$ and a continuum. The diameter of the intense current-carrying core of the channel is thought to range from a few millimeters to a few centimeters and is surrounded by a corona sheath of some meters in diameter. The core conductivity is estimated to be $2 \times 10^4\ \Omega^{-1}\,\mathrm{m}^{-1}$ or a resistance over a 5-km path of only 0.25 $\Omega$!

## 10.24 Physics of Discharges

The classical value of the breakdown potential gradient between parallel plates, at standard temperature and pressure, is $3 \times 10^6\ \mathrm{V\,m^{-1}}$. This decreases linearly with air density, so that the dry air breakdown gradient at 3 km height is about $2 \times 10^6\ \mathrm{V\,m^{-1}}$. Within the cloud it is thought that the breakdown potential is less than in clear air because of the presence of ice particles and water drops. It has been shown in the laboratory that water drops become elongated in a strong field and go into corona. Although the breakdown potential in clouds is not well-known, it is believed to be around $1 \times 10^6\ \mathrm{V\,m^{-1}}$. Potential gradients of $10^5$–$10^6\ \mathrm{V\,m^{-1}}$ were observed by Winn et al. (1974) in 2 of 61 rocket soundings. This implies that breakdown gradients do occur in thunderclouds, but only in rather small regions.

It is important to realize that, once initiated, a discharge may propagate into regions where the potential gradient is well below the breakdown value. Breakdown is initiated when the velocity of free electrons becomes sufficient to ionize by collision, thus forming an electron avalanche. It is suggested that in the case of a cloud-to-ground stroke, this process is initiated in the region between the main negative charge and the small lower positive charge. As the avalanche process grows, thermal ionization and photoionization occur, and the channel becomes a highly conducting arc discharge. The tip of the channel is thus nearly at the potential of the negative cloud charge, and the potential gradient between the tip and the atmosphere immediately ahead may be very large. In this way the channel becomes self-propagating. Studies of laboratory spark discharges have shown the existence of the electron avalanche and also of the so-called pilot streamer, which is very weakly luminous. It has been suggested on the basis of these laboratory results and the form of the electric field changes during a lightning flash that a pilot streamer or pilot leader immediately precedes the stepped leader, but is too weak to show on the high-speed photographs. There is no unambiguous evidence that such a pilot leader exists in natural lightning. Further, there is no generally accepted theory for the pauses in the stepped leader; it is not observed in laboratory sparks. The branching and the characteristic zigzag path is attributed to the nonuniform positive space charge density.

## 10.25 Other Types of Lightning

Sheet lightning is the illumination of a large cloud region without the appearance of the characteristic zigzag ground stroke and results from an intracloud flash. Ribbon lightning results from a strong crosswind that spreads the strokes of a multiple stroke flash. Heat lightning is lightning beyond the audibility of thunder (to be discussed in section 10.27). Bead lightning is a cloud-to-ground flash that appears to break up into segments before disappearing. It may result from a magnetic pinch effect, or it may be that the beads are bright because segments of the channel are seen end-on.

The most intriguing form that lightning takes is ball lightning. This is rather rare, but there are over a thousand reports in the literature and even a few photographs. Ball lightning almost always occurs in or near thunderstorms. It is a bright, more or less spherical object of average diameter 20–30 cm, usually red, but sometimes yellow, blue, or white. The ball may descend from a cloud, often moves horizontally near the ground at a speed of a few meters per second, may enter houses through chimneys or windows, has a typical duration of 1–5 sec, and disappears silently or with a bang. Oddly, most of the balls observed at close range did not give off heat; on the other hand, in a well-documented case, a ball fell into a tub of water that boiled for some minutes. These and other diverse and often conflicting observations have led many to doubt the reality of ball lightning, but a look at Singer (1971) will convince most skeptics. In spite of the efforts of many competent scientists over the years, there is no satisfactory theory of ball lightning. At present it is felt that a ball is composed of a plasma, but there are difficulties in explaining containment and a sufficient source of energy.

## 10.26 Sferics

This curious word is a modified contraction of "atmospherics," which, in turn, is the radio static produced by thunderstorms. Lightning is a very broad-band radiator of electromagnetic power, the radio spectrum extending from 1–10 kHz to more than 10 MHz.[5] Radio direction-finding techniques have been used with considerable success to locate thunderstorm sources, particularly over ocean areas; but the propagation characteristics of the atmosphere and the ionosphere tend to limit the precision of the location of the storm, particularly at the longer ranges.

A somewhat related phenomenon is the "whistler," which is heard in the kilohertz spectral region as a whistling tone of steadily declining pitch. This

---

5. kHz and MHz stand for kilo- and megahertz, respectively; a hertz is one cycle per second.

is thought to be due to low-frequency sferics that penetrate the ionosphere and follow a geomagnetic line of force back to earth at the same geomagnetic latitude in the opposite hemisphere. The waves are dispersive in the magnetosphere, thus leading to the relatively pure tone of decreasing pitch.

## 10.27 Thunder

The almost instantaneous heating of the lightning channel to a temperature of around 30,000°K results in an overpressure of some 10–100 atm (atmospheres). This causes a cylindrical shock wave to form, which decays to an ordinary sound wave within 10 m or so, and this is the thunder. According to Few (1974), more than 80% of the acoustic energy is radiated within ±30° from the normal to a linear segment of the channel. The tortuosity of the channel leads to multiple sources, each at a different distance and inclination from an observer. Few has estimated that a typical three-stroke flash 5 km long generates 3,000 acoustic pulses. In propagating from the channel to an observer, the acoustic wave is subject to attenuation, dominantly at frequencies greater than about 100 Hz, refraction in the atmosphere, and reflection from the surface. The "crack" heard from a close flash owes its characteristic sound to high frequencies that are lost at greater distances. The "clap" is heard when major segments of the channel are more or less normal to the observer's line of sound. The rumbles come from the more distant segments of the channel. The duration of the rumble is a measure of the minimum length of the channel.

The velocity of sound, nominally about 330 m $sec^{-1}$, varies with the square root of the absolute temperature. As is well-known, the distance to lightning in kilometers is approximately one-third of the time in seconds between the flash and the thunder. With the normal decrease of temperature with height a sound wave is refracted upward. This limits the distance at which thunder can be heard from the visible flash below the cloud to 10–15 km. Thunder from intracloud flashes at higher elevations may be heard up to about 30 km, but is usually much weaker. Vertical wind shear also refracts sound waves. In the usual situation of an increase of wind speed with height, sound is refracted downward downwind from the source and upward upwind. It is apparent that the thunder heard by an observer is the resultant of multiple sources along the flash and a complex of effects in the intervening atmosphere. Thunder of a fabric-ripping sound may be heard from nearby leader strokes.

## 10.28 Thunderstorm Charging Processes

Over the years a number of charging processes have been proposed. With one exception, to be discussed in section 10.33, all of these involve microphysical charge separation in which charges of opposite sign appear on precipitation particles and cloud particles or ions. This is followed by gravitational separation of the precipitation particles from the smaller particles and thus an enhanced vertical electric field. Reflecting observations of the charge structure, the viable microphysical charge-separating mechanisms are those that lead to negative charge on the precipitation. The converse may be of importance in the formation of the lower subsidiary positive charge.

The principal microphysical charge-separating processes are (a) the influence or induction process due to the interactions of colliding particles in an electric field; (b) the rupture, fracture, or splintering of liquid or solid particles; (c) the freezing of water in an ice-water mix; and (d) thermoelectric effects in ice. It is not sufficient that the charging process leads to the observed charge structure; it must also be quantitatively adequate to supply the lightning and leakage currents. A typical thunderstorm might produce three flashes a minute, each of 20 C, or an average of 1 A. The leakage current is probably about the same, so the charging mechanism must supply at least 2 A, and often several times as much.

## 10.29 The Wilson Theory

Broadly speaking, the Wilson theory may be placed in the first of the categories given in the previous section. A raindrop in the fair weather field is polarized in such a way that its lower hemisphere is positively charged. As the raindrop falls, it preferentially captures negative ions on its lower surface. If the fall speed of the raindrop is greater than the ion drift velocity, it cannot capture positive ions on its upper surface, and hence the drop acquires a net negative charge. The precipitation carries negative charge down and away from the positive space charge, enhancing the field and regenerating the process.

By taking the upper limit of the fall speed of raindrops aloft as 10 m $sec^{-1}$ and the small ion mobility as $2 \times 10^{-4}$ $m^2$ $V^{-1}$ $sec^{-1}$, the ion drift velocity will become 10 m $sec^{-1}$ in a field of $5 \times 10^4$ V $m^{-1}$. If one prefers to consider large ions, the limiting field exceeds the breakdown gradient. A more serious limitation is the rate of production of ions. The production rate due to cosmic radiation is about $10^7$ $m^{-3}$ $sec^{-1}$, and this would limit the charging rate in a spherical volume of 2 km radius to about $5 \times 10^{-2}$ A. Thus the Wilson theory could be important only if there were a strong secondary source of ions, perhaps by point discharges from drops in a strong field.

## 10.30 The Influence Theory

This theory was proposed by Elster and Geitel (1913), revived by Sartor (1967), and further developed by others. As in the Wilson theory this

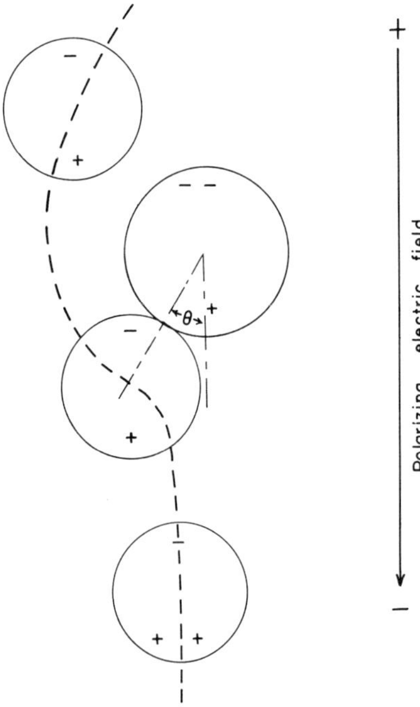


Figure 10.16 An elastic collision between two drops in a gravitational and a vertical electric field.

process assumes an initial field, such as the fair weather field, which polarizes the drops. Charging results when a larger drop collides with and then separates from a smaller drop, as illustrated in figure 10.16. As before, gravitational separation enhances the field and the process regenerates. Sartor (1967) and Mason (1971), among others, have shown that the charge $\Delta Q$ transferred by an event as depicted in figure 10.16 is approximately

$$\Delta Q = 2\pi^2 \varepsilon_0 E r^2 \cos\theta + 4\pi\varepsilon_0 Q/(1 + \pi^2 r^2/6R^2), \qquad (10.25)$$

where $r$ and $R$ are the radii of the small and large drops, $Q$ is the preexisting charge on the two drops, $\theta$ is the angle between the line of centers of the drops and the electric gradient at separation, and it is assumed that $r/R$ is small. The influence effect is represented by the first term on the right of this equation, the second term being concerned with the redistribution of preexisting charge. One uncertainty in (10.25) is the appropriate average value of $\cos\theta$ for the collision-separation encounter. Another is the time of the contact compared with the conduction time constant $\varepsilon_0\kappa/\lambda$, where $\lambda$ is the dielectric constant and $\lambda$ the electrical conductivity of the spheres. Sartor observed charge transfer by a spark without actual contact, but only between drops of several hundred micrometers at large field strengths. When liquid drops separate after contact a portion of the drop remains behind to coalesce with the larger particle. Not much more than a guess has been made of the duration of contact, and the conduction time constant itself depends on the actual conductivity of the always polluted water substance. Obvious difficulties arise in the quantitative use of (10.25) when one or both particles are nonspherical ice forms. Particularly when one or both colliding particles are water drops, only a small fraction of the collisions will be followed by separation. This fraction, which has been variously estimated as $10^{-2}$ to $10^{-4}$, largely determines the rate of charge separation by the influence mechanism in an all-water or a water-ice cloud.

In view of these many unknowns and uncertainties, principal reliance must be placed on experiments. Scott and Levin (1970) measured the charge transfer when natural snow crystals collided with an ice sphere in the presence of a vertical field. Aufdermaur and Johnson (1972) found charge transfer when supercooled water drops interacted with rime in an electric field. Jennings (1975) examined charge transfer due to raindrops interacting with cloud drops in a vertical field. In all three sets of experiments it was established that the charge transfer was due to the influence mechanism and that, in general, it followed (10.25) with some quantitative differences. Aufdermaur and Johnson found no evidence of splinter formation during riming at $-13$°C, but it should be noted that Hallett and Mossop (1974) found splintering only from $-3$° to $-8$°C. Jennings found that charge transfer increased with $E$ as in (10.25) only up to about 15 kV $m^{-1}$ and thereafter decreased toward zero at about 30 kV $m^{-1}$. Jennings suggested that this was due to complete coalescence of all colliding drops as a result of the applied field. It is noteworthy that no such effect was found by Aufdermaur and Johnson at a field of 150 kV $m^{-1}$, suggesting that the droplet separation process from rime is different from that from a water drop. Jennings also deduced values of $f \cos \theta$, where $f$ is the fraction of collisions followed by separation. For $E < 20$ kV $m^{-1}$, he found that $f \cos \theta$ was about $10^{-3}$. This suggests that $f$ was $10^{-2}$–$10^{-3}$, although no determination of $\cos \theta$ was possible.

More experiments are needed to quantify fully the influence mechanism, but it seems clear that it is a viable thunderstorm-charging process. Both in the ice-ice and the liquid-ice cases it appears that the charge separation rate is barely adequate to supply a modest thunderstorm. In the all-liquid case the charging is probably adequate to explain the level of electrification found in subtropical shower clouds by Takahashi (1975), but not in a warm thunderstorm.

## 10.31 Charging by Fracture

Collisions between raindrops usually lead to rupture (see section 8.27). The larger fragments are positively charged, and the negative charge resides on small particles or on ions. Similar effects result when drops splash. This is generally referred to as the Lenard (1921) effect. It is evident that this process would yield a principal dipole with a negative upper charge, but it has been proposed that it is responsible for the low subsidiary positive charge that is often in the rainshaft.

Mason (1971) showed that an average separation of $10^{-5}$ C per cubic meter of water by the Lenard effect was two orders of magnitude smaller than the required thunderstorm charging rate even if all of the drops broke up several times. Matthews and Mason (1964) found that drop breakup in a field of $1.5 \times 10^5$ V $m^{-1}$ led to the separation of a hundred times more charge than in the fair weather field. This suggests that the Lenard effect may contribute significantly to the low positive charge after another process has produced a strong field.

The freezing of raindrops occasionally results in drop fracture; more often the internal pressure is relieved by a jet that may emit small droplets or ice particles, leaving a frozen proturberance or spicule as evidence. Snow crystals collected at the surface often include many fragments of crystals that presumably resulted from collisions with more solid ice forms. All of these fracture processes probably separate charge, but how much is not known.

## 10.32 Charging of Graupel

A number of laboratory experiments have shown that electrification often accompanies the riming of graupel particles. However, the results have been quite diverse, due in part to differences in the experimental conditions. The magnitude and the sign of the electrification seemed to depend on the temperature, the supercooled cloud water content, the presence of ice crystals, and the existence of an electric field.

The laboratory experiments of Takahashi (1978b) have seemingly resolved many of the earlier uncertainties. He found that electrification occurred only when there was active riming and both ice crystals and supercooled cloud water were present. The sign and magnitude of the charging was found to depend on the cloud water content and the temperature, as shown in figure 10.17. When the cloud water content (CWC) is small, the ice crystal concentration is also small and the positive charge weak. When the CWC is very large, the positive charge is again small. Strong charging of the graupel results when the CWC is about 1–2 g $m^{-3}$, a reasonable range for thunderstorms. The charge is positive for temperatures warmer than $-10°C$ and negative at temperatures colder than

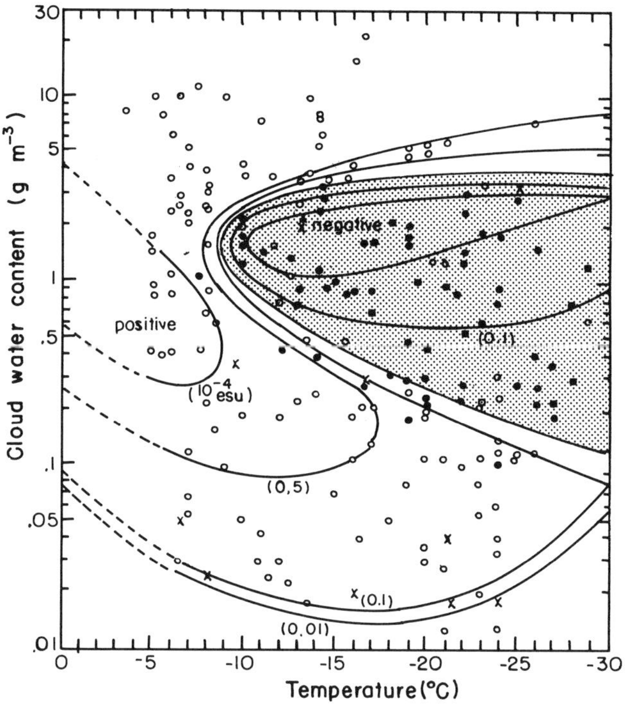


Figure 10.17 Electrification of rime as a function of temperature and the cloud water content as determined experimentally by Takahashi (1978b). Open circles represent positive charge, solid circles negative charge, and crosses no charge. The isopleths show the electric charge transferred by one ice crystal collision in units of $10^{-4}$ esu.

$-10$°C. Takahashi states that high negative electrification of graupel is to be expected in the low-temperature region of the updraft where the ice crystal concentration is large. After collisions with graupel, the crystals will be positively charged and will be carried up into the anvil region. The negatively charged graupel can fall to lower levels in the periphery of the updraft. Takahashi considers that the principal negatively charged region and the low positive charge are due to graupel charging, the former at temperatures below $-10$°C and the latter at temperatures warmer than $-10$°C. As noted, the positive charge in the anvil is attributed to ice crystals that have collided with graupel before being carried into the anvil by the updraft.

Takahashi measured the charge on graupel and found it to range from about $10^{-5}$ to $10^{-4}$ esu in the regions of maximum charging. Simple numerical estimates of the rate of charging in a thunderstorm by the graupel mechanism yielded realistic results. There is some question whether the necessary ice crystal concentrations are found in the low positive charge region where the temperatures must be warmer than −10°C. Takahashi attributes his experimental results on the charging of graupel to thermoelectric effects.

## 10.33 Convective Charging

One proposed thunderstorm-charging mechanism does not fit any of the four categories given in section 10.28. This is the convective process of Grenet (1947) and Vonnegut (1955). It is visualized that the convection responsible for the cloud also carries up the preexisting positive space charge from near the surface to the cloud top. The resulting gradient above the cloud causes negative ions to flow toward the cloud. Instead of neutralizing the convected positive charge, the negative ions attach themselves to cloud drops and are carried down toward the cloud base by downdrafts on the flanks of the cloud. This enhances the field at the ground, finally causing point discharges that supply copious positive ions and regenerate the process. Note that neither precipitation nor ice is required. The principal criticism of this proposal is that the main downdrafts in a thunderstorm do not develop until the mature stage and occur within the rainshaft and not on the flanks. It is also pointed out that the roots of convective clouds are often not close enough to the surface to tap the positive space charge.

Vonnegut et al. (1962) performed a series of experiments in which artificial space charge was generated from an elevated wire about 14 km long that was maintained in corona by a high-voltage source. During periods with small fair weather cumulus, it was found that clouds downwind of the line source became charged and that the sign of the charge could be changed by reversing the polarity of the wire. Flights with an instrumented airplane seemed to show that charge was convected upward into the cloud and that there were regions on the flanks at cloud base of charge of opposite sign, as predicted by the Grenet-Vonnegut process. Suggestive as they are, these experiments were not designed to show that the regeneration due to point discharges and the downdrafts in the small cumuli are not the same as those in thunderstorm cells. Nevertheless, it appears that convection of space charge may well play a part in a future comprehensive model of thunderstorm electrification.

## 10.34 Current View of Charge Generation

The only firm conclusion that can be drawn at the time of this writing is that there is no completely satisfactory theory of thunderstorm charge generation. Mason (1971) has shown that most of the microphysical

Table 10.6
Estimates of the components of the global circuit

| Authority | Conduction current (A) | Point discharge (A) | Precipitation current (A) | Lightning to ground (A) | Sum (A) |
|---|---|---|---|---|---|
| Wait (1950) | +1,620 | −480 | +320 | −320 | +1,140 |
| Israel (1953) | +1,450 | −1,620 | +490 | −320 | 0 |
| Houghton (unpublished) | +1,400 | −1,600 | +400 | −200 | 0 |

charge-separating processes are quantitatively inadequate to separate charge at the rate required in a typical thunderstorm. The only processes that seem able to meet this criterion are the influence or induction mechanism involving the interaction of ice crystals and hail or graupel and the interaction of supercooled cloud drops, ice crystals, and graupel. Even here, it seems that rather favorable assumptions must be made to meet the requirements of a modest thunderstorm. Except for the convective-charging process, which is suspect on other grounds, the proposed processes call for the precipitation to carry the negative charge. Experimental confirmation of this hypothesis would do much to clarify thunderstorm charging.

Further work, both theoretical and experimental, is clearly required. Perhaps two or more processes acting synergistically will be found necessary to explain the charging.

## 10.35 The Global Circuit

This chapter will conclude by reconsidering the global circuit. There are four principal processes that transfer charge between the atmosphere and the earth, namely, the conduction current, point discharges, precipitation current, and lightning flashes to ground. All of these have been discussed, and it is clear that their average values over the globe are quite uncertain. Nevertheless, it is instructive to examine the results of several estimates of these four quantities that comprise the global circuit.

The units in table 10.6 are amperes for the entire earth, and a positive sign indicates positive charge carried to the ground surface. The inclusion of three digits in the first two lines of the table results from a change of units from the original and is not indicative of the precision of the estimates. No special significance should be attached to the closure to zero of the last two lines; this was achieved by selecting appropriate values within the wide range of uncertainty permitted by the data. About the only conclusions

Table 10.7
Conduction currents above thunderstorms

| Authors | Region | Average current (A) | Maximum current (A) | Number of cases |
|---|---|---|---|---|
| Gish and Wait (1950) | Central U.S. | 0.5 | 6.5 | 21 |
| Stergis et al. (1957) | Florida | 1.0 | 4.3 | 25 |
| Imyanitov et al. (1969) | European–U.S.S.R. | 0.1 | 1.4 | — |

that can be drawn are that the conduction current and the point discharge current are of opposite sign and of similar magnitude and that the precipitation current and the charge transfer by lightning are about half an order of magnitude smaller.

Another approach to the global circuit is through the measurement of the current flow above thunderstorms. In this relatively simpler region, if one rules out the possibility of lightning strokes from the cloud to the upper conducting region, only a conduction current need be measured and integrated over the affected horizontal cross section. Such measurements must be made from high-flying aircraft or balloons, and only a few results have been reported; these are summarized in table 10.7.

These data seem to show an increase in average current with decreasing latitude. Since thunderstorms are more frequent in low latitudes than high, it is not unreasonable to estimate that the average thunderstorm current is from 0.5 to 1 A. If the estimate of 2,000 thunderstorms in progress over the globe is accepted, the total current is 1,000–2,000 A, or roughly the same as the conduction current.

In view of the scanty and uncertain data on which tables 10.6 and 10.7 are based, they should not be used to demonstrate that thunderstorms are the generators in the global circuit. Perhaps the strongest argument in favor of the thunderstorm generator is the lack of a plausible and quantitatively adequate competitor.

## Problems

1.
By what processes are small ions and intermediate ions destroyed?

2.
Where does the lower negative charge of fair weather reside?

3.
The near-surface potential gradient is several hundred volts per meter in polluted air, whereas it is only about a hundred volts per meter in clean air. Explain the difference.

4.
Estimate to an order of magnitude the electric power and the thermodynamic power of a typical thundercloud. Show how your estimates were obtained.

5.
Estimate, by adding the times of the separate steps, the total time taken by a typical lightning flash, starting from the leader.

# Appendix: Some Useful Constants and Numerical Parameters

Dimensions of the earth

| | |
|---|---|
| Area of the earth | $5.10101 \times 10^8$ km$^2$ |
| Area of the land masses | $1.48847 \times 10^8$ km$^2$ |
| Area of the oceans | $3.61254 \times 10^8$ km$^2$ |
| Fraction of area of earth covered by oceans | 0.70820 |
| Equatorial radius | 6378.388 km |
| Polar radius | 6356.912 km |
| Mean radius | 6371.229 km |
| Mean distance to sun (=1 astronomical unit) | $1.4968 \times 10^8$ km |

Acceleration of gravity at sea level[a]

| | | | | | |
|---|---|---|---|---|---|
| Latitude (degrees) | 0 | 10 | 20 | 30 | 40 |
| $g$ (m sec$^{-2}$) | 9.7804 | 9.7819 | 9.7864 | 9.7932 | 9.8017 |
| Latitude (degrees) | 50 | 60 | 70 | 80 | 90 |
| $g$ (m sec$^{-2}$) | 9.8107 | 9.8191 | 9.8260 | 9.8305 | 9.8321 |

a. The free air acceleration of gravity decreases with height above sea level by 0.0031 m sec$^{-2}$ per kilometer of height at all latitudes in the troposphere and lower stratosphere. Topography and subsurface inhomogeneities cause local anomalies in gravity that amount to a few parts in 10,000. This is negligible for meteorological purposes, but such anomalies provide one means of deducing subsurface rock structures.

Density of liquid water[a]

| $T$ (°C) | Density ($g\ m^{-3}$) | $T$ (°C) | Density ($g\ m^{-3}$) |
|---|---|---|---|
| −40 | 0.9798* | −10 | 0.9979 |
| −35 | 0.9833* | −5 | 0.9992 |
| −30 | 0.9868* | 0 | 0.9998 |
| −25 | 0.9902* | 10 | 0.9997 |
| −20 | 0.9935* | 20 | 0.9982 |
| −15 | 0.9960* | 30 | 0.9957 |
| −13 | 0.9969 | | |

a. Asterisks indicate extrapolated values.

Density of air[a]

$\rho = 0.348371 p / T_v$ kg $m^{-3}$

a. $p$ is the total pressure in millibars and $T_v$ is the adjusted virtual temperature in °K.

## Thermodynamical Properties

Gas constants

Universal gas constant = 8.31432 J $mol^{-1}$ $°K^{-1}$
Molecular weight of air = 28.964
Molecular weight of water vapor = 18.015
Gas constant for air = 287.05 J $kg^{-1}$ $°K^{-1}$
Gas constant for water vapor = 461.51 J $kg^{-1}$ $°K^{-1}$

Specific heats

Specific heat of dry air at constant pressure = 1,005 J $kg^{-1}$ $°K^{-1}$
= 0.2400 cal $g^{-1}$ $°K^{-1}$
Specific heat of dry air at constant volume = 718 J $kg^{-1}$ $°K^{-1}$
= 0.1715 cal $g^{-1}$ $°K^{-1}$
Specific heat of water vapor at constant pressure = 1,850 J $kg^{-1}$ $°K^{-1}$
= 0.442 cal $g^{-1}$ $°K^{-1}$
Specific heat of water vapor at constant volume = 1,390 J $kg^{-1}$ $°K^{-1}$
= 0.332 cal $g^{-1}$ $°K^{-1}$

Specific heats

| | Specific heat of ice | | | Specific heat of water | |
|---|---|---|---|---|---|
| $T$ (°C) | J kg$^{-1}$ °K$^{-1}$ | cal g$^{-1}$ °K$^{-1}$ | $T$ (°C) | J kg$^{-1}$ °K$^{-1}$ | cal g$^{-1}$ °K$^{-1}$ |
| −50 | 1,738 | 0.415 | −50 | 5,400 | 1.290 |
| −40 | 1,813 | 0.433 | −40 | 4,770 | 1.139 |
| −30 | 1,884 | 0.450 | −30 | 4,520 | 1.080 |
| −20 | 1,959 | 0.468 | −20 | 4,350 | 1.039 |
| −10 | 2,031 | 0.485 | −10 | 4,270 | 1.020 |
| 0 | 2,106 | 0.503 | 0 | 4,218 | 1.007 |
| | | | 10 | 4,192 | 1.001 |
| | | | 20 | 4,182 | 0.999 |
| | | | 30 | 4,179 | 0.998 |

Latent heats[a]

| | Latent heat of vaporization | | Latent heat of fusion | | Latent heat of sublimation | |
|---|---|---|---|---|---|---|
| $T$ (°C) | J kg$^{-1}$ ($\times 10^6$) | cal g$^{-1}$ | J kg$^{-1}$ ($\times 10^5$) | cal g$^{-1}$ | J kg$^{-1}$ ($\times 10^6$) | cal g$^{-1}$ |
| −40 | 2.603 | 621.7 | 2.357 | 56.3 | 2.839 | 678.0 |
| −30 | 2.575 | 615.0 | 2.638 | 63.0 | 2.839 | 678.0 |
| −20 | 2.549 | 608.9 | 2.889 | 69.0 | 2.838 | 677.9 |
| −10 | 2.525 | 603.0 | 3.119 | 74.5 | 2.837 | 677.5 |
| 0 | 2.501 | 597.3 | 3.337 | 79.7 | 2.835 | 677.0 |
| 10 | 2.477 | 591.7 | | | | |
| 20 | 2.453 | 586.0 | | | | |
| 30 | 2.430 | 580.4 | | | | |

a. Temperature of the normal ice point is 273.15°K.

Diffusion coefficient of water vapor in air, coefficient of heat conductivity of air, and dynamic viscosity of air[a]

| | Diffusion coefficient | | Heat conductivity of air | | Dynamic viscosity of air | |
|---|---|---|---|---|---|---|
| $T$ (°K) | $m^2\ sec^{-1}$ ($\times 10^{-5}$) | $cm^2\ sec^{-1}$ | $J\ m^{-1}\ sec^{-1}\ °K^{-1}$ ($\times 10^{-2}$) | $cal\ cm^{-1}\ sec^{-1}\ °K^{-1}$ ($\times 10^{-5}$) | $N\ sec\ m^{-2}$ ($\times 10^{-5}$) | $g\ cm^{-1}\ sec^{-1}$ ($\times 10^{-4}$) |
| −40 | 1.62 | 0.162 | 2.07 | 4.95 | 1.512 | 1.512 |
| −30 | 1.76 | 0.176 | 2.16 | 5.15 | 1.564 | 1.564 |
| −20 | 1.91 | 0.191 | 2.24 | 5.34 | 1.616 | 1.616 |
| −10 | 2.06 | 0.206 | 2.32 | 5.54 | 1.667 | 1.667 |
| 0 | 2.21 | 0.221 | 2.40 | 5.73 | 1.717 | 1.717 |
| 10 | 2.36 | 0.236 | 2.48 | 5.92 | 1.766 | 1.766 |
| 20 | 2.52 | 0.252 | 2.55 | 6.10 | 1.815 | 1.815 |
| 30 | 2.69 | 0.269 | 2.63 | 6.29 | 1.862 | 1.862 |

a. The diffusion coefficients of water vapor in air are for a pressure of 1,000 mbar. For another pressure $p$, multiply by $1{,}000/p$. The other two quantities are independent of pressure in the troposphere and lower stratosphere.

The U.S. Standard Atmosphere of 1962

| Height above sea level (km) | $T$ (°C) | Pressure (mbar) | Density (kg $m^{-3}$) |
|---|---|---|---|
| 0 | 15.000 | 1,013.25 | 1.2250 |
| 1 | 8.501 | 898.76 | 1.1117 |
| 2 | 2.004 | 795.01 | 1.0066 |
| 3 | −4.491 | 701.21 | 0.90925 |
| 4 | −10.984 | 616.60 | 0.81935 |
| 5 | −17.474 | 540.48 | 0.73643 |
| 6 | −23.963 | 472.18 | 0.66011 |
| 7 | −30.450 | 411.05 | 0.59002 |
| 8 | −36.935 | 356.52 | 0.52579 |
| 9 | −43.417 | 308.01 | 0.46706 |
| 10 | −49.898 | 265.00 | 0.41351 |
| 11 | −56.376 | 227.00 | 0.36480 |
| 12 | −56.500 | 193.99 | 0.31194 |
| 14 | −56.500 | 141.70 | 0.22786 |
| 16 | −56.500 | 103.53 | 0.16647 |
| 18 | −56.500 | 75.65 | 0.12165 |
| 20 | −56.500 | 55.29 | 0.08891 |

Surface tension of pure water in air[a]

| $T$ (°C) | Surface tension (erg $cm^{-2}$)[b,c] | $T$ (°C) | Surface tension (erg $cm^{-2}$)[b,c] |
|---|---|---|---|
| −40 | 83.7* | −10 | 77.29 |
| −35 | 82.5* | −5 | 76.44 |
| −30 | 81.3* | 0 | 75.70 |
| −25 | 80.3* | 10 | 74.22 |
| −20 | 79.14 | 20 | 72.75 |
| −15 | 78.06 | 30 | 71.18 |
| −13 | 77.76 | | |

a. This is also the specific free energy of the water-air interface.
b. To convert to surface tension in J $m^{-2}$ multiply by $10^{-3}$.
c. Values with an asterisk were obtained by extrapolation.

Miscellaneous

Speed of sound in dry air at 273.15°K = 331.36 m $sec^{-1}$
Speed of sound at temperature $T$ (°K) = $20.05T^{1/2}$ m $sec^{-1}$
Stefan-Boltzmann constant = $5.6697 \times 10^{-8}$ W $m^{-2}$ $°K^{-4}$
= $5.6697 \times 10^{-5}$ erg $cm^{-2}$ $sec^{-1}$ $°K^{-4}$
Avogadro constant = $6.02252 \times 10^{23}$ $(g\ mole)^{-1}$
Boltzmann constant = $1.38054 \times 10^{-23}$ J $°K^{-1}$ = $1.38054 \times 10^{-16}$ erg $°K^{-1}$

# References


Abbot, C. G., L. B. Aldrich, and F. E. Fowle, 1932. *Annals of the Astrophysical Observatory of the Smithsonian Institution*, Washington, D.C.: U.S. Government Printing Office, pp. 105–110.

Ackerman, T.P., K.-N. Liou, and C. B. Leovy, 1976. *J. Appl. Meteor.* 15:28–35.

Aitken, J., 1923. *Collected Scientific Papers*, Cambridge: Cambrige University Press.

Allen, J. K., 1971. *J. Appl. Meteor.* 10:260–265.

Almeida, F. C. de, 1977. *J. Atmos. Sci.* 34:1286–1292.

Almeida, F. C. de, 1979. *J. Atmos. Sci.* 36:1564–1576.

Anderson, B. J., and J. Hallett, 1976. *J. Atmos. Sci.* 33:822–832.

Ångstrom, A., 1929a. *Beitr. Geophys.* 21:145–161.

Ångstrom, A., 1929b. *Geograph. Ann.* 11:156–166.

Arbel, N., and Z. Levin, 1977. *Pure Appl. Geophys.* 115:864–893, 895–914.

Atlas, D., 1963. *Severe Local Storms.* Meteorological Monograph 5, No. 27. Boston: American Meteorological Society.

Atwater, M. A., and P. S. Brown, Jr., 1974. *J. Appl. Meteor.* 13:289–297.

Auer, A. H., and D. L. Veal, 1970. *J. Atmos. Sci.* 27:919–926.

Auer, A. H., Jr., D. L. Veal, and J. D. Marwitz, 1969, *J. Atmos. Sci.* 26:1342–1343.

Aufdermaur, A. N., and D. A. Johnson, 1972. *Quart. J. Roy. Meteor. Soc.* 98:369–382.

Baker, M. B., and J. Latham, 1979. *J. Atmos. Sci.* 36:1612–1615.

Barrow, G. M., 1964. *The Structure of Molecules*, New York: Benjamin.

Bashkirova, G. M., and T. A. Pershina, 1964. *Trudy, G.G.O.* 165:83–100.

Battan, L. J., 1973. *Radar Observations of the Atmosphere*, rev. ed., Chicago: University of Chicago Press.

Battan, L. J., and J. B. Theiss, 1966. *J. Atmos. Sci. 23:78–87.*

Baum, W. A., 1951. *J. Meteor.* 8:196–198.

Beard, K. V., 1976. *J. Atmos. Sci.* 33:851–864.

Beard, K. V., and S. N. Grover, 1974. *J. Atmos. Sci.* 31:543–550.

Bell, E. E., 1957. "an atlas of reflectivities of some common types of materials," Interim Engineering Report, Contract AF 33(616)-3312, Ohio State University.

Benedict, W. S., and L. D. Kaplan, 1959. *J. Chem. Phys.* 30:388–399.

Bentley, W. A., and W. J. Humphreys, 1931. *Snow Crystals*, New York: McGraw-Hill.

Bergeron, T., 1935. *Proceedings 5th Assembly U.G.G.I., Lisbon* 2:156–178.

Berry, E. X., 1976. *J. Atmos. Sci.* 24:688–701.

Berry, E. X., 1968. *Proceedings First National Conference on Weather Modification*, Boston: American Meteorological Society, pp. 81–85.

Berry, E. X., and R, L. Reinhardt, 1974. *J. Atmos. Sci.* 31 : 1825–1831.

Best, A. C., 1950. *Quart. J. Roy. Meteor. Soc.* 76:16–36.

Bigg, E. K., 1976. *J. Atmos. Sci.* 33:1080–1086.

Bignell, K. J., 1970. *Quart. J. Roy. Meteor. Soc.* 96: 390–403.

Blackwell, H. R., 1946. *J. Opt. Soc. Amer.* 36:624–643.

Blanchard, D. C., 1950. *Trans. Am. Geophys. Union* 31:836–842.

Blanchard, D. C., 1963. "The electrification of the atmosphere by particles from bubbles in the sea," in *Progress in Oceanography*, vol. 1, New York: Pergammon, chapter 2, pp. 71–202.

Blifford, I. H., Jr., 1970. *J. Geophys. Res.* 76:3099–3103.

Blifford, I. H., Jr., and L. D. Ringer, 1969. *J. Atmos. Sci.* 26:716–726.

Bowen, E. G., 1950. *Australian J. Sci. Res.*, Ser. A. 3:193–213.

Bowen, E. G., 1953. *Australian J. Phys.* 6:490–497.

Bowen, E. G., 1956. *J. Meteor.* 13:142–151.

Braham, R. R., Jr., 1952. *J. Meteor.* 9:227–242.

Braham, R. R., Jr., 1964. *J. Atmos. Sci.* 21:640–645.

Braslau, N., 1972. *IBM J. Res. and Devel.* 16:180–183.

Brewer, A. W., and J. T. Houghton, 1956. *Proc. Roy. Soc.* A 236 : 175–186.

Brooks, C. E. P., 1925. *Geophys. Mem.* (Air Ministry, Meteor. Office) 24:147–164.

Brooks, D. L., 1950. *J. Meteor.* 7:313–321.

Brown, E. N., and R. R. Braham, Jr., 1959. *J. Meteor.* 16:609–616.

Brunt, D., 1932. *Quart. J. Roy. Meteor. Soc.* 58:389–418.

Bryan, K., and E. Schroeder, 1960. *J. Meteor.* 17:670–674.

Bucerius, H., 1946. *Optik* 1:188–212.

Budyko, M. I., 1958. *The Heat Balance of the Earth's Surface*, translated by N. A. Stepanova, Washington, D.C.: U.S. Weather Bureau.

Budyko, M. I., 1963. *Atlas of the Heat Balance*, 2nd ed., Moscow: Akad. Nauk., USSR.

Buettner, K. J. K., and C. D. Kern, 1965. *J. Geophys. Res.* 70: 1329–1337.

Burch, D. E., D. Gryvnak, E. B. Singleton, W. L. France, and D. Williams, 1962. "Infrared absorption by carbon dioxide, water vapor and minor atmospheric constituents," Air Force Cambridge Research Laboratories, Research Report AFCRL-62-698, Office of Aerospace Research, U.S.A.F.

Byers, H. R., and R. R. Braham, Jr., 1948. *J. Meteor.* 5: 71–86.

Byers, H. R., and R. R. Braham, Jr., 1949. *The Thunderstorm*, Washington D.C.: U.S. Government Printing Office.

Byers, H. R., J. R. Sievers, and B. J. Tufts, 1957. "Distribution in the atmosphere of certain particles capable of serving as condensation nuclei," in *Artificial Stimulation of Rain*, H. Weickmann and W. E. Smith, eds., New York: Pergamon, pp. 47–70.

Cadle, R. D., and G. Langer, 1975. *Geophys. Res. Letters* 2: 329–332.

Chahine, M. T., 1968. *J. Opt. Soc. Am.* 58: 1643–1637.

Chahine, M. T., 1972. *J. Atmos. Sci.* 29: 741–747.

Chahine, M. T., 1974, *J. Atmos. Sci.* 31: 233–243.

Chahine, M. T., 1977. *J. Atmos. Sci.* 34: 744–757.

Chahine, M. T., H. H. Aumann, and F. W. Taylor, 1977. *J. Atmos. Sci.* 34: 758–765.

Chalmers. J. A., 1957. *Atmospheric Electricity*, London: Pergamon.

Chandrasekhar, S., 1960. *Radiative Transfer*, Oxford: Oxford University Press.

Chapman, S., 1950. "Hydrometeors and thunderstorm electricity," in *Proceedings of the Conference on Thunderstorm Electricity*, Chicago: American Meteorological Society, p. 149.

Cheltzov, N. I., 1952. *Trans. Central Aerol. Observ.* 8.

Choularton, T. W., and J. Latham, 1977. *Quart. J. Roy. Meteor. Soc.* 103: 307–318.

Chylek, P., 1975. *J. Opt. Soc. Am.* 65: 1316–1318.

Chylek, P., G. W. Grams, and R. G. Pinnick, 1976. *Science* 193: 480–482.

Chylek, P., 1977. *J. Opt. Soc. Am.* 67: 1348–1350.

Clark, J. F., 1958. "The fair-weather atmospheric electric potential and its gradient," in *Recent Advances in Atmospheric Electricity*, L. G. Smith, ed., London: Pergamon, pp. 61–73.

Coakley, J. A., Jr., and P. Chylek, 1975. *J. Atmos. Sci.* 32: 409–418.

Cobb, W. E., 1973. *J. Atmos. Sci.* 30: 101–106.

Cotton, W. R., 1975. *Review Geophys. Space Res.* 13 : 419–448.

Coulson, K. L. and W. Reynolds, 1971. *J. Appl. Meteor.* 10: 1285–1295.

Coulson, K. L., J. V. Dave and Z. Sekera, 1960. *Tables Related to Radiation Emerging from a Planetary Atmosphere with Rayleigh Scattering*, Berkeley: University of California Press.

Curtis, A. R., 1952. *J. Roy. Meteor. Soc.* 78:638–640.

Danielson, R. E., D. R. Moore, and H. C. van de Hulst, 1969. *J. Atmos. Sci.* 26:1078–1087.

Davies, R., 1978. *J. Atmos. Sci.* 35:1712–1825.

Davis, P. A., 1971. *J. Appl. Meteor.* 10:1314–1323.

Day, J. A., 1964. *Quart. J. Roy. Meteor. Soc.* 90:72–78.

Deirmendjian, D., 1969. *Electromagnetic Scattering on Spherical Polydispersions.* New York: American Elsevier.

De Luisi, J. J., P. M. Furukawa, D. A. Gillette, B. G. Schuster, R. J. Charlson, W. M. Porch, R. W. Fegley, B. M. Herman, R. A. Rabinoff, J. T. Twitty, and J. A. Weinman, 1976. *J. Appl. Meteor.* 15:441–454.

Diem, M., 1948. *Meteor. Rundschau.* 9, 10:261–273.

Donner, L., and V. Ramanathan, 1980. *J. Atmos. Sci.* 37:119–124.

Dopplick, T. G., 1979. *J. Atmos. Sci.* 36:1812–1817.

Downing, H. D., and D. Williams, 1975. *J. Geophys. Res.* 80:1656–1661.

Draginis, M., 1958. *J. Meteor.* 15:481–485.

Dyer, A. J., 1974. *Quart. J. Roy. Meteor. Soc.* 100:563–571.

Eddy, J. A., 1976. *Science* 192:1189–1202.

Edlén, B., 1953. *J. Opt. Soc. Am.* 43:339.

Elliott, R. D. and E. L. Hovind, 1964. *J. Appl. Meteor.* 3:143–154

Elliott, R. D., and R. W. Shaffer, 1962. *J. Appl. Meteor.* 1:218–228.

Ellis, H. T., and R. E. Pueschel, 1973. *Science* 172:845–846.

Ellis, J. S., and T. H. VonderHaar, 1976. "Zonal average earth radiation budget measurements from satellites for climate studies," Atmospheric Science Paper No. 240, Department of Atmospheric Science, Colorado State University.

Elsasser, W. M., 1942. *Heat Transfer by Infrared Radiation in the Atmosphere*, Harvard Meteorological Studies No. 6, Cambridge: Harvard University Press.

Elsasser, W. M. and M. R. Culbertson, 1960. *Atmospheric Radiation Tables*, Meteorological Monograph 4, No. 23, Boston: American Meteorological Socieity.

Elster, J., and H. Geitel, 1913. *Phys. Zeits.* 14:1287–1292.

Elterman, L., 1966. Environmental Research Paper No. 241, Air Force Cambridge Research Laboratories, Air force Systems Command, U.S.A.F.

Elterman, L., 1968. Environmental Research Paper No. 285, Air Force Cambridge Research Laboratories, Air force Systems Command, U.S.A.F.

Elterman, L., R. Wexler, and D. T. Chang, 1969. *Appl. Opt.* 8:893–903.

Espensheid, W. F., M. Kerker, and E. Matijevic, 1964. *J. Phys. Chem.* 68:3093–3097.

Farlow, N. H., D. M. Hayes, and H. Y. Lem, 1977. *J. Geophys. Res. 82:4921–4929.*

Few, A. A., 1974. *EOS, Trans. Am. Geophys. U.* 55:508–514.

Fitzgerald, J. W., 1972. "A study of the initial phase of cloud droplet growth by condensation: comparison between theory and observation," Ph.d. Thesis, Department of Geophysical Science, University of Chicago.

Fitzergald, J. W., 1978. *J. Atmos. Sci.* 35:1522–1535.

Fleming, J. R., and S. K. Cox, 1974. *J. Atmos. Sci.* 31:2182–2188.

Fletcher, N. H., 1958. *J. Chem. Phys.* 29:572–576.

Fletcher, N. H., 1959. *J. Chem. Phys.* 31:1136–1137.

Fletcher, N. H., 1962. *The Physics of Rainclouds*, Cambridge: Cambridge University Press.

Fletcher, N. H., 1969. *J. Atmos. Sci.* 26:1266–1271.

Flowers, E. C., R. A. McCormick, and K. R. Kurfis, 1969. *J. Appl. Meteor.* 8:955–962.

Foote, G. B. and P. S. de Toit, 1969. *J. Appl. Meteor.* 8:249–253.

Foote, G. B., and C. A. Knight, eds., 1977. *Hail: A Review of Hail Science and Hail Suppression*, Meteorological Monograph 16, No. 38, Boston: American Meteorological Society.

Foster, H., 1950. *Bull. Am. Meteor. Soc.* 31:140–141.

Fowle, F. E., 1915. *Astrophys. J.* 42:394–411.

Fraser, A. B., 1972. *J. Atmos. Sci.* 29:211–212.

Fraser, A. B., 1975. "Meteorological optics," in *The Atmosphere*, R. A. Antes et al., eds., Merrill: Columbus, chapter 9.

Freire, E., and R. List, 1979. *J. Atmos. Sci.* 36:1777–1787.

Friedlander, S. K., 1961. *J. Meteor.* 18:753–759.

Fritz, S., 1954. *J. Meteor.* 11:291–300.

Fritz, S., and T. H. MacDonald, 1949. *Heat and Ventilation* 46:61–64.

Fritz, S., and T. H. MacDonald, 1951. *Bull. Amer. Meteor. Soc.* 32:205–209.

Froehlich, C., 1977. "Contemporary measures of the solar constant," in *The solar Output and Its Variation*, O. R. White, ed., Boulder: Colorado University Press, pp. 93–109.

Fukuta, N., 1969. *J. Atmos. Sci.* 26:522–531.

Fukuta, N., and L. A. Walter, 1970. *J. Atmos. Sci.* 27:1160–1172.

Fulks, J. R., 1935. *Mon. Wea. Rev.* 63:291–294.

Fuquay, D. M., and H. J. Wills, 1957. "The project Skyfire cloud seeding generator," in *Final Report, Advisory Commission on Weather Control*, vol. 2, Washington, D.C.: U.S. Government Printing Office, pp. 273–282.

Furrikawa, P. M., P. L. Haagenson, and M. J. Scharberg, 1967. "A composite, high resolution solar spectrum from 2080 to 3600 Å" NCAR Technical Note 26, National Center for Atmospheric Research, Boulder.

Gagin, A., 1975. *J. Atmos. Sci.* 32:1604–1614.

Gerber, H. E., W. A. Hoppel, and T. A. Wojciechowsky, 1977. *J. Atmos. Sci.* 34:1836–1841.

Gerdien, H., 1905. *Phys. Zeits.* 6:800–801.

Gish, O. H., and K. L. Sherman, 1936. *Nat. Geogr. Soc. Stratosphere Ser.* 2:94–116.

Gish, O.H., and G. R. Wait, 1950. *J. Geophys. Res.* 55:473–484.

Godson, W. L., 1953. *Quart. J. Roy. Meteor. Soc.* 79:367–379.

Goetz, A., and P. Preining, 1960. "The aerosol spectrometer and its application to nuclear condensation studies," In *Physics of Precipitation*, H. Weickmann, ed., Geophysical Monograph No. 5, Washington, D.C.: American Geophysical Union, pp. 164–182.

Gokhale, N. R., and J. Goold, Jr., 1968. *J. Appl. Meteor.* 7:870–874.

Goody, R. M., 1952. *Quart. J. Roy. Meteor. Soc.* 78 : 165–169.

Goody, R. M., 1954. *The Physics of the Stratosphere*, Cambridge: Cambridge University Press, pp. 146–148.

Goody, R. M., 1964. *Atmospheric Radiation*, Vol. 1: *Theoretical Basis*, Oxford: Oxford University Press.

Grenet, G., 1947. *Ann. Geophys.* 3:306–307

Griggs, M., 1970. *J. Geophys. Res.* 73:7546–7551.

Griggs, M., and W. A. Margraff, 1967. "Final report under contract AF19(628)-5517," Air Force Cambridge Research Laboratories, Office Air force Research, U.S.A.F.

Gumprecht, R. O., and C.M. Sliepcevich, 1953. *J. Phys. Chem.* 57:90–95.

Gunn, K. L. S., and J. S. Marshall, 1958. *J. Meteor.* 15:452–461.

Gunn, R., 1950. *J. Geophys. Res.* 55:171–178.

Gunn, R., and G. D. Kinzer, 1949. *J. Meteor.* 6:243–248.

Hale, B. N., and L. M. Plummer, 1974. *J. Atmos. Sci.* 31:1615–1621.

Hallett, J., and B. J. Mason, 1958. *Proc. Roy. Soc.* A247:440–453.

Hallett, J., and S. C. Mossop, 1974. *Nature* 249:26–28.

Hanel, R. A., B. J. Conrath, V. G. Kunde, C. Prabhakara, I. Revah, V. V. Salmonson, and G. Wolford, 1972. *J. Geophys. Res.* 77:2629–2641.

Hansen, J. E., and L. D. Travis, 1974. *Space Sci. Rev.* 16 : 527–610.

Haurwitz, B., 1948. *J. Meteor.* 5:110–113.

Hays, J. D., J. Imbrie, and N. J. Shakleton, 1976. *Science* 194:1121–1132.

Hewson, E. W., 1943. *Quart. J. Roy. Meteor. Soc.* 69:47–62.

Heymsfield, A. J., C. A. Knight, and J. E. Dye, 1979. *J. Atmos. Sci.* 36:2216–2229.

Higuchi, K., 1956. *J. Meteor.* 13:274–278.

Hillary, D. T., E. L. Heacock, W. A. Morgan, R. H. Moore, E. C. Mangold, and S. D. Soules, 1966. *Mon. Wea. Rev.* 94 : 367–377.

Hitschfeld, W., and J. T. Houghton, 1961. *Quart. J. Roy. Meteor. Soc.* 87:562–577.

Hobbs, P. V., 1969. *J. Atmos. Sci.* 26:315–318.

Hobbs, P. V., et al., 1971. "Studies of winter cyclonic storms over the Cascade Mountains (1970–1971)," Contribution from Cloud Physics Group, Research Report 4, Department of Atmospheric Science, University of Washington, Seattle.

Hobbs, P. V. et al., 1974. "The structure of clouds and precipitation over the Cascade Mountains and their modification by artificial seeding (1972–1973)," Research Report 8, Department of Atmospheric Science, University of Washington, Seattle.

Hoffman, K., 1923. *Beitr. Phys. Freien. Atmos.*, 11:1–19.

Hofmann, D. J., J. M. Rosen, T. J. Pepin, and R. G. Pinnick, 1975. *J. Atmos. Sci.* 32:1446–1456.

Hoppel, W. A. and J. H. Kraakevik, 1965. *J. Atmos. Sci.* 22:509–517.

Houghton, H. G., 1950. *J. Meteor.* 7:363–369.

Houghton, H. G., 1954. *J. Meteor.* 11:1–9.

Houghton, H. G. 1968. *J. Appl. Meteor.* 7:851–859.

Houghton, H. G., 1972. *J. Rech. Atmos.* 6(4):657–667.

Houghton, J. T., and J. S. Seeley, 1960. *Quart. J. Roy. Meteor. Soc.* 86:358–370.

Howard, H. G., D. E. Burch, and D. Williams, 1955. "Near infrared transmission through synthetic atmospheres," Geophysical Research Papers No. 40, Air Force Cambridge Research Center, Office Air Force Research U.S.A.F.

Howard, J. N., D. E. Burch, and D. Williams, 1956. *J. Opt. Soc. Am.* 46:334–338.

Howell, H. B., 1969. "Angular scattering functions for spherical water drops," NRL Report 6955, Naval Research Laboratory, U.S. Navy.

Howell, W. E., 1949. *J. Meteor.* 6:134–149.

Hoyt, D. V., 1977. *J. Appl. Meteor.* 16:432–436.

Huffman, P. J., 1973. *J. Appl. Meteor.* 12:1080–1082.

Hummel, J. R., and R. A. Reck, 1979. *J. Appl. Meteor.* 18:239–253.

Humphreys, W. J., 1929. *Physics of the Air*, 2nd ed., New York: McGraw-Hill.

Hunt, G. E., 1973. *Quart. J. Roy. Meteor. Soc.* 99:346–369.

Imyanitov, I. M., B. F. Esteev, and I. I. Kamaldina, 1969. "A thunderstorm cloud," in *Planetary Electrodynamics*, S. C. Coroniti and J. Hughes, eds., Vol. 1, New York: Gordon and Breach, pp. 401–425.

Isaac, G. A., and R. S. Schemenauer, 1979. *J. Appl. Meteor.* 18:1056–1065.

Isono, K., and A. Ono, 1959. *J. Meteor. Soc. Japan.* 37:211–215.

Israel, H., 1953. *Geofis. Pura. Appl.* 24:3–11.

Israel, H., 1970. *Atmospheric Electricity*, vol. 1, Jerusalem: Israel Program for Scientific Translations.

Jacobowitz, H. J., 1971. *Quant. Spect. Rad. Transf.* 11:691–695.

Jacobowitz, H. J., W. L. Smith, H. B. Howell, F. W. Nagle, and J. R. Hickey, 1979. *J. Atmos. Sci.* 36:501–507.

Jacobson, E. A., and E. P. Krider, 1976. *J. Atmos. Sci.* 33:103–117.

Jennings, S. G., 1975. *Quart. J. Roy. Meteor. Soc.* 101:227–233.

Jiusto, J. E., and W. C. Kocmond, 1968. *J. Rech. Atmos.* 3:19–24.

Johnson, F. S., 1954. *J. Meteor.* 11:431–439.

Johnson, J. C., 1954. *Physical Meteorology*, New York: Wiley.

Jonas, P. R., and P. Goldsmith, 1972. *J. Fluid Mech.* 52:593–608.

Joseph, J. H., and N. Wolfson, 1975. *J. Appl. Meteor.* 14:1389–1396.

Joss, J., and E. G. Gori, 1978. *J. Appl. Meteor.* 17:1054–1061.

Junge, Chr., 1954. *J. Meteor.* 11:323–333.

Junge, Chr., 1955. *J. Meteor.* 12:13–25.

Junge, Chr., 1957. "Remarks about the size distribution of natural aerosols," in *Artificial Stimulation of Rain*, H. Weickmann and W. E. Smith, eds., New York: Pergamon, pp. 3–16.

Junge, Chr., 1963. *Air Chemistry and Radioactivity*, New York: Academic Press.

Junge, Chr., 1972. *J. Geophys. Res.* 77:5183–5200.

Junge, Chr., and E. McLaren, 1971. *J. Atmos. Sci.* 28:382–390.

Junge, Chr., and J. E. Manson, 1961, *J. Geophys. Res.* 66:2163–2182.

Junge, Chr., C. W. Chagnon, and J. E. Manson, 1961. *J. Meteor.* 18:81–108.

Kaiser, J. A. C., and R. H. Hill, 1976. *J. Geophys. Res.* 81:395–398.

Kajikawa, M., 1972. *J. Meteor. Soc.*, Ser. 2, 50:577–583.

Kaplan, L. D., 1953. *J. Meteor.* 10:100–104.

Kaplan, L. D., 1959. *J. Opt. Soc. Am.* 49:1004–1007. New York: The Rockefeller Institute Press in association with the Oxford University Press, pp. 170–177.

Kaplan, L. D., and D. F. Eggers, 1956. *J. Chem. Phys.* 25:876–883.

Keeling, C. D., 1978. In *Mauna Loa Observatory. A 26th Anniversary Report.* J. Miller, ed., Washington, D.C.: NOAA.

Kellogg, W. W., R. D. Cadle, E. R. Allen, A. L. Lazrus, and E. A. Martell, 1972. *Science*, 175:587–596.

Kessler, E., 1969. *On the Distribution and Continuity of Water Substance in Atmospheric Circulations*, Meteorological Monograph 10, No. 32, Boston: American Meteorological Society.

Kientzler, C. E., A. B. Arons, D. C. Blanchard, and A. H. Woodcock, 1954. *Tellus* 6:1–7.

Kislovskii, L. D., 1959. *Opt. Spectry.* 7:201–206.

Knight, C. A., N. C. Knight, J. E. Dye, and V. Toutenhoofd, 1974. *J. Atmos. Sci.* 31:2142–2147.

Koehler, H., 1921. Geofys. Publ., 2, Nos. 3 and 6.

Koehler, H., 1925. Meddel. Meteor.—Hydrografiske Anstalt, Stockholm, 2, No. 5.

Koehler, H., 1926. Meddel. Meteor.—Hydrografiske Anstalt, Stockholm, 3, No. 8.

Koenig, L. R., 1971. *J. Atmos. Sci.* 28:226–237.

Koenig, L. R., and F. W. Murray, 1976. *J. Appl. Meteor.* 15:747–762.

Kondratyev, K. Ya., 1969. *Radiation in the Atmosphere.* New York: Academic Press.

Koschmieder, H., 1924. *Beitr. Phys. freien Atmos.* 12:33–53, 171–181.

Kraakevik, J. H., 1961. *J. Geophys. Res.* 66:3735–3748.

Kraakevik, J. H., and J. F. Clark, 1958. *Trans. Am. Geophys. Union* 39:827–834.

Krehbiel, P., M. Brook, and R. McCrory, 1974. *Proceedings Fifth International Conference on Atmospheric Electricity*, Garmish-Partenkirchen, W. Germany.

Kuhn, P. M., 1963a. *J. Geophys. Res.* 68:1414–1420.

Kuhn, P. M., 1963b. *J. Appl. Meteor.* 2:368–378.

Kuhn, P. M., and H. Weickmann, 1969. *J. Appl. Meteor.* 8:147–154.

Kuhn, P. M., H. K. Weickmann, and L. P. Stearns, 1975. *J. Geophys. Res.* 80:3419–3424.

Kuiper, C. P., ed., 1953. *The Sun*, Chicago: University of Chicago Press.

Kumai, M., 1951. *J. Meteor.* 8:151–156.

Kumai, M., 1961. *J. Meteor.* 18:139–150.

Kumai, M., 1976. *J. Atmos. Sci.* 33:833–841.

Kumai, M., and K. E. Francis, 1962. *J. Atmos. Sci.* 19:474–481.

Kung, E. C., R. A. Bryson, and D. H. Lanschow, 1964. *Mon. Wea. Rev.* 92:543–564.

Kunkel, B. A. 1970. "Physiochemical properties of some hygroscopic nuclei," Environmental Research Papers No. 337, Air Force Cambridge Research Laboratories, Air Force Office of Aerospace Research, U.S.A.F.

Ladenburg, R., and F. Reiche, 1913. *Ann. Phys.* 42:181–209.

Lamb, D., and W. D. Scott, 1974. *J. Atmos. Sci.* 31:570–580.

Landsberg, H., 1938. *Ergeb. Kosmich. Phys.* 3:155–252.

Langer, G., G. Cooper, C. T. Nagamoto, and J. Rosensky, 1978. *J. Appl. Meteor.* 17:1039–1048.

Langmuir, I., 1948. *J. Meteor.* 5:175–192.

Latham, J., and C. P. R. Saunders, 1970. *Quart. J. Roy. Meteor. Soc.* 96:257–265.

Lee, A. C. L., 1973. *Quart. J. Roy. Meteor. Soc.* 99:490–505.

Leighton, H. G., and R. R. Rogers, 1974. *J. Atmos. Sci.* 31:271–279

Lenard, P., 1904. *Meteor. Zeits.* 21:248–262.

Lenard, P., 1921. *Ann. Phys.* 65:629–639.

Levin, Z., and B. Machnes, 1977. *Pure Appl. Geophys.* 115:845–867.

Levine, J., 1959. *J. Meteor.* 16:653–662.

Lin, C. L., and S. C. Lee, 1975. *J. Atmos. Sci.* 32:1412–1418.

Linke, F., and K. Boda, 1922. *Meteor. Zeits.* 39:161–166.

Liou, K. N., 1973a. *J. Atmos. Sci.* 30:1303–1326.

Liou, K. N., 1973b. *J. Geophys. Res.* 78:1409–1418.

Liou, K. N., 1974. *J. Atmos. Sci.* 31:522–532.

List, R., 1977. *Trans. Roy. Soc. Canada*, Ser. 4, 15:333–347.

Locatelli, J. D., and P. V. Hobbs, 1974. *J. Geophys. Res.* 79:2185–2197.

Lodge, J. P., Jr., and E. R. Frank, 1966. *J. Rech. Atmos.* 2(2[e] ann.): 139–140.

London, J., 1957. "A study of the atmospheric heat balance," Final Report, under Contract AF19(122)-165, Air force Cambridge Research Center, U.S.A.F., Department of Meteorology, New York University.

London, J., and T. Sasamori, 1971. "Radiative energy budget of the atmosphere," in *Man's Impact on the Climate*, W. H. Matthews, W. W. Kellogg, and G. D. Robinson, eds., Cambridge, MA: MIT Press, pp. 141–155.

Low, R. D. H., 1969. *J. Rech. Atmos.* 4(1[re] ann., No. 2): 65–78.

Ludlam, F. H., 1958. *Nubila* 1: 12–96.

Ludwig, J. H., G. B. Morgan, and T. B. McMullen, 1970. *Trans. Am. Geophys. U.* 51: 468–475.

McDonald, J. E., 1963. *J. App. Math. Phys.* (ZAMP) 14: 610–620.

Mecke, R., 1921. *Ann. Phys.* 65: 257–273.

Magono, C., and C. W. Lee, 1966. *J. Fac. Sci., Hokkaido U.*, Ser. VII, 2(4): 321–335.

Mahata, P. C., and D. J. Alofs, 1975 *J. Atmos. Sci.* 32: 116–122.

Malkus, J. S., 1952. *Quart. J. Roy. Meteor. Soc.* 78: 530–542.

Manabe, S., and F. Moeller, 1961. *Mon. Wea. Rev.* 89 : 503–532.

Manabe, S., and R. F. Strickler, 1964. *J. Atmos. Sci.* 21: 361–385.

Manabe, S., and R. T. Wetherald, 1967. *J. Atoms. Sci.* 24: 241–259.

Manabe, S., and R. T. Wetherald, 1975. *J. Atmos. Sci.* 32: 3–15.

Manabe, S., and R. T. Wetherald, 1980. *J. Atmos. Sci.* 37: 99–118.

Marshall, J. S., and W. McK. Palmer, 1948. *J. Meteor.* 5: 165–166.

Mascart, M. E., 1889–1894. *Traité d'optique*. Paris: Gauthier-Villars.

Mason, B. J., 1954. *Nature* 174: 470–471.

Mason, B. J., 1971. *The Physics of Clouds*, 2nd ed., Oxford: Clarendon Press.

Mason, B. J., and A. P. van den Heuvel, 1959. *Proc. Roy. Soc.* 74: 744–755.

Matejka, T. J., R. A. Houze, Jr., and P. V. Hobbs, 1980. *Quart. J. Roy. Meteor. Soc. 106 : 29–56.*

Matthews, J. B., and B. J. Mason, 1964. *Quart. J. Roy. Meteor. Soc.* 90 : 275–286.

Mauchly, S. J., 1923. *Terr. Mag. Atmos. Elect.* 28: 61–81.

Maxwell, J. C., 1890. *Scientific Papers of James Clerk Maxwell*, vol. 2, New York: Dover, pp. 636–640.

May, K. R., 1945. *J. Sci. Instr.* 22: 187–195.

Middleton, W. E. K., 1952. *Vision through the Atmosphere*, Toronto: University of Toronto Press.

Mie, G., 1908. *Ann. Phys.* 25: 377–445.

Milankovitch, M., 1930. *Handbuch der Klimatologie, Part A*, Berlin: Verlag Bornts.

Misaki, M., and I. Kanazawa, 1969. "Some features of the dynamic spectrum of atmospheric ions throughout the mobility range 4.22–0.0042 $cm^2$/volt sec," in

*Planetary Electrodynamics*, S. C. Coroniti and J. Hughes, eds., vol. 1, New York: Gordon and Breach, pp. 249–255.

Moore, C. B., B. Vonnegut, B. A. Stein, and H. J. Survilas, 1960. *J. Geophys. Res.* 65:1907–1910.

Mordy, W. A., 1959. *Tellus* 11:16–44.

Mossop, S. C., 1970. *Bull. Am. Meteor. Soc.* 51:474–479.

Mossop, S. C., 1976. *Quart. J. Roy. Meteor. Soc.* 102:45–57.

Mossop, S. C., and J. Hallett, 1974. *Science* 186:632–634.

Mossop, S. C., R. E. Cottis, and B. M. Bartlett, 1972. *Quart. J. Roy. Meteor. Soc.* 98:105–123.

Mossop, S. C., A. Ono, and E. R. Weshart, 1970. *Quart. J. Roy. Meteor. Soc.* 96:487–508.

Murgatroyd, R. J., and M. P. Garrod, 1957. *Quart. J. Roy. Meteor. Soc.* 83:528–533.

Nakaya, U., 1954. *Snow Crystals: Natural and Artificial.* Cambridge, MA: Harvard University Press.

Nakaya, U., and Y. Sekido, 1930. *J. Fac. Sci., Hokkaido U.*, Ser. 3, 1:234–264.

Nakaya, U., and T. Terada, 1935. *J. Fac. Sci.*, Hokkaido U., Ser. 2, 1:191–200.

Neiburger, M., 1949. *J. Meteor.* 6:98–104.

Neiburger, M., and C. W. Chien, 1960. "Computations of the growth of cloud drops by condensation using an electronic digital computer," in *Physics of Precipitation*, Geophysical Monograph No. 5, Washington, D.C.: American Geophysical Union, pp. 191–210.

Neilsen, J. R., V. Thornten, and E. B. Dale, 1944. *Rev. Mod. Phys.* 16:307–324.

Nkemdirin, L. C., 1972. *J. Appl. Meteor.* 11:867–874.

Odencrantz, F. K., 1969. *J. Appl. Meteor.* 8:322–325.

Ono, A., 1969. *J. Atmos. Sci.* 26:138–147.

Ono, A., 1970. *J. Atmos. Sci.* 27:649–658.

Orville, R. E., 1968. *J. Atmos. Sci.* 25:839–851.

Orville, R. E., and D. W. Spencer, 1979. *Mon. Wea. Rev.* 107:934–943.

Palmer, K. F., and D. Williams, 1975. *Appl. Optics* 14:208–219.

Paltridge, G. W., and C. M. R. Platt, 1973. *J. Atmos. Sci.* 30:734–737.

Pasternak, I. S., and W. H. Gauvin, 1960. *Can. J. Chem. Eng.* 38:35–42.

Payne, R. E., 1972. *J. Atmos. Sci.* 29:959–970.

Peirce, B. O., 1929. *A Short Table of Integrals*, Boston: Ginn.

Penndorf, R. B., 1956. "New tables of mid scattering functions for spherical particles," Geophysics Research Paper No. 45, Air Force Cambridge Research Center, U.S.A.F.

Penndorf, R. B., 1961. "Atlas of scattering diagrams for index of refraction 1.33," prepared by Avco Corp. for Air Force Cambridge Research Laboratories under Contract No. AF19(604)-5743, U.S.A.F.

Pernter, J. M., and F. M. Exner, 1922. *Meteorologische Optik*. Vienna and Leipzig: W. Braumueller.

Peterson, J. T., and R. A. Bryson, 1968. *Science* 162:120–121.

Pflaum, J. C., and H. R. Pruppacher, 1979. *J. Atmos. Sci.* 36:680–689.

Pietrowski, E. L., 1960. *J. Meteor.* 17:562–563.

Pitter, R. L., 1977. *J. Atmos. Sci.* 34:684–685.

Plass, G. N., 1952. *J. Meteor.* 9:429–436.

Plass, G. N., 1966. *Appl. Optics.* 5:279–285.

Platt, C. M. R., 1972. *J. Geophys. Res.* 77:1597–1609.

Platt, C. M. R., 1976. *Quart. J. Roy. Meteor. Soc.* 102:553–562.

Pound, G. M., L. A. Madonna and C. Scirelli, 1951. Metals Research Laboratory Quarterly Report No. 5, Carnegie Institute of Technology.

Powell, C. F., 1928. *Proc. Roy. Soc.* A119:277.

Randall, H. M., D. M. Dennison, N. Ginsberg, and L. R. Weber, 1937. *Phys. Rev.* 52:160–174.

Rau, W., 1949. *Wetter u. Klima* 2:81–92.

Rayleigh, Lord, 1871. *Phil. Mag.* 41:107–120; 41:274–279.

Remsberg, E. E., 1973. *J. Geophys. Res.* 78:1401–1408.

Reynolds, D. W., T. H. Vonder Haar, and S. K. Cox, 1975. *J. Appl. Meteor.* 14:433–444.

Rider, N. E., and G. D. Robinson, 1951. *Quart. J. Roy. Meteor. Soc.* 77:375–401.

Robinson, G. D., 1950. *Quart. J. Roy. Meteor. Soc. 76:37–51.*

Robinson, G. D., 1958. *Archiv. Meteor. Geophys. Biokl.* B9:28–41.

Robinson, G. D., 1962. *Archiv. Meteor. Geophys. Biokl.* B12: 19–40.

Robinson, P. J., and J. A. Davies, 1972. *J. Appl. Meteor.* 11:1391–1393.

Rodgers, C. D., and C. D. Walshaw, 1966. *Quart. J. Roy. Meteor. Soc.* 92:67–92.

Roosen, R. G., R. J. Angione, and C. H. Klemcke, 1973. *Bull. Am. Meteor. Soc.* 54: 307–316.

Rosen, J. M., and J. Laby, 1975. *J. Atmos. Sci.* 32:1457–1462.

Rosen, J. M., D. J. Hofmann, and K. H. Kaselau, 1978. *J. Appl. Meteor.* 17:1737–1740.

Rossman, K., K. N. Rao, and H. H. Nielsen, 1956. *J. Chem. Phys.* 24:103–105.

Sagalyn, R. C. and G. A. Faucher, 1954. *J. Atmos. Terr. Phys.* 5:253–272.

Sartor, J. D., 1967. *J. Atmos. Sci.* 24:601–615.

Saunders, P. M., 1965. *J. Atmos. Sci.* 22:167–175.

Schaefer, V. J., 1946. *Science* 104:457–459.

Schlamp, R. J., H. R. Pruppacher, and A. E. Hamielec, 1975. *J. Atmos. Sci.* 32:2330–2337.

Schlamp, R. J., S. N. Grover, H. R. Pruppacher, and A. E. Hamielic, 1979. *J. Atmos. Sci.* 36:339–349.

Schnell, R. C., 1977. *J. Atmos. Sci.* 34:1299–1305.

Schnell, R. C., and G. Vali, 1972. *Nature* 236:163–165.

Schnell, R. C., and G. Vali, 1976. *J. Atmos. Sci.* 33:1554–1564.

Schonland, B. F. J., 1928. *Proc. Roy. Soc.* A118:252–262.

Schonland, B. F. J., 1956. *The Lightning Discharge.* Handbuch der Physik 22. Berlin: Springer.

Schonland, B. F. J., 1964. *The Flight of the Thunderbolts*, Oxford: Clarendon Press.

Schuster, A., 1905. *Astrophys. J.* 21:1–22.

Scott, W. D., and Z. Levin, 1970. *J. Atmos. Sci.* 27:463–473.

Sekhon, R. S., and R. C. Srivastava, 1970. *J. Atmos. Sci.* 27:299–307.

Shafrir, U., and T. Gal-Chen, 1971. *J. Atmos. Sci.* 28:741–751.

Shaw, G. E., 1976. *J. Geophys. Res.* 81:5791–5792.

Simpson, G. C., 1927–1928. *Memoirs Roy. Meteor. Soc.* 2:69–95.

Simpson, G. C., 1928–1930. *Memoirs Roy. Meteor. Soc.* 21(3):1–26.

Simpson, G. C., and G. D. Robinson, 1941. *Proc. Roy. Soc.* A177:281–329.

Simpson, G. C., and F. J. Scrase, 1937. *Proc. Roy. Soc.* A161:309–352.

Simpson, J., and V. Wiggert, 1969. *Mon. Wea. Rev.* 97:471–489.

Singer, S., 1971. *The Nature of Ball Lightning*, New York: Plenum.

Skelland, A. H. P., and A. R. H. Cornish, 1963. *J. Am. Inst. Chem. Eng.* 9:73–76.

Smith, E. J., and K. J. Heffernan, 1954. *Quart. J. Roy. Meteor. Soc.* 80:182–197.

Smith, W. L., 1972. *Bull. Am. Meteor. Soc.* 53:1074–1082.

Smith. W. L., H. M. Woolf, and W. J. Jacob, 1970. *Mon. Wea. Rev.* 98:582–603.

*Smithsonian Meteorological Tables*, 1939. Fifth rev. ed., Baltimore: Lord Baltimore Press, p. 240, table 111.

*Smithsonian Meteorological Tables*, 1951. Sixth rev. ed., Washington: The Smithsonian Institution, p. 422, table 137.

Squires, P., 1958. *Tellus* 10:256–262.

Squires, P., and S. Twomey, 1966. *J. Atmos. Sci.* 23:401–404.

Staelin, D. H., A. H. Barrett, J. W. Waters, F. T. Barath, E. J. Johnston, P. W. Rosenkranz, N. E. Gaut, and W. B. Lenoir, 1973. *Science* 182:1339–1341.

Staley, D. O., and G. M. Jurica, 1970. *J. Appl. Meteor.* 9:365–372.

Staley, D. O., and G. M. Jurica, 1972. *J. Appl. Meteor.* 11:349–356.

Stergis, C. G., G. C. Rein, and T. Kangas, 1957. *J. Atmos. Terr. Phys.* 11:83–90.

Stewart, J. B., 1971. *Quart. J. Roy. Meteor. Soc.* 97:561–564.

Stoiber, R. E., and A. Jepsen, 1973. *Science* 182:577–578.

Sulakvelidze, G. K., N. Sh. Bibilashvili, and V. F. Lapcheva, 1967. *Formation of Precipitation and Modification of Hail Processes*, Jerusalem: Israel Program for Scientific Translations, for the National Science Foundation.

Suomi, V. E., and P. M. Kuhn, 1958. *Tellus* 10:160–163.

Suomi, V. E., D. O. Staley, and P. M. Kuhn, 1958. *Quart. J. Roy. Meteor. Soc.* 84:134–141.

Sutherland, R. A., R. D. McPeters, G. B. Findley, and A. E. S. Green, 1975. *J. Atmos. Sci.* 32:427–436.

Takahashi, T., 1975. *J. Atmos. Sci.* 32:123–142.

Takahashi, T., 1977. *J. Atmos. Sci.* 34:1773–1790.

Takahashi, T., 1978a. *J. Atmos. Sci.* 35:277–283.

Takahashi, T., 1978b. *J. Atmos. Sci.* 35:1536–1548.

Telford, J. W., 1955. *J. Meteor.* 12:436–444.

Thorpe, A. D., and B. J. Mason, 1966. *Brit. J. Appl. Phys.* 17:541–548.

Tomasi, E., R. Guzzi, and O. Vittori, 1974. *J. Atmos. Sci.* 31:255–260.

Toon, O. B., and J. B. Pollack, 1976. *J. Appl. Meteor.* 15:225–246.

Toon, O. B., and J. B. Pollack, 1976. *J. Appl. Meteor.* 15:225–246.

Toon, O. B., J. B. Pollack, and B. N. Khare, 1976. *J. Geophys. Res.* 81:5733–5748.

Trent, E. M., and S. G. Gathman, 1972. *Pure Appl. Geophys.* 100:60–69.

Tricker, R. A. R., 1971. *Introduction to Meteorological Optics*, New York: American Elsevier.

Twomey, S., 1959. *Geofis. Pura. Appl.* 43:227–242.

Twomey, S., 1966. *J. Atmos. Sci.* 23:405–411.

Twomey, S., and T. A. Wojciechowski, 1969. *J. Atmos. Sci.* 26:684–688.

Twomey, S., 1976a. *J. Atmos. Sci.* 33:720–723.

Twomey, S., 1976b. *J. Atmos. Sci.* 33:1087–1091.

Twomey, S., 1977. *J. Atmos. Sci.* 34:1832–1835.

Twomey, S., and J. Warner, 1967. *J. Atmos. Sci.* 24:702–703.

Twomey, S., H. Jacobowitz, and H. B. Howell, 1966. *J. Atmos. Sci.* 23:289–296.

Twomey, S., H. Jacobowitz, and H. B. Howell, 1967. *J. Atmos. Sci.* 24:70–79.

Uman, M. A., 1969. *Lightning*, New York: McGraw-Hill.

Unsworth, M. H., and J. L. Monteith, 1975. *Quart. J. Roy. Meteor. Soc.* 101:13–24.

Vail, G., M. Christensen, R. W. Fresh, E. L. Galyan, L. E. Maki, and R. C. Schnell, 1976. *J. Atmos. Sci.* 33:1565–1570.

Van de Hulst, H. C., 1957. *Light Scattering by Small Particles*, New York: Wiley.

Vernekar, A. D., 1972. *Long Period Global Variations of Incoming Solar Radiation*, Meteorological Monograph 12, No. 34, Boston: American Meteorological Society.

Volmer, M., and H. Flood. 1926. *Zeits. Phys. Chem.* 119:277.

Volz, F. E., 1959. *Arch. Meteor., Geophys., Biokhm.* B10:100–131.

Volz, F. E., 1960. "Some aspects of the rainbow and the physics of rain," in *Physics of Precipitation*, H. Weickmann, ed., Geophysical Monograph No. 5, Washington, D.C.: American Geophysical Union, pp. 280–286.

Volz, F. E., 1969. *Appl. Optics* 8:2505–2517.

Vonnegut, B., 1947. *J. Appl. Phys.* 18:593–595.

Vonnegut, B., 1955. "Possible mechanism for the formation of thunderstorm electrification, "*Proceedings of the Conference on Atmospheric Electricity, Portsmouth, NH*, Geophysical Research Paper No. 42, Cambridge: Air Force Cambridge Research Laboratories, pp. 169–181.

Vonnegut, B., C. B. Moore, R. G. Semonin, J. W. Bullock, D. W. Staggs, and W. E. Bradley, 1962. *J. Geophys. Res.* 67 : 3909–3922.

Wait. G. R., 1950. *Archiv. Meteor. Wien* A3 : 70–76.

Waldram, J. W., 1945. *Quart. J. Roy. Meteor. Soc.* 71:319–336.

Walshaw, C. D., and C. D. Rodgers, 1963. *Quart. J. Roy. Meteor. Soc.* 89 : 122–130.

Wark, D. Q., and D. T. Hillary, 1969. *Science* 165:1256–1258.

Warner, J., 1969. *J. Atmos. Sci.* 26:1049–1059.

Wegener, A., 1911. *Thermodynamik der Atmosphäre*. Leipzig: J. A. Barth.

Weickmann, H., 1949. Bericht. Deut. Wetterdienst., U.S. Zone, No. 6, Bad Kissingen.

Weickmann, H., 1957. "The snow crystal as aerological sonde," in *Artificial Stimulation of Rain*, H. Weickmann and W. E. Smith, eds., New York: Pergamon, pp. 315–325.

Weinstein, A. I., 1970. *J. Atmos. Sci.* 27:246–255.

Wexler, R., and D. Atlas, 1958. *J. Meteor.* 15:531–538.

Whelpdale, D. M., and R. List, 1971. *J. Geophys. Res.* 76:2836–2856.

Whipple, F. J. W., 1929. *Quart. J. Roy. Meteor. Soc.* 55:1–17.

Whytlaw-Gray, R. W., and H. S. Patterson, 1932. *Smoke: A Study of Aerial Disperse Systems*, London: Arnold.

Wieland, W. 1956. Eidg. Tech. Hochsch., Zurich, Promotionsarbeit No. 2577.

Wiener, C., 1900. *Nova Acta Halle* 73 : 1–240.

Wilhelmson, R., and Y. Ogura, 1972. *J. Atmos. Sci.* 29:1295–1307.

Willett, J. C., 1975. "Fair-weather electric charge transfer by convection in an unstable planetary boundary layer," Ph.D. Thesis, Department of Meteorology, Massachusetts Institute of Technology.

Willson, R. C., and J. R. Hickey, 1977. "1976 rocket measurements of the solar constant and their implications for variation in the solar output in cycle 20," in *The Solar Output and Its Variation*, O. R. White, ed., Boulder: Colorado University Press, pp. 111–116.

Willson, R. C., S. Gulkis, M. Janssen, H. S. Hudson, and G. A. Chapman, 1981. *Science 211:700–702.*

Wilson, C. T. R., 1897. *Phil. Trans. Roy. Soc.* A189:265.

Wilson, C. T. R., 1899. *Phil. Trans. Roy. Soc.* A193:289.

Wilson, C. T. R., 1922. *The Observatory* 45 : 393.

Winn, W. P., G. W. Schwedo, and C. B. Moore, 1974. *J. Geophys Res.* 79:1761–1767.

Woo, S. E., and A. E. Hamielic, 1971. *J. Atmos. Sci.* 28:1448–1454.

Woodcock, A. H., 1953. *J. Meteor.* 10:362–371.

Work, D. Q., and M. Walk, 1960. *Mon. Wea. Rev.* 88 : 249–250.

Wormell, T. W., 1927. *Proc. Roy. Soc.* A115:443–455.

Wormell, T. W., 1939. *Phil. Trans. Roy. Soc.* A238:249–303.

Yamamoto, G., 1952. Scientific Reports, Tohoku U., Ser. 5, Geophysics 4: No. 1.

Yamamoto, G., 1962. *J. Atmos. Sci.* 19:182–188.

Yamamoto, G. and G. Onishi, 1953. Scientific Reports, Tohoku U., Ser. 5, Geophysics 4: No. 3.

Yamamoto, G., M. Tanaka, and T. Aoki, 1969. *J. Quant. Spectr. and Rad. Trans.* 9:371–382.

Yamamoto, G., M. Tanaka, and S. Asano, 1970. *J. Atmos. Sci.* 27:282–292.

Young, K. C., 1974. *J. Atmos. Sci.* 31:1749–1767.

# Index

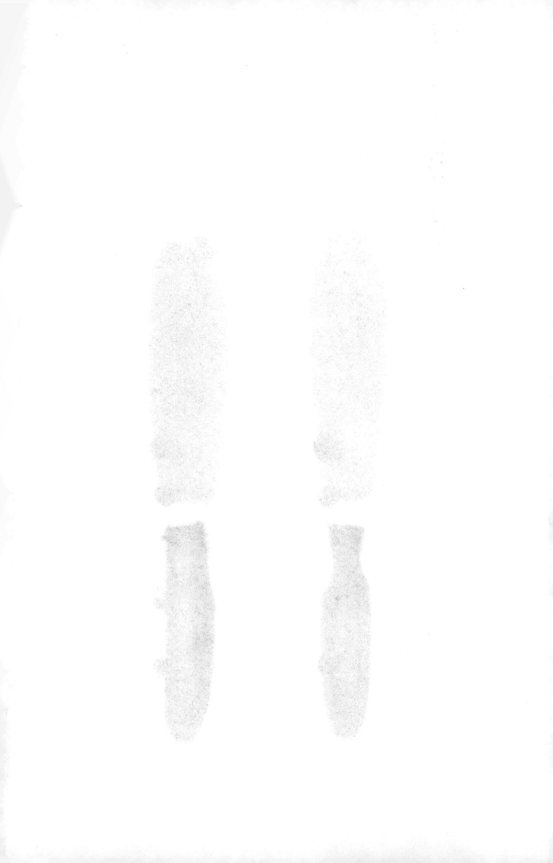

DATE DUE

| | |
|---|---|
| DEC 14 1994 | |
| MAY 1 2003 | |
| FEB 22 2004 | |
| JUN 07 2005 | |
| OCT 01 2008 | |
| NOV 22 2008 | |
| | |
| | |
| | |
| | |
| | |
| | |
| | |
| | |
| | |
| | |
| | |